CAMBRIDGE LIBRARY COLLECTION

Books of enduring scholarly value

History

The books reissued in this series include accounts of historical events and movements by eye-witnesses and contemporaries, as well as landmark studies that assembled significant source materials or developed new historiographical methods. The series includes work in social, political and military history on a wide range of periods and regions, giving modern scholars ready access to influential publications of the past.

William Whewell, Master of Trinity College, Cambridge

William Whewell (1794–1866) was born the son of a Lancaster carpenter, but his precocious intellect soon delivered him into a different social sphere. Educated at a local grammar school, he won a scholarship to Cambridge, and began his career at Trinity College in 1812; he went on to be elected a fellow of Trinity in 1817 and Master in 1841. An acquaintance of William Wordsworth and a friend of Adam Sedgwick, his professional interests reflected a typically nineteenth-century fusion of religion and science, ethics and empiricism. Published in 1876, and written by the mathematician and fellow of St John's College, Isaac Todhunter (1820–84), this biography combines a narrative account of Whewell's life and achievements with extracts taken from his personal correspondence. Volume 2 contains a selection of his correspondence with scholars including Herschel and Lyell, revealing much about the conflicts, debates and friendships that shaped nineteenth-century academic life.

Cambridge University Press has long been a pioneer in the reissuing of out-of-print titles from its own backlist, producing digital reprints of books that are still sought after by scholars and students but could not be reprinted economically using traditional technology. The Cambridge Library Collection extends this activity to a wider range of books which are still of importance to researchers and professionals, either for the source material they contain, or as landmarks in the history of their academic discipline.

Drawing from the world-renowned collections in the Cambridge University Library, and guided by the advice of experts in each subject area, Cambridge University Press is using state-of-the-art scanning machines in its own Printing House to capture the content of each book selected for inclusion. The files are processed to give a consistently clear, crisp image, and the books finished to the high quality standard for which the Press is recognised around the world. The latest print-on-demand technology ensures that the books will remain available indefinitely, and that orders for single or multiple copies can quickly be supplied.

The Cambridge Library Collection will bring back to life books of enduring scholarly value (including out-of-copyright works originally issued by other publishers) across a wide range of disciplines in the humanities and social sciences and in science and technology.

William Whewell, Master of Trinity College, Cambridge

An Account of his Writings

VOLUME 2

EDITED BY ISAAC TODHUNTER

CAMBRIDGE
UNIVERSITY PRESS

CAMBRIDGE UNIVERSITY PRESS

Cambridge, New York, Melbourne, Madrid, Cape Town,
Singapore, São Paolo, Delhi, Tokyo, Mexico City

Published in the United States of America by Cambridge University Press, New York

www.cambridge.org
Information on this title: www.cambridge.org/9781108038546

This edition first published 1876
This digitally printed version 2011

ISBN 978-1-108-03854-6 Paperback

WILLIAM WHEWELL, D.D.

MASTER OF TRINITY COLLEGE, CAMBRIDGE.

LETTERS.

WILLIAM WHEWELL, D.D.

MASTER OF TRINITY COLLEGE, CAMBRIDGE.

AN ACCOUNT OF HIS WRITINGS

WITH SELECTIONS FROM HIS LITERARY AND
SCIENTIFIC CORRESPONDENCE.

By I. TODHUNTER, M.A. F.R.S.

HONORARY FELLOW OF ST JOHN'S COLLEGE.

VOL. II.

London:
MACMILLAN AND CO.
1876.

𝕮𝖆𝖒𝖇𝖗𝖎𝖉𝖌𝖊:

PRINTED BY C. J. CLAY, M.A.
AT THE UNIVERSITY PRESS.

Dear Morland,

In order that you may reap the full benefit of all the philosophy I intend to bestow upon you and have some distinct idea of what I mean, I will tell you the situation in which you are to suppose me addressing you. After spending an idle day, I came into my rooms here about eleven in the evening, and having contrived to lay hands upon Miss Edgeworth's Patronage I made myself a fire and some tea and gravely sat down to read it. Having finished one volume of the four, I find it to be about 3 o'clock and not feeling the least inclination to put myself to bed I do not see what I can do better than make up a packet of cogitations and reflections for your edification. If you have ever sat up till early morning reading, you may possibly have some idea of that flow of spirits which one feels a little after midnight, when the propensity to sleep is totally vanished and one finds a kind of—a sort of—I know not what manner of alacrity and fermentation of the animal spirits, which I suppose if one had been asleep would have evaporated in the form of a cloud of nonsensical dreams, without any such benefit to society and to literature as is likely to be derived from my waking contemplations at present. But in order that you may have distinct ideas upon the subject which all metaphysicians make to be of so much importance, (from which observation you may possibly fancy that I am at present

reading metaphysics) it may not be amiss to set forth what
I mean by an *idle day* at Cambridge; seeing that there are as
many different modes of idleness as there are of action. I awoke,
therefore in the morning, and looking at my watch, found that it
was 10 minutes too late to go to Chapel, with which disappointment I was not very much mortified, as I did not think myself
after that obliged in conscience to get up immediately. I therefore took unto me Locke's Essay, and read in bed till eleven
o'clock—got up and breakfasted—grew out of humour with Locke
—went out—called on a great mathematician—driven out by his
private pupils coming—called on another great mathematician—
found him with a tooth just drawn—Thompson's Chymistry in
one hand and spitting blood into a bason in the other—asked me
to dine at his rooms· as his mouth did not allow him to go into
hall—lounged at the Booksellers—dined at 5—a great quantity
of mathematics, puns and nonsense talked—(N.B. He was a
Johnian)—Eight o'clock we began to read Tom Jones, four of us
reading a chapter alternately—much deep Philosophy solved—
and so on; upon various parts of which day's work one might
very sagely philosophize—"But what," you cry, "is all this to
Miss Edgeworth?" Now, my dear Morland, consider what a very
high rank Patience holds among the Christian virtues; consider
how very necessary it is to get habits of self-command, of restraint
upon the passions and above all that most wicked passion *Curiosity;* and peradventure before I get to the end of this sheet
I may tell you something about it, though you are to reflect that
I have myself only read the first volume, and that therefore
I cannot be supposed to know how many marriages there are at
the end, which of course is the first object of solicitude. Nor
indeed if I could have given you an outline of the story, would
you I suppose at all thank me for it, if ever the book itself should
fall into your hands, at least I would at this moment most willingly take a glass of the waters of Lethe, for my evil stars led me
to the Edinburgh Review where I found an abstract of the plot
whereof it now much repenteth me—Well, but to give you my
opinion of what I have seen; though the plan of the tale appears
not very original but rather *novelish,* what is of much more im-

portance, the Characters, and those pleasing parts, the Dialogues, appear supported with Miss Edgeworth's usual style of ability. To be sure it may be said of it that all of it is not equal to the best parts of her former works; but as Johnson said on a similar occasion "There must be some parts which exist for the sake of others—a diadem cannot be one entire diamond—the jewels must be held together by some substance less precious," to which I have often been tempted to add "that it is very well if that substance be gold." The *moral* which this fair lady appears desirous to inculcate is one which always has appeared to be her favourite principle, Decision and Firmness of Character and Self Dependence in opposition to that dependence on others, that System of 'Patronage,' which is the order of the day. By the bye— entertaining not the least doubt, as I certainly do not, of the wisdom and weight of this fair moralist's observations, and feeling all the inclination that any soul can feel to act upon those principles which she has so well illustrated, is it not an infinite nuisance that one is never likely to be put in any situation where one might make the experiment upon one's self and try what latent powers and hidden capabilities the mind possesses? Is it not, in short, enough to put one in a boiling rage to think that one must live the dormouse life of the inhabitant of a college instead of going amongst other two-legged beings in the wide world to see whether we have got any thing superior to instinct about one; any soul—and if we have, to find what colour it is of? Though to be sure I think a person in any situation may, if it contributes to his satisfaction, contrive to lead himself into temptation in order to experiment on the strength of his own mind. For instance here at this place there is no small degree of courage required, if a person get into certain situations in society, to avoid running into expense—not merely for the sake of dissipation; but because it is difficult without incurring certain expenses to keep up that connection with persons of genius, learning, and eminent character, which one would not like when once begun to let drop for impediments of this sort. I suppose you begin to perceive by this time that though I do not myself appear disposed to sleep I have a tolerable knack at reasoning other people into a state of slumber.

In fact I have written such a parcel of stuff that, large as this sheet is, I do not know where I must put the tail of my letter

> Summi plenâ jam margine libri
> Scriptus et in tergo necdum finitus.

Moreover I expect that you will write to me again and as soon as may be and then, as I shall have read the whole of this book, if you are not satisfied with this I shall be able to give you a more satisfactory account of it.

Your Local Militia business is certainly what would be called here *an infinite bore;* in truth it is a serious vexation and many are the philosophical theories and sublime, that I could sport for your consolation; only looking forwards to the day when I shall see you another Xenophon, I think that we ought rather to rejoice.   *   *   *   *

Believe me to be Dear Morland Ever yours w. w.

Ap. 6th.

Babylon is fallen, that mighty city. The Allies entered Paris by capitulation on the 30th. Did I not suppose that you would be laetified by this news before this could reach you I should be inclined to say something more about it and to heap still higher this running over bushel of nonsense. Adieu.

CAMBRIDGE, *June* 15, 1814.

DEAR MORLAND,

Your cautions against vanity were, I can assure you, by no means unnecessary; the adversary assails me very strongly; I am at the present moment in the greatest perplexity how to return pretty modest answers to half a dozen flaming complimentary letters that I have received. But really you are very inconsistent and whilst you are giving me ghostly advice you have no more mercy than to be one of those to increase my embarrassment. I had a great wish to *stay up* to escape the congratulations of my Lancaster friends who I find are disposed to give me credit for much more than I have a claim to; and actually talk of my fair fame resting on a foundation which &c.—and of being the first young man of the age, and of immortal honours, and several other fine phrases which are enough to turn the head of any mortal alive.

Now I wish you would contrive to make them understand that this said prize and many others of equal value must be got every year by some one or other, and that every year some one or other must be first at every college; and that therefore the quantity of fair fames and immortal honours thus acquired will be so great that it will be no small burthen to posterity to give them due credit, and that posterity will very wisely solve the difficulty by not troubling her head with them in the smallest degree, and that even in this generation these fair fames &c. after buzzing for a few months will terminate their ephemeral existence.

I intended to send you a metaphysical letter in answer to your preceding one, but I was prevented first by the bustle of the examination and then by a fortnight of idleness and dissipation which has followed it. It is now I dare say out of my power, for as I only absorbed a quantity of metaphysics for the examination I gave it all out again in answer to the questions, so that I have scarcely a particle remaining. This may appear strange to you but it is perfectly well understood here. For instance, in the examination for A.B after passing a most severe examination in the profoundest parts of Mathematics the men go into a place where they are stuck up in a box, and a man stuck up in a box opposite asks some very simple question such as "Quid est circulus?" and to which a man's mathematical knowledge is so much exhausted that he always answers "Nescio."

I am puzzled to make out what "little society" it is that you have instigated to buy Metaphysical books; and also how Paley's Philosophy finds its way into the list. I fancy you will like the study of the science of the Mind, as it is a subject where when you once set off a reasoning you may pass the bounds of space and time and travel on to all Eternity without coming to any conclusion, or rather may arrive at a dozen contradictory conclusions all equally certain. But if you want to see the science in its triumph over common sense, with all its train of conclusions that *admit no answer and produce no conviction*, read Bishop Berkeley's Principles of Human Knowledge and his Dialogues on the existence of Matter, where he proves beyond all contradiction that our belief of the existence of external objects on the evidence of our senses is equally as fallacious as when in a dream we deduce the same conclusion from the same evidence.   However after what the Scotch Metaphysicians, Reid, Stewart, &c. have written it must be allowed that Metaphysics is a very entertaining and also a very useful study.

After indulging in a little idleness since the examination I am setting to work again.   I find it more than ever necessary to read in order to come as near as may be to the expectations that my friends entertain.

I have just sent my poem to the press and find from the bore of copying, correcting, &c. that a person cannot meddle in any degree with authorship with impunity.   I shall come down in a short time laden with verse.

*   *   *   *

Dear M. Yours sincerely W. WHEWELL.

TRIN. COLL. *Aug.* 10th, [1815?]

DEAR MORLAND,

Though our correspondence may not be the most punctual in the world it is very regular in some points.   For instance every epistle has for its exordium an apology for not writing sooner and for its peroration a very urgent request for an immediate answer.   I have I am afraid more than common occasion to begin according to these established rules, and certainly I have the greatest inclination in the world to conform to them in ending.   Till I looked at the date of your letter I was not aware that I had been such a capital delinquent.   One evil consequence of this vile practice is that I hardly know what to say in reply to observations which two months ago I could have answered with the most profound wisdom.   I no longer know, and you no longer care what I saw at London, or what the Dean of Peterborough said to Mr Lingard.   With regard to the first of these points I have only an indistinct recollection of seeing a multitude of houses, each to the best of my remembrance occupied by its respective inhabitants, except some of the largest which are in the hands of people to whom you give sixpences and shillings that you may say you have been in St Paul's, Westminster Abbey, etc. In a week I began to be tired of seeing a world where every body but myself seemed to have something to do, where society seemed made up into sets without leaving any vacancy which I could fill, and accordingly I left them, knowing that they would mind their business just as well if I were in my dressing gown at Cambridge. I very soon by the help of philosophy overcame the mortification of finding that I was not, so far as I could discover, a necessary part of the system of things; though as the immense bustle in the streets and the grand aspect of St Paul's seemed for a moment to have been created merely to excite pleasing sensations in my mind, you must allow it would have been very flattering if the wheels of business and of the great clock had stood still when I withdrew my presence. You are quite mistaken in supposing that there is no account of a voyage or travel to London. Besides all the innumerable lists of Guides to, Pictures of, Vade-mecums in, London—a description of a journey to the Metropolis and of introduction into Society there is become an ingredient in a novel as necessary as an amour.

By the bye, there lay upon the bookseller's counter yesterday a book entitled "A view of London," which is one of the gravest pieces of absurdity which I have seen for some time. The author, who calls himself a Beneficed Clergyman in some part of Suffolk, seems to have led one of those happy tranquil ecclesiastical lives, in which a journey of fifty miles is the greatest event that happens between the beginning and end thereof; at least if one may judge from the space he devotes to the undertaking of removing any surprise you may chance to feel at such an improbable adventure. He begins—"Innumerable are the inducements of business and pleasure which may draw the stranger to our far famed Metropolis,—*which* has nearly doubled its population within the last two hundred years." How the latter part of the sentence is connected with the first (except indeed by the word *which* which is a never failing connector) no unbeneficed person I am afraid will ever find out. But I have not the least intention of writing a critique upon this man, *who* is as loyal and as cautious of giving offence as his bishop could possibly desire, as his book does not contain above 40 pages.

What are you reading now? This is the time of life when you and I ought to be reading something, systematically I mean; I suppose Divinity is your subject—but do you not intend to squeeze something else into your Cyclopædia? It is an observation that I never made or at least never put into words till the other day, that here I have the means of getting at almost every book that ever was written. This is obvious enough, nevertheless it tended to excite some scientific enthusiasm. But some people bid us beware of the Demon of universal knowledge, and I suppose some people are wise. I wish exceedingly you would take to studying some science or other.   *   *   *   *

Believe me Ever Yours W. WHEWELL.

TRIN. COLL. *Aug.* 10, 1815.

DEAR GWATKIN,

When I got your epistle I was meditating to send you a very vehement philippic against people who let you hear so little of them that you begin to imagine they must have migrated to some other part of the solar system. I was rejoiced to hear that you had only been describing a trajectory over some parts of the earth; and I am afraid I have procrastinated my essay on procrastination till I have lost the privilege of making it. From what you say I suppose you will account for this by supposing me lying on the sofa all day reading novels in my dressing gown. (N.B. I have a sofa.)   But my indolence is not precisely of that character. *Our* vocations are much more numerous than you suppose—N.B. upon "our." Of the small number of men who are here every body knows every body, and thence it comes that they have a great quantity of time and amusements in common. Depict upon your intellectual retina Wilkinson, Slegg, Powell, Wollaston, Reed, Whewell, shooting swallows, bathing by half dozens, sailing to Chesterton, dancing at country fairs, playing billiards, tuning beakers into musical glasses, making rockets, riding out in bodies, and performing a thousand other indescribable and incomprehensible operations, and you will have some idea of the means which are used to keep *ennui* at arm's length—and hitherto with tolerable success.

That wizard Michael Slegg is becoming a man of this world— You may have observed him to be possessed with the desire of universal knowledge, but I dare say you did not expect that this demon of universality (beware of the foul fiend!) would ever put him upon the back of a great cantering horse. But so it was. Some demon whispered "Slegg you ought to ride," and to ride he accordingly began, to the very great astonishment of every passer by, for his horse, though a very quiet one, happened to have at that moment some singular fancies, in which Slegg to gain the animal's confidence indulged him—such as sidling up against the door of a house, running his head against a house side, walking up narrow passages, and at length, which inconvenienced his rider most of all, trotting—in consequence of which Slegg came to the ground before he got out of the streets of Cambridge. S. however has persevered most manfully and in spite of repeated descents still threatens to become an accomplished cavalier. The other day his beast got the dominion over him and deposed him in such a way as to collect almost all the inhabitants of Barnwell about him, who appeared somewhat disappointed that he was not as they said "all smashed to pieces." He was however no worse, though he thought it advisable to get himself phlebotomized.

Among all this you will suppose that mathematics do not go on very well—better perhaps than you imagine but certainly not so well as they ought to do. I wish you were at Cambridge— I have not got any body to talk mathematics to—Nobody that would care a fig if I were to tell them that when Force varies as $\frac{1}{r^7}$, and Velocity equals that from infinity, a body describes a lemniscate with two loops, when Force varies as $\frac{1}{r^9}$ a curve with three loops, when Force varies as $\frac{1}{r^{11}}$ a curve with four loops; and so on. I scarcely ever hear the Senate House mentioned except by

Whittaker, who always asks whether I am reading. Wilkinson leaves Cambridge tomorrow and I rather think is going N. to stay for some weeks. Perhaps you may recollect a man of your college who took his degree a year or two back of the name of Thomas Briarly—do you know whether he is yet alive and in England? Since he left Cambridge there has not been anything heard of him. Communicate what intelligence you can concerning him as it is hoped he may yet possibly be in existence. At present I suppose you are "teaching the young idea how to shoot"; are any of the young ideas that you have got likely to make good shots? Will any of them bring down a high wrangler? In Higgin I suppose you will not have a pupil who is bigoted to some theory or notion of his own. And Whitcombe? what manner of mathematician,—or (what is a much more important question) what manner of wrangler will he make? Expound to me concerning these things—I would add if I thought you would pay any attention to it—*very soon.*

Yours truly, W. WHEWELL.

CAMBRIDGE, *Dec.* 15, 1815.

DEAR MORLAND,

I ought in all conscience sooner to have answered your request for something intelligible on the subject of my last letter, which so far as I can recollect must have been very particular nonsense (not a very uncommon occurrence). I believe it conveyed to you certain yearnings after the whole circle of the sciences, certain ecstatic aspirations after universal knowledge, certain indefinite desires to approximate to something like omniscience. It is most certain that the mind of man is given to such morbid generalizations, but woe betide the wight who either expects that such a spirit will support him through all the details even of one science, or that without the accuracy of detail he may content himself with the amplitude of general views, the magnificence of extensive vacuity. But though not much good would be likely to come of me if I were to remain in such an all-reading, all-learning mood for ever, I am much rejoiced that you seem to have formed some designs of plucking a few of the apples of the tree of knowledge. Unless indeed, in the transitory nature of all human things, and more especially of all human emotions, the enthusiasm of literature has gone by and you are settled into a contentedness with such a share of human learning as may in the common jostling of sublunary matters tumble into your cup. But this is precisely what nobody who intends to lead a literary life, and such a clerical life is, ought to be. All knowledge is valuable for its own sake, and, independently of the value of the object, the time spent in acquiring knowledge is worthily, because happily, spent. And by studying systematically both the satisfaction of the pursuit and the value of the end is increased. As to the precise system of attack, and points to be attacked, it is a matter of some difficulty to decide. To recommend mathematics, or even a very extensive study of natural science, would be to give you a subject which it requires more time than you have to spare to read to any purpose. But do by all means read so much mathematics as to get an idea of mathematical reasoning; that is to say, get an Euclid and read the three or four first books. It almost seems to me that a man who has no idea of mathematical reasoning has no idea of reasoning at all, certainly no proper notion of the powers of the human mind and of the processes by which it may arrive at truth, no conception of those faculties by which the dominion of human knowledge has been extended, is extending, and is hereafter to be extended. If you do get this same book, do not expect to be struck with astonishment by the glare of discovery, do not even expect to see the full value and bearing of each proposition as you go on; that is, in plain words, expect to find a system of reasoning which you may very likely think dull, but of which, if you go on with perseverance, you will come to see the force. If you do not after this feel any particular call to mathematics (I do not think you need be under any great apprehensions) your own taste is a much better judge for you what to addict yourself to than that of anybody else. Though you are married to the church, you may be "married to immortal verse," or to any other of the immortals of the same species, without offence to the statutes of polygamy. But as people have always a tendency to give advice which they do not follow, I should say read systematically and accurately. This is so exceedingly wise an epistle that it would be a pity it should be lost, and yet now that I have written it I begin to wonder where it will find you. I wish you would find time to write again, which you may do as you cannot by any chance be so busy as I am or ought to be, for behold the end of my undergraduateship is at hand. You will hear in a few weeks the event of all things. Do not be too much surprised if a friend of yours should be four or five places lower than you might wish to see him.

Dear M., yours truly, W. WHEWELL.

BURLINGTON QUAY, Aug. 10, 1816.

DEAR MORLAND,

It is a longer time than usual since we have addressed one another; and now that I am domesticated in this out of the way place you will scarcely know where to direct your imagination if in any idle mood it should wander towards me, unless I give you indications of my existence. Upon looking back at the time which has passed since I saw you, it seems perhaps longer than it might otherwise do, from the changes that have taken place in it. We naturally conclude that an event which has changed the colour of our existence has occupied a time proportional to its importance; and however sudden the blow may be it seems as if the parts of life before and after it were separated by a wide interval. I have been called to Lancaster since I saw you, and without the pleasure, as it would under any other circumstances have been, of meeting you there. You were snatching, I suppose, a draught of your native air in the short pause which the bustle and hurry of life now and then allows you. I was obliged to proceed immediately to commence my summer operations here. I am here as idle as a man well may be who is compelled to confine his attention six hours a day, and, so far as I can perceive, stand an admirable chance of being more ignorant at the end of the vacation than the beginning. There is scarcely anything but the habit of being in literary society which is at the same time sufficiently strong and sufficiently constant to act as a stimulus to literary exertion. The abstract desire of accumulating knowledge is too vague and general to stick close about us, and the prospect of future advantage in the way of fortune or fame too distant and uncertain to keep us going, so that you perceive it is quite inevitable that I must be idle. You have no idea of the satisfaction it gives me to form theories such as this to account for my indolence, because it half persuades me of what I know to be false, that in other circumstances I should do anything or to any purpose. This same human life is a very strange business—in which we are led from step to step only by forming designs which are never to be executed and hopes which are never to be fulfilled; in which we amuse ourselves and tire our neighbours by talking of intentions which perish as fast as they rise, and out of which we shall go, having purposed everything and done nothing; just when we were going to do what we should never do if we were to live for ever.

I do not know whether this train of moralizing ever gets into your head, but it amuses me very much to observe all the inconsistencies and absurdities of man, taking for granted, as moralizers have from time immemorial been privileged to do, that every body is as absurd and inconsistent as myself.

I hope all your concerns spiritual and temporal go on as you wish, that you have no lack of audiences for your sermons or of sermons for your audiences; that you and the vicar are upon what he calls good terms, and that you do not find more than usual tedium in teaching the young idea how to shoot.

It is so long since I wrote to you before that my good wishes have, as you perceive, accumulated to a more than ordinary magnitude. I would send you more of them but I have a man writing at my elbow to whom I have been every half sentence introducing into his head ideas that nature never intended should find their way there.

I suppose at the present moment our occupations are much the same. Let me hear from you soon and believe me

W. WHEWELL.

TRINITY COLL., CAMBRIDGE, *Feb.* 2, 1817.

"Yet once more, O my Morland, and once more" my future Bachelor of Divinity, I come to commune with you on the subject of our last two or three letters, and as this is a matter of no small importance let us consider it threefoldly. I shall therefore discourse on the topics with which Milton's devils amused themselves—viz. Fate, Freewill and Foreknowledge. First of Fate—is it predestined that you shall become a member of Trinity? Second of Foreknowledge—if so let me know before the nineteenth of February as the term divides then. Third of Freewill—if you do not I shall conclude I am free to enter you at some small college. By entering before the division of the term you will keep the present term and your name will then remain upon the books till in the fulness of time you shall come to see the glories of Cambridge. At all events find time to write to me before division and inform me whether Mackreth has got you a letter from Hudson. I wish your coming hither were not at such an immeasurable distance, for though I do not at present see the probable fulfilment of your prayers that I may not waste my sweetness on the desert air of Cambridge all that time, yet it is not to be expected with any degree of confidence that you will find me here ten years hence. But I have no doubt that when you come you will be sufficiently amused by the animals you will find here. A very perfect specimen of the class of them called senior fellows departed this life, or more properly speaking and which excites much greater interest here, vacated his fellowship yesterday. An old man between 80 and 90 who must have been in college ever since the middle of last century, and consequently had outlived all his early connections with any society except his college society without forming new ones. All our fellows who stay here live to a great age—we have one or two of his coævi still left. I breakfasted this morning with a French abbé (speaking wondrous bad English and exhibiting a wondrous quantity of beard, dirt and foul linen—but no matter for that) with whom I had a stout battle about the catholic religion. He accuses you clergymen of the church of England of misrepresenting from ignorance or from malice the doctrines of his Church. I did what I could to drive him into some inclosure of fixed principles, but he always contrived to slip through the paling. However he talked so perseveringly that when the dinner bell rang I was obliged to take advantage of a fit of coughing which he had to slip in a good morning.

Truly yours W. WHEWELL.

TRINITY COLL., CAMB., March 6, 1817.

DEAR HERSCHEL,

From what you said in one part of your letter I hoped to have seen you before this, sailing through the courts of St John's with all the pride and ample pinion that bear a Master of Arts hither at every contested election, and in consequence I was prepared to be not only in charity but in good humour with the whole system of electioneering, canvassing, voting, contravoting, and outvoting which predominates here. I do not know a stronger instance of the power of Providence in thus 'from evil still educing good,' and of the ingenuity of man in educing evil from good—than the way in which one of the parties contrives to make such an establishment as this university little else than an arena of petty, interested, idle, gossiping and malicious college politics, and the other manages to produce from this same paltry frame of college politics events which increase the sum of human happiness so much as the reappearance of one's friends upon this terrestrial theatre—or more properly senate-house. You must allow me to be out of humour with the university, because I have not yet recovered my chagrin at finding when I began to look about me and rub my eyes at the beginning of last term that both you and Jones were vanished. It was, to use a familiar illustration, a feeling like that of coming to yourself after a dose of nitrous gas. I cannot tell you the real intellectual loss which I feel it—but amongst other things I have got no soul to talk functions to—nor in fact anything else that is worth talking. By the bye, we had Bromhead here a little while back, who was as usual absolutely overflowing with theories, more particularly mathematical. The rapidity and extent of his generalizations is absolutely overpowering. They expand before you like a canister of gunpowder which should explode while you held it in your hands. He certainly has got a very long way into new views of his subjects, but from what I have seen of him I cannot for the life of me tell whether it is from the lightness or the firmness of his tread. Exceedingly pretty the way in which he gets at Fagnani's theorem, and what is still better gives you the naked principle of the thing. I have been trying to generalize it but the ulterior forms demand some pitiful restrictions. I am not certain that I shall not soon have the honour of appearing in your capacity of translator and annotator of French mathematics. Rose came to me the other day and told me he thought of translating Lacroix's Application of Algebra to Geometry. I said, carelessly enough, "I will write you some notes." The book is a good book and might be made very useful here, but Rose is not the man to be concerned in such a thing : he knows very little about it, and I suspect will not be the more manageable on that account. If you can suggest anything that you think will be useful pray do. You have I suppose seen Peacock's examination papers. They have made a considerable outcry here and I have not much hope that he will be moderator again. I do not think he took precisely the right way to introduce the true faith. He has stripped his analysis of its applications and turned it naked among them. Of course all the prudery of the university is up and shocked at the indecency of the spectacle. The cry is "not enough philosophy." Now the way to prevent such a clamour would have been to have given good, intelligible, but difficult, physical problems, things which people would see they could not do their own way, and which would excite their curiosity sufficiently to make them thank you for your way of doing them. Till some one arises to do this, or something like it, they will not believe even though one were translated to them from the French. Every thing here is just as it ever was, that is all the world is eaten up with pupils. Whittaker has six, so have I, besides which I am or ought to be reading for a fellowship. I was in London during the vacation and learnt that Jones had intentions of visiting Cambridge this term—learnt it from one who saw Jones in a situation where it would be whimsical to see him, playing at whist, *utpote* Parson of the Parish, with his farmers and their daughters. There are few things that would give me so much pleasure as to see you and him; therefore my dear H. I pray that some election or other which I hear people talking about may bring you here. Yours truly W. WHEWELL.

TRIN. COLL., *14th Sept.* [1817].

MY DEAR ROSE,

You did a most praiseworthy action when you called up your sympathy for a man about to be examined and wrote to me. I am glad to hear again of an old and favourite scheme, and to see it tending more and more to its fulfilment. I am glad to see my friends about to exert their talents when I know that they want only to be exerted in order to be valued. Now as to the manner. You still seem to have a partiality for the form of a Review, which is no doubt in these times as accommodating a form as we could take, and will admit as many subjects treated in as many manners as could be wished. But is there—I ask merely for information, because I know nothing of the politics of literature—is there an opening for one? Would you annex yourself to any party in politics or in religious opinion? If you would not, do you think that a review with no such seasoning would go down? People read reviews at present to spare themselves the trouble of reading original books and forming their own opinions; and of course, from the importance of the subjects, opinions about church and state are those for which there is the highest demand. This would of course be expected at first, and the book would first have to get over people's disappointment at not finding it what you never intended it to be, before it began to get credit for what it was. No doubt the system upon which reviews are in general conducted has many and huge faults, but would you be more likely to cure them by falling into their ranks? Would you not be something like an army surgeon who is put in uniform for the purpose of healing the wounds that the rest make, but generally looks rather awkward in his regimentals?

Do you think in short that there is a sufficient number of people to support a review merely because it was conducted with sound knowledge, good taste, right feeling, and reasonable good writing? Would it not be the tendency of a review to force you upon personalities, to make you sacrifice truth and fairness to wit or paradox, and in short to plunge you at once in all the *tricks* of a literary life? Again, does not a review at least for a long time, though it seems to exert only the shadowy majesty of an anonymous critic, yet in fact derive its weight from the known character of its conductors; and would there not be a strong prejudice to be overcome against a few young men, just emerging or not emerged from the university, who should set themselves up to direct public opinion? This might be overcome by perseverance and talents, but still it would be for a long time very uphill work. Besides, so far as I can discover, there is nothing to mark and particularise your plan; your publication would at first seem merely to be lifting its buzz among the crowd of ephemerals that every day brings forth, distinguished only by its name and the colour of its cover. To think of giving it any very marked connexion with Cambridge may arise perhaps from a narrowness of view, but it would at least have the effect of *individualizing* it more completely and of pointing out in some measure at least the nature of the work, as rather undertaken from a strong and, if you will, an *academic* attachment to science and literature, than from a love of literary gossiping. There is another consideration which ought to have some weight, and that is the effect upon the mind of such habits of writing as a review would require. There is a very great difference both between the books you would read and the way in which you would read them if you were catering for a review, and if you were getting knowledge for your own purposes. You would have to read a number of worthless books, and to read many good ones with very little enjoyment or profit. The perpetual habit of watching the sensations that a book makes will not by any means tend to make them more vivid or distinct; and there must be a very wide difference between analysing the effect which a writer produces upon you when it is so strong as to give you a curiosity to do it, and the feeling it your duty to perform such a process upon every author who comes in your way; and the everlasting drain of your ideas, the drawing off your feelings before they have time to settle and clear, the pouring out every thing that comes into your head as fast as it collects, must be, and I think you may observe from the instances of young and prolific writers, is one of the things the most impoverishing to the mind. And on subjects of science and information the case will be much the same. In a set of men so few and so young as the proposed reviewers it could not be expected that they would have sound information and extensive reading on all the subjects that would come under their cognisance. In many cases they would have to get up an article, to scrape together information in a very hasty and indiscriminate way, and most likely to do it ill at last. Not to speak of the probability of their opinions and views of subjects changing completely. For my own part I will not answer for my opinions upon any subject out of the circle of demonstration for two months. I find them changing perpetually, and should be by no means certain if I advanced an opinion upon a subject in one number that I should not be disposed to retract it in the next. For my own part indeed I should feel much inclined not to attempt to write anything at all for the next year or two, for there are so few subjects on which I am not on the very verge of ignorance that it would be of much more use to myself, and I am sure to the world, that I should try to learn than to teach. There is no occasion to say that I speak this in the most profound sincerity of heart, but as it is difficult to analyse such feelings so as to discover how much indolence is mixed up in them, I should be willing to engage in any project where I could suppose myself likely to do any thing at all. I do think however that it would be advisable to wait a little before publishing, if it were only to collect some materials. I do not much like the plan of living from hand to press; writing just enough to keep the first number on its legs and leaving the rest to take care of itself. There are many reasons why, if a review were to be written, I should like to have half a year devoted to collecting materials, that is, to writing articles sufficient for three or four numbers. It would be well, because we should most of us I think be unpractised in writing, and it might be some time before we got a style that we should wish to write and could write with ease. I think it very likely that in that respect at least the first essays would be the worst; now there is no reason under the sun why we should send our exercises into the world with all their unavoidable imperfections on their head. Besides, by writing and communicating articles we should each see something of the spirit in which the rest expected to have the work conducted. Without this half a dozen articles might meet together all looking different ways and giving each other the lie in every page. Not to speak of the very disagreeable case of panting and toiling after time in vain, having to seek for articles when you ought to be publishing them; and prowling about in the vain hope of dragging some unfortunate author to your den. I have written so far because it is highly desirable that in such a matter we should know one another's views. You will consider all my objections only as queries to let me more fully into your plans. If the plan be carried into execution, and if it be supposed that I can assist in it, I shall have satisfaction in contributing my share if I can make out what it is. But I shall find great difficulty in fixing on any subjects which are not better pre-occupied. On scientific subjects, to which perhaps I have the strongest bias, I know nothing; and where Herschel is concerned, if I did, I could be of no use. The same is true, in at least an equal degree, on subjects of general information. If I had a year or two of leisure I fancy I could read to some purpose. But pray let me know how you have arranged so far. Who are your half dozen? Are matters so managed that there is room for me to be one of them? For the periodical time I think certainly with you that a month would be too short; two months might perhaps be long enough and would have the advantage of not bringing us into competition with the quarterly reviews, and upon the whole seems to have as much for it as any scheme. George Peacock talks of a six months' review; upon this hint I suggested a *secular review*. Spinetto has been trying to collect a body of Cambridge reviewers.

He proposed to Peacock that he and Bland should take the mathematics, which did not at all quadrate with George's notions. I believe the thing has fallen through. I have thought frequently of something like a magazine or periodical collection of essays upon all subjects, scientific, literary, spectatorial, or any other. It would give us more liberty than any form. If its circulation at Cambridge were a matter of much importance, I have no doubt that we might annex to it a sufficient quantity of Cambridge mathematics neatly done to make it sell here—an odd expedient in the way of bladders to make it swim, but I think it would answer. The remainder of the publication which should be much the largest part might, I do not doubt, be so written as to do much good here and elsewhere. I cannot tell whether you will find anything in all that I have said, but let me know, and consider my situation, you know now reading for an examination

Affigit humo divinæ particulam auræ.

I hope when I have got my troubles here over to be able to think more freely. If, which is not improbable, I feel a tendency to gulp some other than Cambridge air I will go to see Jones, and I shall be able to arrange matters much more to my own satisfaction. Whittaker is with Jones at present.

I have not made any agreement with Deighton. I have left the matter in a state of suspense. When do you come up to Cambridge? If it be soon I will leave it so till you arrive, for I acknowledge, my dear Mr Editor, that I have much more confidence in your powers of managing booksellers than in mine.

I am glad Mr Rose has determined to send your brother here. I shall rejoice to see one of a family all of whom I recollect with so much pleasure, and by all of whom I should wish to be kindly remembered. For the sake of consistency in your kindness, my dear Rose, let me hear from you soon. Paynter talked of writing to you to-day.

Keen is here and I believe reading for a fellowship—to which that we may all come may, &c., &c.

My dear Mr Editor believe me ever most truly yours,

WILLIAM WHEWELL.

TRINITY COLLEGE, CAMBRIDGE, *Oct.* 16, 1817.

MY DEAR JONES,

I have not often been more severely mortified and irritated than I have been to-day on finding the tricks fortune has been playing me the last week. The freedom from all immediate cares and occupations which the end of an examination brought with it induced me to realize a project of a visit to you which has long been floating before me in the region of desirable possibilities. A series of blunders brought me within a few miles of you for two days and sent me hither to learn it. My first piece of ill luck was the not seeing Whittaker before he set off to you: in consequence of which I was ignorant of your locality. In fact I left Cambridge a day or two after Whittaker, under the impression—got I cannot tell how—that you were at Tunbridge: whither I wrote to you and where I was proceeding to seek you if I had not met with somebody in London who informed me that you were not there. I then found Musgrave who told me you were at Petworth—and to Petworth I proceeded on Saturday last. Conceive my disappointment when, upon alighting from the coach and making rapid enquiries about you, mine host informed me that Mr Jones went to London on Thursday. He introduced some episode or other about another gentlemen in spectacles who was with Mr Jones, but as I could not possibly know that your eyes had failed you or imagine from the familiar way in which he mentioned your name that he had mistaken Whittaker for you, I paid little attention to him. It will do me much good if you will write me word that you have broken the knave's head or got your clerk to do it, if you think it inconsistent with the character of a minister of peace to use your hands for any purpose except to give effect to a sentence. To add to my good fortune, next day was Sunday and it was not possible to return to town by any common conveyance. I had nothing to do but to gird my practical philosophy about me and amuse myself as I could. I went on to Arundel that evening—rode over to see Chichester cathedral next day—went on Monday to look at the castle and at the Roman pavements at Bognor, and in the course of Tuesday and Wednesday conveyed my body to the great pyramid here where it will now perhaps rest for some centuries longer.

I had soon satisfied myself that it was to no purpose to seek you in London, but it was not till I got here that I learnt from Whittaker what a blind and blundering journey I had had. My vexation was great in proportion to the pleasure I had so narrowly missed. I had trusted to revive many old and acquire many new ideas; and more especially just now when I have cleared away the obstacles that stood between me and the speculations about which we used to talk, I had anticipated much edifying discourse upon the past the present and the future. Such a meeting would have been no small object in a life which, as Mad. Neckar says of hers in a letter which I recollect your admiring, is a succession rather of ideas than of events. It grieves me to the soul to have missed all this, but I am so completely discouraged by the disappointment that I have not the heart to begin to anticipate over again. But to make me some amends overcome your uncorresponding disposition and write to me. Among many other things I wanted to talk to you about the Review which Rose says is again labouring into existence. I made many objections, but the amount of them was that I do not think we have *strength* for it—including knowledge, reading, time, habits of writing—and, which would be a great matter in ensuring the success of a laudable undertaking, weight in the literary world. For my own part if I am not deceiving myself a year or two would give me much more confidence in what I could do. At present I do not believe I should be worth anything—if however you carry the project into execution and can find anything for me to do I will exert all the powers I have. What does Herschel think of it? I shall have Rose and Whitcombe here in a day or two and we will talk of it. I have got several projects for mathematical works. I think I could at least be useful but I have not put any of them in a tangible shape yet. Saw Slegg in town. He is much as usual but I fancied I discovered an energy in his character which I was not aware of before. With motive and a friend who was acquainted with his domestic circumstances I should not despair of seeing him an active man even yet. I am very angry with our seniors for condemning Higman to another year of academic trifling. With the kindest recollections of the pleasure and more than pleasure that

I have owed to you and I hope shall again, believe me sincerely yours—W. WHEWELL.

June 19, TRINITY COLLEGE, CAMBRIDGE.

DEAR HERSCHEL,

          *   *   *   *

    I was in town soon after you and was to have gone with Babbage to Sir Joseph Banks's when I was seized with a most critically unfortunate illness. Jones writes me word that he has almost blown off his arm with an unlucky powder flask, and asks me to go to see him which I think I shall do. I understand from Peacock that you are untwisting light like whipcord, cross-examining every ray that passes within half a mile, and putting the awful question "polarized or not polarized" to thousands that were never before suspected of any intention but that of moving in a straight line. On any cloudy day that you have got I should be very happy to hear what the sunshiny ones produce. In a few weeks I remove to Barmouth in Merionethshire where I stay till October.

                Always truly yours W. WHEWELL.

CAMBRIDGE, *June* 19, 1818.

DEAR JONES,

    After losing sight of you so long it was with no common satisfaction that I received your assurance of your still being in the world and out of humour with it: for I should almost have doubted the former fact if it had not been confirmed by the latter. My pleasure was in some degree alloyed by not receiving this testimony under your own hand, and by the history you give of the cause of this flaw in the evidence; but after so much good I am willing to hope more: and therefore I trust that flesh wounds will cicatrize, arteries anastomose, and that your hand will be ready to receive a friendly shake in a fortnight. My summer campaign does not begin till the beginning of July, and as the pages of existence which I have before me at present are rather insipid, I am much tempted to turn back a few leaves and read with you some of the annals of your Cambridge days. We cannot recall the

Cambridge of that time completely, but I have no doubt I shall find much more of it in Sussex than there is left in Cambridge. You are, you say, much the same, and I have not been acted on by any society I have since seen as I was by that, and we shall therefore, notwithstanding the eternal flux of all things corporeal and intellectual, easily recognize each other and slide back a couple of years without a jolt.

I am going to lead my flock of pupils to graze at the foot of Cader Idris. On my road to seat myself in the giant's chair I shall proceed in the first place to London, and for the satisfaction of seeing the best of all Welshmen first I will set off a few days sooner and diverge into Sussex. I will not stake my reputation for punctuality by telling you the exact day when I hope to see you, but it will be within a day or two of Sunday the 28th, possibly on Saturday the 27th, that I may have the satisfaction of seeing and hearing your reverence in the pulpit—though, on recollection, with your wounds you will not be able to wrestle with the evil one. At all events that day is as likely as any other. I am almost tired of this vile alternation of grinding in Cambridge and out of Cambridge, but as yet there is not much else to be done. It has kept me still lamenting, as you have known me lament, over my ignorance and the small probability of its being dissipated. There is one thing however which I am doing for which if you have a drop of Celtic blood in your veins you ought to reverence me. I have begun to read Welsh and have already made some fierce attacks on Taliessin Ben Bevidd and Llywarch Hên : I now anticipate with much eagerness the time when I shall hear a genuine *ll* and *ngh* uttered by a Cymro throat.

I was either at London or at Oxford when your letter arrived here : and I have not answered it immediately, not out of any feeling of justifiable revenge for your long silence but from a sluggishness of the correspondential faculties which does not allow me to be very vindictive towards such inaction. I have, I am ashamed to say, hitherto forgotten to give your message about your parcel to Whittaker, but I will remedy that omission before I sleep.

Ever truly yours, W. WHEWELL.

TRINITY COLLEGE, *June* 24, 1818.

MY DEAR ROSE,

*   *   *   *

With respect to Butler it is long since I read him. At the time however I thought there was more close reasoning, fighting the ground inch by inch, fairly taking hold of a difficulty and squeezing it hard till its bulk dwindled to nothing, than I had often seen. You are to consider however that it is a book of negatives. Its object is not to prove, but to remove the presumptions against, natural and revealed religion. In the Chapter you mention he does not undertake to establish a future state, but merely to shew that the persuasion against it which is perhaps the first impression produced by the phenomena of dissolution, &c. is unfounded on any sufficient reasoning, and this, you seem to allow, his analogies do for him. This is I think his plan all the way through; he never ventures to the positive side of the line; content if he can expel the enemy, he does not attempt to make conquests; he clears the ground and then leaves revelation and other arguments to erect the building. He attacks *dis*believers, but has very little to say to mere *un*believers. In consequence of this his book has rather a sceptical effect; he disposes of so many arguments that have a good sound look at first, that he leaves you with an idea that reason is much more lucky in finding flaws than in mending them, and that you may disprove with much more certainty than you can prove.

*   *   *   *

CAERNARVON, *Aug.* 21, 1818.

DEAR JONES,

In the article of letters my pen is not that of a ready writer, otherwise I should have employed it sooner to inform you that I am here and to remind you of your plan of visiting Snowdon. I should not easily forgive myself if I had neglected this till you had put yourself on the Barmouth road, for that would be, to say the best of it, giving you much superfluous Welsh travelling. I did not at all succeed in my plan of getting thither. I made divers unsuccessful attempts to penetrate through S. Wales, but being repulsed both at Bristol and Worcester I was obliged to betake myself to Shrewsbury, which is, it appears, the only point at which there is any chance of entering N. Wales. Within the space of two days after I reached Barmouth I was joined by nearly the whole of the party, and it being resolved upon investigation that Barmouth was not a proper local habitation for the summer, we performed a nomadic expedition hither, where we are at present comfortably established. Among other advantages Caernarvon has that of offering much greater temptations for your expedition than Barmouth or almost any other place. The chain of hills which extends from Conway to the hook of Caernarvonshire is crossed by four or five passes all of which afford exceedingly fine scenery and almost all within half a day's journey of us. The top of Snowdon is an easy day's excursion and you may take a horse and ride within a mile or two of the top. I have been there but once. While I stood there great shapeless clouds and heavy mists came rolling up the inland side choking up the valleys and blotting out the hills; when they reached the point where I stood they paused leaving the other half of the view round me clear and sunny, so that I seemed to stand in the middle of the first day of creation, with all on one side, chaos, and all on the other, world. You get from Shrewsbury to Bangor by the coach; which is only nine miles hence—and the road I believe through some of the finest parts of Wales. I can easily provide you with a bed at my lodgings as they were calculated for a family, and you will see cromlechs and mountains and lakes in abundance; and when you get to the top of Snowdon you shall shew your contempt of the world in any way you like.

I am cramming seven pupils but living comfortably enough withal. One of them is going to leave me. One advantage of having my hands so full of employment is that it allows me to dream of undertakings metaphysical, philological, mathematical and others which I would execute if I had time. In the mean while nothing prospers but Mechanics. I have additional reasons for wishing to get that afloat as soon as possible. Monk has written to offer me the mathematical lectureship which I have accepted; so that there is hope of doing some good, though I shall be obliged to talk with some moderation of Wood[1] for the first year till there is a better book extant. Pray let me hear from you soon and know when you are coming. I want much to know how you go on in mind and body—how your lame hand is, how Kenty's business and his pamphlet proceeds, and how moral sentiments and political economy prosper. I have not done much in the way of Welsh, but I am in a fair way to be convinced that Madog ap Owen Gwynedd discovered America about three centuries before Columbus.

Dear Jones, always truly yours, W. WHEWELL.

TRINITY COLLEGE, *Nov.* 1, 1818.

DEAR HERSCHEL,

Your account of your optical experiments came most welcome to Caernarvon where it found me employed in grinding down several specimens of senior sophs, who however were much less promising than your pieces of nitre, as they will not take a very high polish nor are they likely to exhibit any brilliant phenomena. Since I came here I have been looking at the *Philosophical Transactions* with the intention of comparing it with your account, but a host of new occupations which I have not yet learnt the art of reducing into a small compass of time have not left me much leisure for such studies. I can perceive however that you are treading close on the heels of Brewster and, so far as I can make out, that Brewster has got a long way ahead of Biot in the race of discovery which has been going on for some time. I cannot help considering you as extremely fortunate in finding, just when your mathematical and physical knowledge had made you want such an opportunity of applying it, so rich a field of discoveries where it is hardly possible to observe much and accurately without finding something new. Facts about the minute properties of light are accumulating beyond measure: and if you go on it is impossible that some of the general laws which are enveloped in them should not fall to your share. Brewster's

---

[1] Dr Wood's *Elements of Mechanics.*

papers are very beautiful, but they always seem, especially the theoretical part, to want some of that mathematical distinctness which might be given them, and in consequence his theories, that of the elliptical rings for instance, are not satisfactory without more attention than I have yet been able to give them. Your accounts of what you are doing will always be very grateful, for I have an exceeding curiosity to see to what this stream of new facts is tending, and what degree of simplification we may expect in the general laws at which we shall finally arrive. There is another point of view which occurs to us lookers on, who, not making a single experiment to further the progress of science, employ ourselves with twisting the results of other people into all possible speculations mathematical, physical, and metaphysical. What are the final causes of these phenomena and their use in creation? It is true, as Bacon says, that speculations about final causes are like vestal virgins, pretty, dedicated to Heaven, and barren : yet still it is difficult to believe that the laws which produce all this multitude of phenomena should exist if it were not that some good purpose is or may be answered by them, or that they are necessarily comprehended in other laws which act a more prominent part in the system of the universe. Thus if our accuracy of observation had increased before the discovery of gravitation, so as to discover all the moon's inequalities, we might in the same way have been puzzled *why* they existed. I acknowledge however that such disquisitions are more fitted for those metaphysical prancers who 'shew all their paces, not one step advance', than for you who are riding forwards as if the devil were after you. Have you made anything of your red and green conchoids yet? There has not, I think, much been done in the determination of a curve from its different forms when its constants change, which is a problem which your images seen through films immediately suggest. I have to thank you for your Isoperim[1]. Woodhouse who, though he is muddy-headed about other matters as well as Isoperimetricals, generally does good by his books, has just published one, which was much wanted, on Phy-

[1] The article on Isoperimetrical Problems in Brewster's *Edinburgh Encyclopædia*.

sical Astronomy. From the little I have seen of it, it seems as deficient in neatness and symmetry as well may be; but that will not prevent it from being very useful. The great point is to have detached from the formidable mass of generalities of the writers on Celestial Mechanics parts of a moderate size and tractable shape. He has put a quantity of new mathematics in circulation in the university, as in his former books, and people will be found who will work it up into better forms. The subject is so interesting that if the book be readable it will be read. Even Vince had constantly people making attacks upon him though he was always found to be impregnable. Accordingly I consider this as an epoch in Cambridge mathematics. In a year or two it will become a senate-house book—especially as W is known to have no liking for the ultra-analysts—and I should not be surprised if in a short time we were only to read a few propositions of Newton as a matter of curiosity; it would however as yet be treason to breathe such an idea to most people. You see that I talk to you about these matters taking for granted that you still retain some interest for your old plan of reforming the mathematics of the university. I have it now more in my power to further this laudable object by the situation I have taken of assistant tutor (i. e. Mathematical Lecturer) here. Whatever may be the disadvantages of the office this is one of its advantages. I shall have a permanent and official interest in getting the men forwards—I shall have an opportunity of directing their reading— and I shall write books (good ones of course) and be able to put them in circulation. By using such powers wisely but discreetly much may be done. I have the first volume of a system of mechanics, containing statics and a little dynamics, ready for the press. It consists in a great measure of a classification of problems. To serve as an elementary book introductory to the higher applications of mechanics is one object, and another is to establish the science on simple and satisfactory principles, which I do not think has been done. By the by I wish you would tell me upon what principles or proof you suppose that it is assumed in French books of mechanics, that velocity communicated by pressure is as pressure directly and mass moved inversely. It is generally done very quietly without any hesitation by taking the accelerating force f, and multiplying by m the mass, which they say gives mf the force producing motion. Our moderators are Gwatkin and Peacock; in G. I have great confidence, he is cautious but reasonable, and he has been reading a good deal of good mathematics. Whittaker is employed in confuting a man of the name of Bellamy, who has been translating the scriptures out of Hebrew though certainly not into English. I believe he is very soon to put forth his red right arm. I shall be obliged to you to tell me how you found Jones, for it is hopeless to expect to hear immediately from his Reverence. When I saw him it seemed very unlikely that he should ever have the use of his right thumb, which must nip some of his political and metaphysical lucubrations in the bud—for he will not venture to dictate treason and what is worse. There are few things which will give me more pleasure than to hear from you whenever you have half an hour to spare. Your experiments I shall rejoice to see both on their account and yours. I find in the people here very little sympathy to keep alive the ardour for physical knowledge. Pray have you not begun to torture calves' feet jelly or to subject glass to the *peine forte et dure* and to the hot iron? In the case of jelly what is it, in the name of the principle of sufficient reason, which determines the neutral and depolarizing axes? Believe me, Dear Herschel, ever sincerely yours W. WHEWELL.

[CAMBRIDGE. *Feb.* 25, 1819]

MY DEAR HARE,

*　　*　　*　　*

It is no doubt a circumstance calculated to make one distrust one's feelings and opinions very much to find that character depends so much upon situation. There seems to be scarcely any alternative between seeing objects tinged with false colours by viewing them through the medium of professional habits or favourite pursuits, and seeing them perfectly colourless and uninviting by establishing yourself in a vacuum of prejudices. Your profession[1] has a greater tendency than others to efface the

---

[1] Mr Hare was at this time studying *law*.

simplicity and energy of the mind because it is generally cultivated more exclusively. The quantity of reading and attention
which it requires, and still more the quantity of trifling and common place which are necessary, operate as a very heavy window
tax upon the intellect; and in that case of course it is the sky-
lights which are first shut up. But I should conjecture the change
in * * * to be probably owing to another cause. I have
told you before, and I believe still, that your principles of poetry
are incompatible with the mental habits of men existing in a
state of society like ours, and though a man whose age has fixed
his character past change may remain a lake poet among towns
and cities, a young man on whose character the seal is yet warm
and soft will most likely have the impression obscured or oblite-
rated by being jumbled in a bag with others. Certainly the
writer of that review seems to look back with an eye of longing to
the critical days of Addison. And after all there is no doubt that
those came much nearer the method of the critics of antiquity
than anything that now goes for criticism, which you ought to
allow to be some praise. But we are not very likely to agree on
these points, and you have the advantage of believing that there
is certainty to be obtained in matters of taste. You must allow
that science is a much more satisfactory study: your knowledge
there is undeniable and its accumulation eternal and imperishable.
You know what truth is and you are sure that when you possess
it no change of feelings can prevent your holding it fast and
reaching it to future ages. It is true that besides mere Reason
there is much in the spiritual nature of man, that his reasoning
powers are but a small portion of his existence, but all the rest of
him is so mysterious and unaccountable, so capricious in its operations, appearing in general to acknowledge the authority of
reason and yet eluding her grasp when she attempts to bring it
before her judgment seat, that I can make nothing of it and could
almost find in my heart to forswear speculating about it, and to be
content to feel and love the beautiful, or what seems so to me,
without knowing why or caring wherefore—Adieu. This is nonsense, but fortunately it is not the first you have heard me talk.

Yours always, w. w.

[*September,* 1819.]
[BRIGHTON] *New Ship Inn,* 10 *o'clock.*

DEAR JONES,

The Nancy is safe at the bottom of the Channel along with every ounce of our luggage. We are safe here and quite unincumbered with superfluities. If you can lend each of us a change of linen (stockings, shirt and cravat) and Sheepshanks a black coat besides, you will be rewarded in this world or the next for your charity to such poor shipwrecked souls.

Yours truly, W. WHEWELL.

TRIN. COLL., *Sept.* 26, 1819.

MY DEAR HARE,

I believe Sheepshanks wrote you word from Brighton that we were just going to France. So I thought at that time as well as he; and nevertheless at present I am not spending my mornings at the Louvre and my evenings at the Théâtre Français, but writing mechanics and indulging occasionally in the Barnwell theatricals. The explanation of this is simply that we have been shipwrecked. We sailed about 9 in the evening that S. wrote to you, and before we had been out two hours by great mismanagement ran foul of another vessel, so as almost to stave in our own. We instantly got out of it, and by still greater absurdity allowed it to go down without saving a single article of our baggage. We were landed at Brighton the next morning by the vessel which had thus disposed of our unfortunate packet; and making what refitments we could, with the assistance of Jones who is resident there, we next day wended our way dolorous and discontented towards our English homes, untravelled and unportmanteaued. Our persons were never I believe in any danger and I was rather mortified to be the victim of a wreck so tame as ours was. Except that after we had left our packet the two vessels worked and rolled together very furiously, and at last the mast of the packet came thundering down with a great crash, there was nothing picturesque in the affair. We lost every thing even to our hats and Sheepshanks's smart frock, and were of course quite disabled from proceeding.   There may perhaps be some things which Undine's relatives may never before have had an opportunity of seeing; for instance an air-jacket of S.'s which could not even save itself, and some sheets of my mechanics which were to have been finished at Paris.   However there is good in everything; I believe I should have done better to have taken your advice to stay here and finish my book, and now I have no alternative.   I should have made great confusion by spending this month out of England, but while the voyage was in prospect I resolutely shut my eyes upon the future.   I am now likely to be out by the end of October.   If I believed in omens as much as you do, I should suppose that Fortune had taken upon herself the care of my offspring; for besides sending me here to watch over its birth in spite of myself, she is going to make me moderator, which will do something towards ensuring it a favourable reception.

Still I cannot help being mortified that when, after talking so long of going abroad, I had condensed my intentions into an act of volition, my project should be dispersed by a puff of wind, and I left without any addition to my threadbare wardrobe of ideas, and with a sensible diminution in my more valuable wardrobe of bodily vestments.   I believe Sheepshanks suffered more in the latter respect than I did; as for his suit of ideas it is a sort of buff jerkin that is a most excellent robe of durance, and he may wear it another season without any inconvenience.

*   *   *   *

I was very much obliged to you for the sight of the books you sent me.   They were all works which I had much wished to see; nevertheless I did not find time to read any but Schlegel's pamphlets on Provençal literature before our disastrous journey. I brought them to town with me to leave them for you, but hoping to find time to read the other two I have brought them back hither.   We are all here busy or eager about the fellowships. The examination has been moderate.   I should think the probability is now for Malkin, Lefevre, Ellis, Stainforth, Gambier. Adieu.   I hope your friends the Germans will speculate upon politics and morals to some purpose.   At all events they have the consolation of having a right side and a wrong, which is more than we can boast who have only to choose between an illegal and an irrational part.

Ever affectionately yours W. WHEWELL.

TRINITY COLLEGE, *Oct.* 3*rd*, 1819.

DEAR MORLAND,

If I pretended to any thing like regularity of correspondence I should generally have to begin a letter with an apology for the violation of it; and very often, as at present, my repentance itself would need to be repented of, for my apology comes so late that it requires to be apologised for itself. Nevertheless it might have happened that I should have reserved all the valuable communications which I have to make to you till I should see you at Christmas, if it had not been for one circumstance; which is that I shall in all probability not see you at Christmas. I have just accepted the office of one of the moderators, which involves me in employments that will make it impossible for me to leave Cambridge for any length of time till Easter; and as that is a long time to hear nothing of you I have resolved without further delay to ask you how you do. I should have been glad to do this in person, but I shall, besides my college employments, be engaged in the duties of presiding in the disputations in the Schools and examining the candidates in the Senate House. Perhaps you may recollect in Cumberland's life and I think also in Kirke White's, an account of the disputations which I have mentioned. They are one of the most marked relics in their form of the ancient discipline of the University. They are held between undergraduates in pulpits on opposite sides of the room, in Latin and in a syllogistic form. As we are no longer here in the way either of talking Latin habitually or of reading logic, neither the one nor the other is very scientifically exhibited. The syllogisms are such as would make Aristotle stare, and the Latin would make every classical hair in your head stand on end. Still it is an exercise well adapted to try the clearness and soundness of the mathematical ideas of the men, though they are of course embarrassed by talking in an

3—2

unknown tongue. You would get very exaggerated ideas of the importance attached to it if you were to trust Cumberland; I believe it was formerly more thought of than it is now. It does not, at least immediately, produce any effect on a man's place in the tripos, and is therefore considerably less attended to than used to be the case, and in most years is not very interesting after the five or six best men : so that I look for a considerable exercise of, or rather demand for, patience on my part. The other part of my duty in the Senate House consists in manufacturing wranglers, senior optimes, &c. and is, while it lasts, very laborious.

I might perhaps have surprised you with a letter from Paris, but my first essay in travelling was peculiarly unsuccessful. Perhaps you may have heard one way or other that the Nancy packet of Brighton sunk on its passage, and that I took the prudent measure of getting out of it before that took place. Never was shipwreck more expeditiously or simply transacted than ours. I was sleeping quietly on the deck when I was roused by a furious crash caused by our running plump against another ship. In an instant I found myself getting into the other vessel, helping the sailors to pull ropes, talking bad French to console some women who were exceedingly frightened, witnessing the mast of the packet tumble down, and in a short time sailing away and leaving everything I had had with me at the bottom of the Channel. We were landed at Brighton next morning, where fortunately a clergyman, who was a Cambridge acquaintance of mine, was residing, who helped a little to refit me and my companion, and as soon as I could I returned to Cambridge carrying my all with me, disburthened of many superfluities and some necessaries, and doomed to form plans of peregrinations for another year.

I am much edified by your loyal manifesto at Lancaster which is so harmless that I do not see why it should not be signed every fortnight. I am only afraid that you have lost the chance of being thanked for it by Lord Sidmouth by not making it strong enough, for it does not even go to the length of saying that if there be a meeting in the Market Place at Lancaster you will treat the people as the magistrates of Manchester did their townsmen. There are several names which I am surprised not to see there—why does not the vicar sign it? Is it that his loyalty is of so superior a cast that it is not right to shew it in the same way with other people? And why is your name wanting? Is it possible, that you do not see the necessity of cutting down people who though they are standing very quietly may very possibly at some time, or at least whose sons and grandsons may sometime, do some mischief to somebody; and who have probably got sabres in their walking sticks and great guns in their pockets? I hope I am not talking treason, but that is a point which in these times nobody can be sure of; the air is so tainted with it that a man breathes it before he perceives it. If it be so, I beg your pardon and hope you will burn this letter before you read it that it may do no harm.

Dear Morland, yours very truly, W. WHEWELL.

July 1, 1820.

My DEAR HARE,

*   *   *   *

The small quantity of festivity which our commencement generally exhibits will be subject to additional deduction in consequence of the death of our Master of which you have most probably heard. His interment in our Chapel takes place on Monday. Perhaps too you have heard that Kaye is his successor in his Bishopric. It is said that when Kaye received the intelligence this morning, without any previous expectation of its nature, he handed the letter to a friend who was sitting with him, observing that "he had always thought the best plan for advancement to be to stay quietly at home and do your duty." I am glad of the choice, for I do not know of any body upon whom it could have fallen more creditably to us and the givers of it. The disposal of the Mastership is of course a matter about which much is speculated and little known. The general opinion gives it to Wordsworth. If this turn out so, he shall invite his brother here and you shall come and meet him, and we will be the most poetical and psychological college in the universe, though certainly some of us are bad materials for such an edifice.

I am extremely glad to receive you in print. I certainly think you might be more worthily employed than in translation, but life has room for one such work, and especially when it is to diffuse valuable ideas and feelings, as I know you intend this to do (though I dare say you will not like my phrases), besides being very delightful to read, which from your comparison of this with Undine I expect to find it. I only hope that you will not become one of a body of septuagint translators of German novels for the conversion of the heathen. Your preface I perceive abuses unfortunate people who are puzzled with the connexion between the mind and the soul as I used to be ; and who try by anatomizing to discover the way in which the flesh and muscles of the moral man act upon the "wordy skeleton" of reason. One thing occurred to me which from what you say I think I may venture to tell you. You have succeeded very well in naturalizing Sintram, if you have made the work look as unlike a translation from the German, which it is, as you have made the preface look like a translation from the same language, which I suppose it is not.

*   *   *   *

Yours affectionately, W. WHEWELL.

TRIN. COLL., *July* 16, 1820.

DEAR JONES,

I was rather surprized to find that you were at Paris, having expected to see you *in transitu* provided our road had been through Brighton. Sheepshanks and I are preparing to take wing, but it will be nearly a fortnight before we reach Paris; and having almost given up our original and more philosophical plan of staying a good part of the summer in that capital, and of making ourselves thoroughly acquainted with its state, moral, political, literary and culinary, we have almost determined upon the commonplace John Bull proceeding of getting over as much ground as possible in a given time. We have accordingly been speculating upon the possibility of staying a very short time in Paris, passing through the west of Switzerland, thrusting our heads into Italy, perhaps to Milan, perhaps even to Venice, returning through the eastern parts of

Switzerland, and perhaps making a second and rather longer stay at Paris on our return. We are not quite resolved upon the outline of our tour and still less on the detail, and the temptation of seeing and speculating with you on the whims of the Parisians would be almost sufficient inducement to persuade us to add a week to our first sojourn there, if we cannot persuade you to go with us and luxuriate among the scenery of the Alps and the marble walls of Venice. But I do not think we can take up our habitation in your suite of apartments according to your proposal, which would from your account be very desirable if it were possible. I shall however hope to find you in Paris about the 28th, and shall be well pleased to follow your guidance if you have nothing better to do than give it us. It is with great regret that I think of abandoning the idea of a month's residence in Paris and only with the reserved hope of indulging in something of the kind at some future period. I am very glad that it is a place to put one in good humour with England, for God knows there is need enough of such a place. However when your children, the French, are grown to years of discretion and we are sunk in the decrepitude of age, they will have the advantage of us.

I am very sorry that you have got your mouth out of order. Is it a consequence of too violent and persevering efforts to pronounce the _u_ properly? I hope nothing more serious, for to be confined to your room in Paris must be provoking enough. If there be no good reason to the contrary, your being at the Hôtel de Boston is a very good reason for our taking up our abode there while we stay. We shall, if our present intentions hold, take the Dover passage, as we have no time to admire the Rouen road.

I wish we could persuade you to take our more extensive ramble. You would be the better for it all your life afterwards, and you may take the length and breadth and height of the French constitution at some future time when the edifice has settled into a more permanent form. I trust we shall be able to come to some plan of compromise when we see you. Always most truly yours,

W. WHEWELL.

LAUSANNE, *Sept.* 1st, 1820.

DEAR MORLAND,

I have been resolving ever since I left England to take a shot at you from some part or other of my path, by way of proving to you what a trifling obstacle distance is in the way of correspondence, if idleness and bad habits were not in operation. A rainy day here, where the only things to be seen are the outdoor views, has left me without any excuse for not executing my resolution. What kind of communication you are likely to have from such inspiration is your concern—perhaps you will be very much disposed to wish, from selfish as well as benevolent feeling, that the weather had continued fine. If I were disposed to give you any account of the more recent part of my travels, I do not know exactly how I should set about it. Scenery, as has been observed a hundred times, and found by experiment still oftener, is not easily made to live in description and look green in black and white. The country about Geneva and along the banks of the lake to this place is almost as beautiful as any thing can be. A lake much larger than Windermere, and of a very fine shape, forms the floor in the centre. Then the walls that rise from this are hills as large and picturesque as those about Windermere on one side, and on the other slopes clothed with vines and country houses in the best English taste, with neat towns here and there. But this is not all. Beyond the first range of mountains which I have mentioned you see the higher summits of the Alps clothed in their perpetual snow, looking with a cold and wintry eye from an immense distance upon the laughing prospect about you, or sometimes catching the warm rays of the setting sun when he is so low that they pass above every thing near you. Add to this that the lake is as superior to others in its material as in its form and accompaniments—the deep blue colour and bright transparency of its water are so striking as at first to appear almost unnatural. I am afraid this will not give you any idea of the luxuriant majesty of the prospects here, but at all events it may help to shew you how very easily in attempting any thing of the kind one slides into nonsense à la Mrs. *Ratcliffe*.

I dare say that you would prefer it if I could give you some account of men and manners, as the phrase is; having like Ulysses visited the cities of many men, you may fancy I ought to know their minds. But of course a progress so rapid as ours does not admit of much speculation of that kind, or rather admits of nothing but speculation. Without letters of recommendation, time, and a complete knowledge of the language, you can get but a little way into the national character and manners. It is no doubt much easier to go to a certain extent in any country than it would be for a foreigner in England. They enter into conversation much more easily and fluently, at least in France, though even there not at the rate I had expected. But the greatest advantage that way is their *tables d'hôte*. In all the great inns (except in Paris) almost all the people who are at the inn dine and generally sup together. To be served *en particulier*, as they call it, is rather unusual. Hence you get much better dinners, much more company, and if you set properly to work much more information than you could any other way. Ladies who happen to be at the hotel come to table along with the rest, so that it is a good deal different from an English ordinary. In Paris it is nearly as it is in London. People go to the Restaurateurs by ones, by twos and by threes, and eat what they like; and for the same money you may certainly get a much better dinner in their Metropolis than ours, including their wines, some of which are exquisite and would make an impressien even upon you who do not like claret. My companion, another fellow of our college, will not allow that their meat is so good as ours to begin with, but in general the cook so completely gets dominion over the original flavour that that is, comparatively speaking, a matter of small importance. You will perhaps think it very scandalous that I should talk so much about eating and drinking, but as it is a matter which every day requires our attention and about which a good deal of the little money and little French we brought with us is expended, you need not wonder at its making an impression. In other matters you are obliged to take some pains to get information, but there it comes of itself. I was three Sundays at Paris and wished to hear a French sermon or two, but the people of whom I made my enquiries were so stupid or ignorant, and I was so unfortunate, that I have only heard two fragments of which I could only understand a very little, though I could see that one of the preachers was much more animated and familiar than any I ever saw in England. But the Church in France is at present ill provided for and little considered : the revolution sucked up all its revenues and it has only been restored on a very economical scale. Neither the feeling nor the wealth of the country would allow of more. A letter from home in a foreign land is always a comfort; if you write to Berne by the middle of this month your letter will be there before me. I am going into the heart of Switzerland to return thither. Yours most truly,

W. W.

August 4, 1821.

DEAR JONES,

It is very provoking that the only time when you are in the humour and in possession of time to pay Cambridge a visit should be when I am intending to leave it immediately. I had fixed upon Monday to begin a migration northwards : but if your visit were more certain and longer I should be tempted to wait for it here. I have been in college about a fortnight, principally for the sake of writing a little of that bad metaphysics called mechanics; and having perpetrated enough of that for the present, I begin to be impatient of wasting fine summer weather here, knowing by last year's experience that it is not at all wise to put off a journey among mountains to the end of the autumn. The mountains which I have a design upon at present are those of Cumberland and Westmorland. I have long intended to make the regular Cockney tour of the lakes. I suppose they are very pretty, and coming almost from the borders of them I am ashamed of never having seen but one of the number, though every day of my life when I am at Lancaster I see the heads of the mountains in whose laps they lie. I dare say they are not so magnificent as the valleys and lakes of Switzerland but I expect they are of a character which will not set you upon making comparisons, so it is no matter. I shall travel, so far as I know at present, alone and with divers sources of enjoyment or at least employment. You have no idea of the variety of different uses to which I shall turn a mountain. After perhaps sketching it from the bottom I shall climb to the top and measure its height by the barometer, knock off a piece of rock with a geological hammer to see what it is made of, and then evolve some quotation from Wordsworth into the still air above it. He has got some passages where he has tumbled the names of those hills together till his verses sound like the roaring of the sea or like a conjuration which would call the spirits of them from their dens. I wish you could come with me but I suppose that is impossible. I believe you have good taste enough to like for a time the company of hills and rivers and such respectable characters better than that of the foolish and knavish and miserable people whom you must see if you travel to extend your knowledge of human nature and so forth. Not but that we will do the latter too sometime or other. But to end where I began. The consequence of this plan is that except you can come down here by the Fly on Saturday I shall, I fear, have little chance of seeing you at present. I mention *the Fly* because on Saturday I shall be returning to Cambridge in it the last stage, from a visit on that road: but there are now coaches hither at almost all times of the day. The Fly in question sets off I think from the George and Blue Boar at 10 in the morning which is a practicable hour.

I was obliged to resist Rose's blandishments to entice me to Brighton, because we returned by Oxford where I staid a few days with much satisfaction. They are more violently geological there than we are here and in fact all but the geological people were out of the University. I am really glad that at last one science, and that a most important and interesting one and (whatever you may think) with indisputable claims to be called a science, has not only been vigorously cultivated in the Universities before it had been prosecuted out of them for 100 years, which is in itself a new fact, but has owed and is likely to owe its substance and value in a great measure to academical professors. Laugh at me, if you choose, for my attachment to this our Diana of the Ephesians, the University, but the novelty of the circumstance makes it really curious. Perhaps you can tell me *why* they are so slow in abandoning what is obsolete every where else and adopting what is established every where else, but the fact is so or nearly so. Like very obstinate artillery-men they stand by their guns even after they are spiked. However we hope to alter a good deal of this in time.

I never had an opportunity of telling you how much I was pleased with the book of your Romeo and Juliet friend. I read it all the way from Brighton to London and was in perfect delight with his account of the origin of language—of course you will understand because it agreed with my own—and it is moreover exceedingly clear and well developed. I wonder if he did it himself. His account of accent I am not so well satisfied with, and about English versification he seems to be all wrong.

I am glad you got the ale and liked it. The brawn cost 1. 10. 7½. One way or other I will hope to see you by next Christmas. Cannot you manage to come here sometime next winter ? Somebody told me you had finally made your Welsh bargain. I wonder if it be true. If I were writing to any body else in similar circumstances I should say let me know if it be so, but I am afraid there is no conjuring a letter from the vasty deep of your inactivity. They will not come when we call them. Remember me kindly to Rose and likewise to Townsend.

Always affectionately yours W. WHEWELL.

N.B. There is nobody in the University. Peacock is in Paris, and every body else in divers places.

TRIN. COLL., *May* 12, 1822.

MY DEAR HARE,

I have just parted with Blackstone who goes away to-morrow morning. He has been here since yesterday week and has been, I believe, seeing sights with a most laudable energy and perseverance. What I have seen of him has made me regret that I could not give him more of my time, and he has been so much liked by those of our Cambridge men in whose company we have been, that I hope he has carried away a favourable impression of us on his side. I rode over with him to Wimpole, but as the family were at home we were not allowed to see the inside of the house, so that we had nothing but the ride and the sight of the country for our pains, which however at this time of the year are well worth the while.

I am going to trouble you with two commissions which I hope will not be very inconvenient, though they will both require some despatch. The first is to ascertain when a person must be in London to keep the next term (sometime in June) at the Middle Temple and also to obtain the same information with respect to Lincoln's Inn. I have to send notice of this to a person at Paris whose motions are to be regulated by it, so that it would be desirable to have it by return of post if possible. My other commission is a selfish one. I have never heard Catalani and should be grieved if she quitted the stage while that continued to be the case. I would willingly make the exertion of going to London one day and returning the next, if there were a fair chance of listening to her. Now I do not know what faith to put in her pretence of not issuing tickets beyond the number her room will hold, but if I am not precluded by that I should be greatly obliged to you to procure me a ticket for Wednesday next. If such a thing be to be had I will run the chance of finding a place in the Union Coach to London by 8 o'clock and in the Argyll rooms afterwards. If you find anything hopeful in this plan, have the goodness to send my ticket to the care of mine host of the White Horse, Fetter Lane, who I suppose may be depended on for a charge so weighty. I am afraid I shall hardly see you, as I must return by the earliest coach I can find the next day.

I am sorry you do not like the law and the prospects which it offers, especially if you resolve to take to nothing else. There is one way in which I think you might do something *tüchtig* and which I hope you will soon have an opportunity of choosing.

I mean infusing good principles of taste and scholarship and, if you like to call it so, of philosophy, into the rising generations of academic youth. I do not venture to say more, but I hope that you will receive shortly an explanation of what I mean and that you will decide as I could wish. I should be very sorry if you voluntarily continued under that hard stepmother, the law, when you might take refuge so honourably and usefully with your *alma mater* who likes you so much better. At any rate I hope we shall meet in Cambridge during the vacation. I think all my plans of travelling are blown up or will give place to my other plans of working here, so that I shall be here during some part of the summer in all probability.

I have heard from Sheepshanks who says the *Miserere* &c. are not so good as an English oratorio and that the only good things in the holy week are the illumination of the dome and the fireworks at S. Angelo. He is too *poco curante* a person to be implicitly followed on those matters, so if you choose we will suppose him wrong, though I have received the same account from another of our fellows who has also been at Rome this spring. On the whole he seems to be as well pleased as one should expect him to confess. He is going to leave his sisters at Rome and to proceed to Naples himself.

I must finish in haste and for want of time leave poor Locke at your mercy, who however little deserves your abuse because he has done only what he and all other reasonable metaphysicians pretend only to do, namely give an account of the operations of his own mind and analyse them to their elements.

*   *   *   *

Ever yours truly W. WHEWELL.

We are just going to bring forward another plan for the improvement of the examinations, which I have great hopes will pass.

July 17, 1822.

MY DEAR HARE,

You have determined as well as if you had previously drunk a gallon of ale or whatever other liquor it was with which the ancient Germans used to ballast themselves when they had to balance weighty matters in their minds. In modern times we go upon the same principle except that we eat instead of drink—but in both cases I suppose the object is similar. The stomach and the head are like the inside and outside of a coach, and if you give all the loading to the latter you are in some danger of an overturn. But by whatever process you arrived at your decision I am heartily glad of it. You would have had the question proposed to you some time ago, but that Brown is very dilatory and the Master very busy. However it is all very well and you have time plenty to look forwards to the commencement of your illuminatory course in October. I augur all great and good things from our combined operations and I think you will find that, whatever there may be of wearisome and disagreeable in your office there is a great deal that is animating and gratifying. To be sure, we will be very rational and persevering and regular but at the same time we will have an ambition to be something more. I am here and shall remain for the present. So come down and we will concert measures for doing the greatest possible quantity of good. I suppose you have heard or will hear from the Master certain stipulations which he chooses to make with every body concerned in the tuition. They are nothing of any consequence except so far as they indicate his disposition to introduce a somewhat stricter and what he considers, a more paternal superintendence than has prevailed of late. I must not urge you to come without telling you that we have had some alarm here about a return of our fever. Some Johnians have died, though none of them in Cambridge. At present it has disappeared entirely—at least the only two invalids were recovering fast some days ago. Sheepshanks's letters find him at the Temple. I wish you could find him and bring him here. There are several of us in residence, Peacock, Thorp, Higman's new lecturer, George Waddington &c. You may of course find rooms here now at half a day's warning. I do not know if you will be able to secure good ones next term. We eat venison and turtle on Friday and dine at four o'clock—if any additional argument be requisite to bring you here I am quite at a loss.

Ever yours truly W. WHEWELL.

DEAR JONES,

I have sent you the Encycl. along with the two pamphlets
which I mentioned and which I suspect to be worth nothing.
The library books must be returned about October to avoid fines;
but I will let you know the time more accurately when it
approaches. Now touching *metaphysics,* I am almost surprised at
the fierce burning of your indignation against the poor word. It
is no doubt true that people apply it to the speculations of others
when they want them to appear abstruse and unsubstantial, and
to their own when they would have them seem profound and
philosophical, but if there were no such word as the one in question they would find some other to answer those laudable purposes,
and I do not think that 'metaphysical' in itself carries much
weight with it. As to its genealogy it is certainly, I believe, to be
traced up to Aristotle, or rather his transcribers. His φυσικά are
much the same as our physics as to subject, and the μετὰ τὰ
φυσικά were so called either because they were put *after* the
physics in the books, or were *beyond* them in the order of speculation. Your proper metaphysician is very angry if you maintain
the former hypothesis. But the word itself was unknown to
Aristotle, and in the books which are headed by it he appears
somewhat at a loss for a name for his science. He generally calls
it "The science of the causes and principles of things", which you
will allow to be a good science. Of the book itself I can tell you
nothing except that it is a great deal about substance and accident
and category and the opinions of philosophers on the universe,
and all the abominations which arise from erecting abstract terms
into principles. Harris's *Philosophical Arrangements* is said to be
a good deal taken from it. And this is the origin of those fatal
four syllables which have come down even to our times, spreading
darkness and offence on their course, like a column of smoke from
the pot of some magician. It is perhaps less easy to determine,
in all the various and vague senses which float about people's
minds when they use it, what is the distinct and precise meaning
to which they may most naturally be considered as approxima-

tions. This is all that you can do in such a case, like determining the brightest point in a nebula. Sometimes you can separate your misty spot into two or three more distinct ones. But I think certainly if you will not allow 'metaphysical' to have its application to the philosophy of mind you will find no *precise* use for it. In general it seems to mean either that which depends upon the examination of our intellectual powers and properties, or that kind of reasoning, depending upon the legerdemain of abstract terms, which the origin of the term would attach to it. In the first sense I do not know any one word to substitute for it, and paraphrases are bad things.   *   *   *   *

By the *metaphysics of mathematics* I mean the examination of the laws and powers of the mind on which their evidence depends, the analysis of their principles into the most simple form and, if you choose, the history of their development. It is not easy to stick to the distinction between this and the *logic* of the science; but the latter examines the accuracy of your mode of deducing conclusions from your principles, and the former your way of getting your principles. The *metaphysics of language* is equally intelligible, for the mutual influence of mental operations and signs is to be called metaphysical, if the word is to have any reputable signification. But what your wiseacres mean by the Metaph. of Polit. Econ. I cannot tell; for there are no peculiar principles of observation or deduction employed in that science— they may as well talk of the metaphysics of chemistry. The thing which I suppose leads them into error is that some abstract terms are necessarily introduced in your science, and some even referring to the moral and intellectual qualities of man, and that blockheads have thought themselves usefully employed in increasing the number. I have no objection to your calling those people metaphysical in the scurrilous meaning of the word who, like Say and Hauterive, 'confound the language of the nation with longtailed words in *osity* and *ation*'. But never mind if other people call you so—being well assured that you and Malthus belong not to the *metaphysical* but to the *ethical* school of Political Economy. I have no time to write more. I am printing at a great rate, but my book is growing monstrously too large. If you would come here we could discuss these matters with great effect. Sheepshanks is here, and Peacock going out but returns shortly.   Adieu.

Ever yours W. WHEWELL.

Oct. 17, 1822.

DEAR HERSCHEL,

I am afraid you are by this time tired of hearing of Mount Rosa and discussing the question as to what part of her body you crawled upon, but I believe I promised you the outline which I made from the opposite side of the Valais of her and her neighbours, and I have accordingly sent it on the other side of the sheet[1]. The view was drawn at the head of the pass of the Gemmi, and principally for the purpose of giving my memory a stronger hold upon one of the most splendid scenes I ever saw. The whole range from Mt. Rosa to Mt. Blanc was as bright as snow and sunshine could make it, except the little streak of cloud, which I have indicated towards the middle. The outline, so far as I recollect, is tolerably correct as to the general features, and the one which I have called Mt. Rosa is certainly what the guides told me was Mittaghorn so far as I understood their German. You see it appears a good deal insulated and of a respectable comparative bulk. Mt. Blanc cannot be seen without going higher than the road. We climbed about half-an-hour to the east of the top of the pass to get to the point of view. Of course you will not attach any importance to the opinions of the people about the place, but I learnt from some persons who were there this spring and willing to make the ascent of Mt. Rosa, that there men were quite persuaded you could not have been to the top in consequence of the shortness of the time which you employed. They maintain that no less a time is requisite than Mt. Blanc requires. I suppose they think their dignity injured by your doing it in a less period.

You have of course heard from Peacock of Airy, a pupil of

---

[1] The letter contained a pen and ink sketch of a portion of the Alps, including Mt. Rosa, the Matterhorn, and Mt. Blanc.

his, and certainly a man of very extraordinary talents. The reports about Babbage's machine have, it seems, excited him to attempt something of the same kind. He and another man have made a machine to solve cubic equations, but besides this he has, so far as I can make out, invented something a good deal in the way of Babbage's contrivance. He is not here at present, but his friend tells me that it has got toothed wheels, working in one way for the differences and in another for the digits. If it be a similar invention to that, it is probably an independent one; for I do not know any way by which he has got any lights about Babbage's affair. I will enquire into it as soon as he returns.

I cannot conclude without wishing to offer, as far as I can without exciting too many painful reflections, my condolence on your recent loss. Few persons have got so many sources of consolation; and it is much to know that he saw as much as he could hope to see of your progress with pride and pleasure, and that he could anticipate the illustrious name he delivered to you becoming more illustrious in your hands. These are feelings which must soothe at least the complaints of natural affection. Excuse me for having said so much, and believe me,

Dear Herschel, your sincere and faithful friend, W. WHEWELL.

[*The next two letters are to* Dr MONK, *Dean of Peterborough, after-wards Bishop of Gloucester.*]

June 2, 1823.

MY DEAR MR DEAN,

I am on the point of leaving Cambridge for a considerable portion of the vacation, and before I quit the venerable walls of Trinity for a time, I venture to trouble you with a few lines. I should have thanked you sooner for my pleasant visit to Peterborough if I had not hoped that you would take for granted my sense of your kindness, and if Peacock had not promised to tell you that it is still a favourite subject of recollection. Since that time I believe I have to claim your congratulations on my admission into the Tuition. By an arrangement which has lately been made, Mr Brown and I are to conduct the affairs of one *side* jointly for the ensuing year, after which he retires entirely from the office. I shall do my best to keep up its former glories.

I am sincerely sorry that we are no longer to have your assistance in the capacity of Greek professor. The notice of your resignation was brought to the Master to-day while I was with him. We all naturally feel much interest as to who is to be your successor in an office which it is so difficult to fill. Dobree and Evans refuse to be candidates[1]. Ward is talked of but his intentions are not known. Scholefield will, I believe, certainly be a candidate, and as the Johnians will of course have one, it is said Hastings Robinson is to be theirs. It appears to me that Hare is a person much more likely to make a great scholar and to fill the office worthily than any of the others, and I have done what I could to persuade him to offer himself. He is fully sensible of the disadvantages under which he lies in making his claims on the University, and I am afraid his merits are hardly sufficiently known as yet. If you agree with me, which I should hope might be, from the opinion which I have heard you express of him, I have no doubt that you might produce a great effect on the minds of some of the electors by stating your opinion. The Bishop of Bristol, I am persuaded, would be much influenced by it. I do not know whether it is necessary to say that I am convinced, as I believe every body is who has known Hare lately, that he would be as valuable for his temperate and sober judgments in practice, as for his knowledge of Greek, which, if I could presume to judge, I should suppose to be uncommonly extensive and accurate. And I think there is no person who is so likely to feel the responsibility of the situation and to exert himself vigorously and perseveringly to cultivate the knowledge to which it refers. High as your opinion may be of his qualifications, I have no doubt that if he should succeed he would more than justify it. If I am fortunate enough to agree with you I shall think myself happy, and certainly to Hare it may be of great consequence, as the nature of his career through the University has hardly been such as to give him anything to refer to as a proof of his capability.

---

[1] Mr Dobree was elected.

I shall leave England in less than a week. My immediate destination is for the abbeys and cathedrals of Normandy, after which my route is rather uncertain. In September I take my state as Tutor. Pray present my best respects to Mrs Monk, and believe me

Ever most truly yours W. WHEWELL.

The subscription for the building goes on very well. The sum at present is about £4200, including the £2000 from the College.

Oct. 17, 1824.

MY DEAR MR DEAN,

I should have addressed a letter to you before this time, if it had not been that I hoped to acquire for you more definite information on the subject of some of your commissions than I have hitherto been able to obtain. In consequence of King's absence from the University during the vacation, I had no means of acquainting myself with the state of his library. I have now learnt that he does not possess either Ernesti's Cicero or the Oratores Græci. If you will give me your instructions with respect to the matter which you formerly mentioned I shall be most happy to execute them. I have not been able yet to extract from our Junior Bursar (Macfarlane) the rent of your last rooms: but as the Master will, I believe, inform him that an order of the seniority was made such as you mention, he will perhaps allow himself to be convinced before long. The alterations in progress and in project here are so numerous that I hardly know how to give you an account of them. Our new court goes on prosperously towards its completion. We expect to be able to inhabit three staircases of it in a week or two, and we have had as much difficulty in electing bed-makers as we had in electing the four new Fellows. With the latter election I hope you are satisfied, as, I think, they have taken the four best men. The new buildings at Corpus are going on with equal splendour, and still more at King's. But what is doing bears no proportion to what is hoped to be done, as you probably have heard. The alterations contemplated, of course with various degrees of confidence, extend almost from Peterhouse

to St John's.   The proposed site of the future "Pitt Press" is the
ground opposite St Botolph's Church, where the University have a
good deal of land, and where, I learn, as member of a Syndicate
for the purpose, we have an opportunity at present of buying
more.   The houses which narrow the street on each side of
Catharine Hall, will, I hope, be removed before we have done, and
the opposite side of the street to Rutlidge's corner is to come down
in the course of next year.   When the old houses at King's are
demolished, this will lead us to St Mary's by a fine open street, and
then, *en passant*, we will reform the portico and knock off the balls
from the steeple.   Then we come to the great debateable ground
of Caius.   I hardly know what to expect as the end of Bankes's
project about that.   When people return to the University and
meet at Syndicates, we shall see how they have made up their
minds upon the subject.   There will be great opposition to any
plan of raising the requisite sum, and great difficulty in finding
an eligible site for the new college, but perhaps both these
obstacles may be got over.   If it be placed at the end of Trum-
pington Street, opposite Addenbrooke's (which by the way is also
much beautified), it will extend still farther our line of improved
building.   Besides all these plans we have to remove the site of
the Botanic Garden; and if, as some propose, we make the present
ground into a large square with a good market in the centre, we
may turn it to a very profitable speculation, and may perhaps
finally convert that part of Cambridge into the commercial and
shop-keeping quarter of the town; and by thus diminishing the
value of the houses near St Mary's, enable ourselves to get some
additional openings in the Academical quarter.   You will think,
my dear Mr Dean, that all this is very visionary, and so perhaps it
is; but even this is moderate compared with a plan which I heard,
at a Syndicate, proposed and approved by some of the gravest
heads of our Colleges; which was no less than to turn the course
of the Cam so that it shall follow the present carriage-road behind
the Colleges; while that road is to be relegated still farther off.

After cutting away, however, all that is inexpedient or impracticable in our plans, there is much that will certainly take
place.   The drainage, about which I think you took some interest,

has answered completely as far as it has been carried. It would be much more serviceable if it were not that the Master and Bursar of St John's steadily resist all attempts to drain Garlick-fair-lane, which, you may perhaps recollect, is a perfect 'cave of the smells.'

Perhaps most of what I have told you is not news, but it is too early in the term to have much either of University or College novelty to communicate. Our admissions on all the sides are as large as the Master allowed them to be, and I think our arrivals are likely to be very numerous. Lord Fordwick, son of Earl Cowper, comes here on Tuesday as a pupil of mine. We begin lectures on Wednesday in the 'Seven against Thebes.'

I do not know that we are likely to have much change in college shortly. Macfarlane does not leave us till the end of this year. The senior Bursar, Judgson, is, I am sorry to say, in a very infirm state of health. It is doubtful whether he will be replaced by Clark or Musgrave; the latter is favoured by the Master, and the former by a majority of the Seniors.

Pray present my best respects to Mrs Monk. I should be happy to be recalled to the remembrance of Miss Monk, of the continuance of whose good opinion I am very ambitious, but I am afraid that is not to be expected. I understand that Miss Hughes intends to pay her respects to Miss Monk. I hope there will be no jealousy between the two lady-cousins, so that the same Deanery may contain them. I should also be very glad if you would give my remembrances to Dr Bentley, and inform him that a great number of his friends here are very impatient for his appearance among them. I hope, Mr Dean, you can answer for its not being long before he gratifies them. As I know that all your friends here recollect you with great kindness, I will venture to present their remembrances, and am, Dear Mr Dean,

Always very truly yours W. WHEWELL.

I think you requested me to ascertain if your subscription to the University had been paid. I cannot learn that it has been received either at Mortlock's or Deighton's. I shall be happy to make any other enquiry which you will suggest on the subject.

June 25, 1825.

MY DEAR JONES,

When I got to London after leaving you, I found letters
informing me that Henslow had been appointed Botanical Professor
by the King[1], and that consequently the Mineralogical Chair would
be vacant next term (in October). I came here immediately and
began to canvass for the office, and so far I have met with all
possible encouragement and with no rival. I conceive therefore
that I have not much chance of failing to be elected, except what
may arise from the usurpations of the Heads as to the form of
election. I have been to several of them to request expressly that
they will not elect me *their way*, and still hardly know what to
expect. One consequence of this plan is a change of purpose
respecting my travels. My present intention is to go to Freiberg
and Berlin which seem to be the best Mineralogical schools in
Germany and especially given to crystallography. I have not yet
determined when I shall migrate, but perhaps in the middle or
end of July. I shall not therefore be able to send you any information about the tenure of your Tyrolese or your Grisons property,
but I shall be happy to learn for you what is the price per night
of goblin labour in the Harz forest, and what rent is paid to the
king and queen of fairy land, for each square foot of the enchanted
ground there. If I understand right I shall be near Dresden, and
if you have therefore any Saxon commands I am your man.

*   *   *   *

Give my best respects to Mrs Jones. I hope your bees are well.
It did not occur to me while I was with you that you had got
them to make experimental enquiries upon the prosperity of
communities, and that you were concocting political economical
speculations when I fancied you were merely feeding your eyes
with the honeycombs which were hereafter to supply your breakfast table.

Yours ever W. W.

---

[1] See the *Report* of the *Cambridge University Commission*, 1852, page 62.

MY DEAR HARE,

It was not part of my plan, as perhaps you know, to bring my mineralogical materialism into this enchanted ground, but so it is that here I am. And as I have been looking for some time for leisure to write a letter to certify to some one in your most doubting island that I am in the land of the living and upon the surface of the earth, without success, I am reduced to the necessity of taking an odd half hour here to inform you that I am here, in the land of witches and spectres, and the fourth half-thousand feet above the level of the sea. I have just been witnessing a lordly sunset and taking a sketch of the witches' altar, which stands on the shoulder of the hill just above the moss where those ladies love to have their visitors ; and am going to bed in a few minutes for the purpose of rising before the sun. If I succeed in getting out of the Harz tomorrow, I shall travel, as fast as German posts will carry me, over Leipzig to Freiberg; where indeed I ought to have been before this time. I do not however regret the digression that I have made. I had the most beautiful passage from London to Hamburg that can be imagined. If I am to give you a history of my journey I ought to tell you that by incredible exertions, after leaving Cambridge on Friday, I got into the steam-boat on Saturday morning with a passport which I hope the Prussians will be quite satisfied with, as I paid £2 7*s.* to Mr Canning's secretary for his signature thereto. The weather was so superb that though I was four days at sea, I ate, drank, slept and talked with as much comfort as I ever performed those important functions on land. I left Hamburg the day after I reached it, and was jolted to Brunswick in a manner which I did not think the frame either of the human body or of any piece of coachmaker's work could stand. It was principally a wish to change the character of the motion to which I was subjected, that made me leave the plain for the mountains when I got to Brunswick. I have in consequence been moving about the Harz for the last four days with great satisfaction. It will perhaps give me something less of Mr Mohs's company at

Freiberg, but it is still in a mineralogical point of view a very proper proceeding. These regions are, or rather were, very rich in both metals and crystallisations of other kinds, and it is at any rate something to be able to say 'I have seen and sure I *ought* to know.' At Clausthal I introduced myself to a Mr Bauersachs, who is teacher of mineralogy there, and a very worthy and kind old man, who shewed me such carbonates of lime as English eyes never saw. I introduced myself to him, as I have sometimes taken the liberty of doing in similar cases, calling myself Professor; and I have, in consequence, had the gratification of hearing myself called Professor by my postilion ever since. There is an omen for you! I made a mistake as I have done before, in despising these mountains of 3000 feet and taking for granted that my portmanteau could perforate them in all directions. The result has been that I have had to-day to describe the whole northern semicircumference of this knot of mountains from W. to E., in order to have a chance of dropping into the Leipzig road the day after to-morrow. I left my waggon at the outgoing of one of the valleys which diverge from this point, and took a guide to come here with me and to carry my shaving-box. When I requested to have such an attendant, I looked to have been provided with some horrid black-bearded villager with an utterly unintelligible dialect of Deutsch. I was therefore agreeably surprised to find that I was to be accompanied by a young woman with the prettiest small features and morning-red cheeks that I have seen in these parts. She was, I suppose, much like Mimili, but though something of a coquette, not quite so much as that meritorious lady. As my German does not yet run very freely it is lucky that she was very talkative, and with a soft voice and a laughing eye, she went on telling me stories about witches and fairies and herself and her mother and the *condition* at Hamburgh to which she was going at *Michaeli*. She came all the way bare-headed, with thin shoes and a black apron, so that you may imagine there is no great difficulty of ascent in this journey. I must not go on talking of Louise till I forget one of my principal objects in writing to you. I dare say you have not got any of my correspondents who are difficult to dispose of; but if there be anything which requires immediate or personal consideration, pray write to me at Dresden, *Poste Restante*. I believe I shall be in time for letters there, but at any rate I will take care to have them forwarded. I wished to write to Pierce Morton, and will do so if I have any chance of hitting him before his departure. I suppose Lodge and Sheepshanks set off the day but one after I did. Considering the great hurry in which I came off I do not recollect anything very particular which I had forgotten, but I shall be very thankful for all news of any body at Cambridge. So good night. Yours affectionately, W. WHEWELL.

DRESDEN, *Aug.* 15, 1825.

MY DEAR ROSE,

I have just got your kind letter and am mortified to find that I have given you the trouble of making up a packet for me. I hoped by writing as soon as possible to spare your time, which, I take it, has got demands enough upon it. I have in other respects every reason to rejoice that I came away just when I did, for I found Mohs just preparing to leave Freiberg for Vienna where he spends his vacation, and at his invitation I have resolved to join him at that capital, where I shall stay, I conceive, till near the end of September; so that all communications there will be very welcome. Previous to receiving your letter I had met with your friend Hase who accosted me with enquiries after you on finding that I associated Cambridge with my name in the Gallery of Antiques. When I told him that you had been preaching against some of the German divines, he was very desirous of knowing what class you had attacked. I told him that I did not know their usual distinctive appellations, but that I supposed he would understand me if I called them *Rationalists;* which he appeared to do, and said that you were a great *Supernaturalist.* I shall convey your message to him before I leave this place, for Böttiger is, he informs me, absent from the city, having gone to some bathing place for his health. I shall be very glad to have it in my power to do anything to assist your researches, and will make enquiries of any ghostly or otherwise likely persons. So far as I have seen of their public service, it appears to consist of nothing but psalm-singing in the loudest possible voice; and their songs are I dare say like Sir Piercy Shafton's, excellent but somewhat of the longest. If you wish for any reasonable quantity of this interminable bawling I shall be happy to import it for you. I am now on the point of quitting Protestant Germany, for I think I shall set forwards for Vienna to-morrow; but I shall certainly return by this place, and if you address any further commissions to me *Poste Restante* here I shall take them on my return. Some of my principal employments at Vienna will be seeing mineralogical collections and sitting at the feet of Professor Mohs while he discusses them, but I hope to find time for other things. I had begun to consider the advisableness of leaving Berlin till another time and returning by another route, but I think your letter will determine me to include it in my homeward track. I saw Hermann at Leipsic who enquired after you; and I had the satisfaction of hearing a lecture of his on the art of Criticism. I do not know whether I shall be able to give as good an account of the German schools of Mineralogy, as you have given of their Theology, but there is much to be said thereupon. They have got a Natural History system of my science, as they have got a Naturalist scheme of yours: and a very dangerous heresy it is considered to be. I am afraid however that I may not bring back my faith as untainted as you have done: for I find my mineralogical *supernaturalismus* giving way in some respects. It may perhaps be possible to bring about a union between the two creeds, which I hope will not be such a horrible thing in science as you hold it to be in faith. I have been here about a week, but have employed half that time in seeing the Sächsische Schweiz, which, to be sure, is not Switzerland, but is a very pretty assemblage of deep rocky dells and bold cliffs hung with woods. If you see or communicate with Jones I shall be obliged to you to inform him that I am gone to Vienna. I get into the Prague dilly to-morrow. I hope it is a different business from the Hamburg and Brunswick one; for if you were not submitted to the operation of removal in that vehicle, which I suppose you were not, you have no idea of the quantity of jolting and pummeling that the human frame can bear or that coachmaker's work can be made the instrument of. In pure weariness I turned aside and rambled some days among the Harz mountains. So your Museum and your sermons are not yet out. I suppose both will be extant on my return. Remember me to Mrs Rose, and if any other of my friends come in your way tell them that I remember them all the more kindly for being at such a distance from them.

Ever truly yours W. WHEWELL.

Oct. 1825.

MY DEAR JONES,

* * * Send me your Political Econ. and I will make time for it. Do not imagine that I am not aware of my own folly in supposing I can understand the matter with such snatches of attention as I give it; but I always persuade myself it will be otherwise by and bye, and then one's vanity is such a comfortable thing. I have got hundreds of mineralogical maggots in my head which I found in Germany and which may crawl into daylight hereafter—but now mind this my injunction. Do not go and conspire with Peacock or any body else to tell our friends that I am bewildered with German philosophy, as you once raised an outcry with the accusation of *a priori* metaphysics. If you do so you may easily give people an impression, which you will not be able to remove when I have convinced you, as I certainly shall at the first opportunity, that everything which I believe is most true, philosophical, and inductive. Another injunction I also would give you. Do not imagine I am doing all for the material sciences with my mineralogy. If I do not fail altogether, it will be seen that this is one of the very best occasions to rectify and apply our general principles of reasoning; and my science shall, without ceasing to be good and true mineralogy, be also a most profitable example of that higher philosophy of yours which legislates for sciences. Remember also that we have got to do something for that same philosophy one day. I have got hold of some good ends of speculations which I think will wind out well. The sad thing is that for the next month or two I shall hardly have half an hour to spin them in. My tutorship hangs about the neck of my theories in a wonderful manner—I mean in the way of a millstone and not of a mistress. Nothing is done or known about my professorship. I think people seem disposed to try some way of amicable arrangement with those rascally usurpers, the Heads. At present we are all agog about another professorship—the Greek. Rose and Hare are considered the most likely candidates. Besides these there are Waddington, Walker, Hastings Robinson and Scholefield. To be decided on Saturday[1]. I had rather see Rose Divinity Professor than Greek, but shall be content in any probable event. The candidates utter Latin prelections about Prometheus the day after to-morrow. I hope you liked Normandy and saw plenty of traces of Duke William. I have much enjoyed my summer. Could not learn much for you about the Prussian cultivators, for my stay in Prussia was very brief. They all agree that agriculture is much improved since the alteration, but I could not hear distinctly in what way. I was told that one result had been scattering the peasantry more in farm-houses over the country instead of leaving them collected in villages. They are a coarse substantial-looking set of people, apparently above want. I have got the laws on the subject coming amongst my books. What is the news of my horse? I shall want him as soon as he can be got here if he is *in rerum naturâ*. Pray let me know this soon. I found Peacock busily employed in comparing the numeral words of the Nanticocks and the Mandingoes, and proving that people talked decimal notation in central Asia—all for the Encyclopædia Metropolitana. For the present good bye. It has played the devil with my employments being detained nine days on the salt sea between Hamburg and Harwich, and I must try to work my way to daylight. Always yours, W. WHEWELL.

---

[1] Mr Scholefield was elected.

MY DEAR JONES,

I intended to have written to you by young Attree, but finding that he does not go to Brasted immediately I think it better to adopt the shortest possible line from me to you, which is in such cases a line of posts. I was somewhat surprised at not hearing from you, but having made up my mind that you went upon the principle of writing the smallest possible number of letters I waited till I had time to ask you again. I hope some day to have an opportunity of convincing you that I have ten times as much reason to be angry, and weary, and dissatisfied with myself and my life as you have; and this I trust will be a great consolation to you. It is a right which I every now and then exercise most abundantly, and I confess that I am somewhat jealous of seeing it exercised with equal freedom by persons who seem to have got a claim to it so much inferior as you have. But of this another time. I guessed that something had prevented the production of that imaginary quadruped my horse, and when I got your letter I had already provided myself with a beast, who has not got all the beauties and perfections that I stipulated for, but carries me very well.   *   *   *

I am sorry you have promised not to come and vote for Lord Palmerston. I shall think the worse of the University if he is turned out upon that eternal No-popery cry, which I do not think impossible. I cannot, however, persuade myself to care very much about the matter, and am almost sick of hearing of that or any subject of politics. Principally, I believe, because I can find no general principles at all to my liking, and therefore cannot have the pleasure of applying them. If I could get rid of my tiresome occupations here and find time for some glimpses into your world of moral speculations, I should at least have the pleasure of theorizing. However, when your book comes I will begin. I should, on all accounts, like to see you during the vacation. Are you likely to be in London between the 1st and 12th of January? If you are not I should almost be tempted to come to Brasted for a day, though I shall be much pressed for time. I am going to

Sir John Malcolm's for a day or two, and shall then be in Cambridge till the end of the month.   We have just elected Richardson as arbiter in our dispute about the professorships, and have therefore a chance of having the matter decided in a finite time. Adieu.   Yours ever W. WHEWELL.

TRINITY COLL. CAMBRIDGE, April 11, 1826.

DEAR HERSCHEL,

I believe you know that Airy and I intend to try, this summer, to swing a pendulum somewhere in the bowels of the earth, getting as near as we can to the centre.   I understand from him that the Royal Society and Board of Longitude are disposed to give us what assistance they can in the way of allowing us the use of such instruments as they possess.   We are of course desirous to ascertain as soon as possible to what extent this assistance will go, and for this purpose I shall, I believe, come up to London on Thursday.   I shall, I hope, be able to find you in the course of the day, if not sooner, at any rate at the meeting of the Royal Society. I should be very glad if you would learn, as far as any opportunity comes in your way, what implements we are likely to obtain.   The most important for us are *two clocks*, which I understand are to be had, belonging to the society; and one, or better two, invariable detached *pendulums* which I am told are in the possession of Lieut. Foster and Capt. Hall, and may probably be disengaged.   Also a few good *chronometers*, and, what is very essential, a *double tent* for observing, which it appears may perhaps be had.   If, besides, we could have *another tent* it would be all the better for our comfort.   Another very desirable thing would be, if attainable, one or two such *artillerymen* as you had: and if this favour were likely to be granted us, we should be glad if possible to have these persons *before our operations begin,* to superintend the conveyance of our apparatus and the arrangement of our preparations.   Other matters which we should wish to borrow from the Society, if it possesses and could lend them, are a measuring chain of approved accuracy, a transit instrument, a small theodolite, some barometers, and various thermometers.   I mention all these articles, because upon the amount of what the Royal Society could do in providing us with them would depend the endeavours which we should make to supply them from other quarters. And the more accurate the information I can get on Thursday and Friday the better. I take the liberty of applying to you about it, supposing that you will be willing to do what you can to learn what the kernel of the earth is made of. I should also rejoice to know anything about the Cornish mines, because, though the Ecton in the borders of Derbyshire seems to be the best adapted for our purpose, it is still easy to change our plans if any other seemed likely to be better.

I shall be in London about four on Thursday, and shall be obliged to return the next day.

Ever truly yours W. WHEWELL.

UNDERGROUND STATION, DOLCOATH MINE,
CAMBORNE, CORNWALL, June 13, 1826.

MY DEAR HERSCHEL,

As I have to sit here, about half-way between you and the centre of the earth, for many hours waiting for coincidences, it occurs to me that I may as well give you some account how our experiments go on, as I gave you a reasonable share of trouble in the course of our preparations for them. So far as extraneous circumstances go, nothing could be better than the opportunities we have here. I am sitting in a small cavern 1200 feet below "the grass", as the miners say, as comfortable as possible; peering ever and anon by means of two telescopes through a boarded partition into a larger den, where the Royal Society clock and Foster's pendulum are oscillating with due gravity and without betraying by any irregular motions the surprize they must necessarily feel to find themselves in such a situation. The way to and from the superior air is laborious, as you may easily imagine a quarter of a mile of vertical ladders through narrow passages in the rock to be; but it does not offer any other difficulty, and our chronometers being properly packed in a cage of wood, rope, and leather, travel up and down apparently without much dis-

turbance, appended to the neck of a miner who attends us.   I do
not know if I ever explained to you exactly the way in which we
propose to obtain our result.   It differs considerably from any way
of dealing with pendulums which has yet been practised, inasmuch as it is altogether independent of transits and takes nothing
for granted but the uniform rate of our chronometers for a few
hours.   The process is this.   Pendulum P is swung below with
clock R.   Hence its rate of oscillation with respect to R is determined for that time. R is compared with seven chronometers
before and after the observation.   Hence we have the rate of
oscillation of the detached pendulum P compared with each of
the chronometers.   In the same way, a pendulum Q swung at the
other station (above) at the same time, is, by means of a clock S,
also compared with the chronometers.   Hence we have

$$\frac{\text{oscillation of } Q \text{ above}}{\text{oscillation of } P \text{ below}} = a.$$

In the same way we have, changing the places of the pendulums,

$$\frac{\text{oscillation of } P \text{ above}}{\text{oscillation of } Q \text{ below}} = b.$$

And hence, since the ratio of the oscillations of P above and
below is the same as of Q, we have

$$\frac{\text{oscillation of } P \text{ above}}{\text{oscillation of } P \text{ below}} = \sqrt{ab}.$$

All which is easy to work by logarithms.   There are of course
corrections to be applied for arc and temperature; the latter is
easy and is at this lower station pretty steady.   The arc is more
difficult to manage.   It has generally been assumed that the arcs
decrease in geometrical progression.   It is not true in fact, and is
altogether false in theory, for that would make them to have
their decrements as the squares of the arcs.   We have hitherto
applied a correction founded on this latter principle, but the
progression does not seem to agree with experiment, and we are,
I suppose, to infer that there is either at the knife-edge, or in the
air, some retarding force besides that which varies as the square
of the velocity.   I cannot yet foresee at all what conclusion we are

likely to arrive at. Our operations are necessarily slow. The time and labour of going up and down between here and the surface are such, that it is impossible to communicate with the observer above when you are once here, and any mistake or defect in your equipment here loses the day. The stands of our instruments have also given us much trouble and vexation. Each clock and each pendulum refuses to accommodate itself to the stand of the other, which perverseness it cost us loss of time and patience to comply with. Our apparatus has been let down a vertical shaft the whole 1200 feet deep very safely, though if you were to see the fury with which the iron bucket, dangling at the end of its quarter of a mile of chain, dashes itself against the sides of its pit, you would think it rather a hazardous experiment. The thunder of this vessel in its ascent and descent is one of the sounds which visit my solitude at intervals. Another indigenous noise is the deep dull gong-like bang of the explosions where they are blasting, of which I hear many in an hour, some of which come with such a puff as makes me tremble for the tranquillity of my pendulum. We have once exchanged the place of our pendulums, but have not carried our calculations so far as to foresee the result in any measure. We have had no transit observations for various reasons; one is that we have hardly time. The underground observations can seldom be made in less than 8 or 10 hours, and often take more, which leaves little of the day for any thing else. But another cause is, that Mr Parkinson packed one clock in such a manner as to put it *hors de combat*. The pendulum was broken at both ends, viz. the screw and the spring: and when we had got this repaired, we found that one of the rubies in the pallets was both broken and displaced, so that all the efforts of the clock to *escape* were unavailing. We have been obliged to give up all hope of this clock and think ourselves very fortunate that we had a third.

I have written to beg off about the election, so I shall not see you there. Let people look to the consequences of walking any longer upon an earth of which neither Whig nor Tory knows the density. Adieu.

Ever yours W. WHEWELL.

MY DEAR HERSCHEL,

I wrote a little while ago to you to give you some account of the way in which we were conducting our experiment in Cornwall. I am sorry now to tell you that our proceedings came to an abrupt termination last Monday, by the accident of one of our pendulums falling vertically through about 1000 feet, on the way from the upper station to the lower. By some means or other, which we cannot give more than a conjectural account of, the bucket, in which the pendulum box was packed with a quantity of small reeds, took fire, while Foster's pendulum was journeying upwards in it, and in consequence both this pendulum and a gridiron one belonging to one of the clocks of the R. S. fell nearly the whole depth of the shaft. This happened in the afternoon, and in the course of the evening of the next day various small fragments of our apparatus were brought to us, all hammered and twisted in the most violent manner. I had seen the packages ascend from the bottom of the shaft and had exulted in the slow and steady way in which they ascended, hanging quite free of the sides. The fire must have broken out about three-quarters of an hour from that time, and the conjecture of our miners is, that it was occasioned by a candle-snuff thrown into the shaft by some labourer working near it.

I am sorry that our pendulum has proved the action of subterraneous gravity in such an unexpected manner, for I think we were going on very well and should have come to a satisfactory conclusion. We had made some alterations in the method of comparing our clocks, which would I think reduce the errors within the requisite limits, and we have ascertained, I expect pretty correctly, the ratio of a vibration of Hall's pendulum above to one of Foster's below. We wanted the ratio of their vibrations at the same place, which we should have had by swinging them side by side, and observing either their coincidences with one another (which I think would have been very practicable) or their coincidences with the same clock at the same time. It is now impossible to obtain this latter ratio. If the knife-edges had not been altered we might have had it from their previous history, but this can now be of no service. If you see Kater you will perhaps ask him if this is not the case, though I conceive there can be no doubt of it.

I am more and more persuaded that the method is a good one, that the place was admirably adapted to it, and that we could have obtained our result if it had not been for this sudden manœuvre which the earth's attraction has played off against us. But more of this another time. I am in London only for an hour on my road from Cornwall to Cambridge, where I am obliged to be immediately. I shall return to town soon and hope to see you then.

Ever truly yours W. WHEWELL.

BRASTED, KENT, Sep. 9, 1826.

MY DEAR AIRY,

I have put off writing to you till I could say something about the matters referred to in your letter, most of which have gone on but slowly. But to tell you what is to be told about them. Imprimis, your paper about Laplace is printed and on its road to you, if it has not yet reached your hands. I took the opportunity of Herschel's visiting Paris to send it by him. He is gone to geologize and volcanologize and so forth in Auvergne. I requested him to give copies of your memoir to Laplace, Poisson, and such other persons as might be capacious thereof; thinking this would answer your purpose. He will probably send you word when he has finished with it. Babbage also is at present at Paris with his wife; but whether you will find him there in the beginning of October, which I suppose will be about the period of your migration, is more than I can tell you. Your other paper has lost some time in consequence of not being very clever in following my motions, which have of late been subject to no law of continuity. But two days ago I sent up to town the whole corrected with the additions; and as soon as it is ready I will forward it to you. If however it lingers much longer it may be doubtful whether it will be worth its while to seek you at

Orleans. Had you not better send me word when you move to Paris, and how I can send it so as to meet you there? Of course there are *Bureaux Restans* and such devices, where it may stay till you fetch it. I had not read it before but I am greatly pleased with the mathematics of it, and grieved that the degrees and pendulums are so insubordinate. I am rather amused with the number of people that you have managed to contradict lately, for besides Laplace who, I hope, has heard of you by this time, and Young and Kater, to whom I will convey your contradictions when I have an opportunity, I find that you have occasion to resist Ivory, Lambton, and Sabine, with some others in your paper. I did not very well understand Young's reasoning in the article to which you refer in the Journal of the Royal Institution, but when I have an opportunity I will compare his result with yours. I suppose your investigation about the law of reappearances will be proper to accompany the account of our operations. For an account, Young says, we ought to write. But I do not understand on what supposition of the law of decrease in the arc you are to calculate it. I have not been thinking much or doing anything about the revision of our luckless pendulums. The results are not comfortable to look at, nor the history to reflect upon. However, so far as I have revised, I maintain 4 seconds to be right. I have begun within these few days to draw up a short account of our proceedings, that there may be one ready when the R. S. meets. I will not at present make any agreement to repeat the experiment, but I should in fact be very desirous of doing so with fitting time and circumstances. I do not know whether any pendulum has been ordered which may dangle in the place of the lamented F. I had another project for a different apparatus for which however I fear the materials will fail. It is of this manner. If you have two noddies with springs of different lengths and strengths and with different weights they may both be made to vibrate seconds or nearly. Let one have a little tail and the other a little disk and observe their coincidences. Then this apparatus, taken underground far enough, will shew a difference in the coincidences which may serve to determine the variation of gravity. It would, as you see, save comparisons, be very portable, and if your springs were steady in their elasticity and persevering enough in their nods, I do not see why it should not be as accurate as the other way. But touching this *nous verrons*.

* * * *

I have not been much at Cambridge since Commencement. For some weeks I have been living in this beautiful valley which runs for near a hundred miles through Kent between the Chalk hills and the Green-sand range. It is now full of hops and *hoppers*, which are as good to look at as any thing which you will see in the way of vintage. To be sure, you are not in a vine country.

If you see Biot, get what *renseignemens* you can about his way of observing coincidences. And I recommend to your serious consideration the propriety of writing to me directions to send your Phil. Trans. to Paris. I should be glad if you would write at any rate. Adieu.

Yours truly, W. WHEWELL.

Oct. 13, 1826.

MY DEAR JONES,

So you are not come, and I suppose it is every day more uncertain whether and when you will. I wanted to see you on various accounts, but at present particularly I want to consult you about a matter in which I think you will feel some interest. In short another professorship affair. This is the case. The Lucasian professorship of mathematics (Barrow's, Newton's and Milner's) is expected to be shortly vacant, Turton, who holds it, having taken a college living with which it is not tenable. It is for a mathematician far the most desirable we have, though not very rich in money (perhaps £200 a year). Now I doubt whether to apply for it (the Heads are the electors), but one main doubt is whether Herschel would offer and would succeed. Do you know where he is or how a letter can reach him soonest? I am afraid he will not return home for some weeks yet. The only certain candidate I hear of is French. Peacock's intentions I do not know, for I have not thought of the matter till this moment. King will not come forward if Peacock or I do; but will, I think, if we do not. Of course he will also give way to Herschel; and I should hope that if he were to offer himself nobody would oppose him but French. I think French would, and being one of the Heads, I am afraid his chance would be better than it ought to be. But at any rate Herschel's name and reputation would make the absurdity more glaring than in the case of any of us. I might influence some. I think therefore he has no bad chance. I think too he would like it and would fill it well, and I am sure we should all be glad to have him connected with the University. If therefore you can devise any way of letting him know immediately the state of things and getting his resolve we will lose no time. The Heads and they alone are the persons to be applied to.

If I were certain that he would not be a candidate I should be tempted to propose myself, for, as I think I have told you, I should like the professorship much. At the same time I had rather not declare myself if I were sure that the thing was desperate, and I should thus escape the ridicule of asking for professorships indiscriminately. I have not consulted with Peacock because I have not made up my own mind, and it would be an additional reason for declining it that I might not again stand in his way if he advances. I shall write to our Master immediately (who is at Brighton) asking his advice &c., that is, trying to ascertain whether he will support either Herschel or me cordially. If anything occurs to you which can be done let me know. I confess that I have a horrible dread that the Heads will favour one of themselves, and perhaps King might have a better chance than I should. French has long been known to be giving his time to Divinity and Hebrew.

I have no time to talk of other matters. Moreover the straggling van of the freshmen is already upon me. But even in this indecision and difficulty my best remembrances wait upon Mrs Jones.

Ever yours W. WHEWELL.

It just occurs to me that I could make very grand lectures on the principles of induction in mixed mathematics that I have talked to you about. This would be good : but better still would be—Herschel for ever.

Oct. 13, 1826.

MY DEAR HERSCHEL,

I wish I knew where a letter would find you soonest, because what I have now to say requires expedition. The matter is this: the Lucasian professorship is soon to be vacant, and I believe the greater part of people in Cambridge would naturally think you fitter for it than any body else that can be named. The electors are the Heads of Colleges. I do not yet hear of any candidates except Dr French the Master of Jesus; whether he would have the grace to retire if you came forwards I cannot tell, but I think any body else would. We, that is every body whose sentiments I can pretend to understand, would be delighted to have you fixed in official connexion with the University, and I have no doubt you would come among us and lecture till our hair stood on end. But the provoking thing is that there is not at present any way of getting at your declaration of your being desirous of it. In case this should reach you where you can write to the Heads better than wait till your arrival here to do so, I will give you their names which it is not likely you should recollect. Dr Barnes, Peterhouse, Dr Webb, Clare Hall, Dr Turner, Pembroke Coll., Dr Davy, Caius Coll., Dr Thackeray, King's, Dr Godfrey, Queens', Dr Procter, Catharine Hall, Dr French, Jesus Coll., Dr Wood, St John's, Hon. G. N. Grenville, Magd., Dr Wordsworth, Trin., Dr Cory, Emman., Dr Chafy, Sid., Dr Frere, Downing Coll.[1] Write to all these people and ask for their support, and you have done all that is requisite. I have not seen any of the people of your College and indeed know so little of them now that I can hardly tell how to ascertain whether they have thought seriously about suggesting this to you. But I hope you do not doubt that it would be a good thing that the business should be brought about.

---

[1] Three Heads are omitted, who were perhaps absent from Cambridge at the time; namely, Dr Le Blanc of Trinity Hall, Dr Lamb of Corpus C. Coll., Dean of Bristol, and Dr Kaye of Christ's Coll., Bishop of Bristol and afterwards of Lincoln.

I write in great haste and have said, I think, all that is requisite to let you know the state of things: so adieu for the present.

Yours ever, W. WHEWELL.

Oct. 18, 1826.

MY DEAR JONES,

Every word that you say is perfectly true, and what is more, I had made out nearly the whole of it before I got your letter. What I was not so confident about, that Herschel would not come forward, was confirmed to me by a letter which I had from him this morning, in answer to one which I sent, *viâ* Lady Herschel, at the same time that I wrote to you. I told King, as soon as I had made up my mind, that he had nothing to do but make the most of his hand, which I suppose he would proceed to do. I did not then know that Peacock will certainly stand, which he to-day told me he should. Nor does King know it, and it is quite as well, for his withdrawing in favour of Peacock will be better than his not advancing. But besides all this Babbage is making application, and has written to people here on the subject. He has no chance whatever, and it is mere extravagance, at least as appears to me, his taking up the thing. I do undoubtedly believe that he would be a good Professor *now*, but it is too much to expect that our Heads should understand not only his merits, but the varying shape of them as time and circumstances may have modified it. They, good easy men, will never get as far as the first step in such an investigation. I think Peacock is the most desirable man now forthcoming; I suppose Airy will not think of opposing him, for though he would be a better professor, it would be ungracious in him to fight Peacock; and after all it makes no difference, for French, if he be a candidate, will undoubtedly be elected. So there is an end of that matter, and so you see, much as I should wish for you, I cannot claim your company merely to listen to my vacillation. I have been for the last two days employed in receiving and establishing my freshmen, which is I suppose what people call business, and is undoubtedly a process for which I cannot flatter myself that I am peculiarly fitted. I grieve much to hear of your cold of which Attree also told me— get better as soon as you can; and then you cannot make a better use of your convalescence than to come here. I think we are not growing worse. You know that George Waddington and Sheep-shanks are establishing themselves here, and just lately Thirlwall has resolved to give up the law and take up his abode among us. This is good every way. I hope your speculations go on increasing in length, breadth and depth, as they should do. When you get anything finished which I can look over—that is any part at all— send it me here. Remember me to Mrs Jones.

Yours always, W. WHEWELL.

TRIN. COLL. CAMBRIDGE, *Nov.* 19, 1826.

MY DEAR ROSE,

When I got your sermon a little while ago and read it over, I partly intended to write to you for the purpose of remonstrating with you as to one or two things more hard than was necessary which you have said of my friends the experimental philosophers. However my intention was swept away, with several similar ones, by a current of other business, and I do not think I ought to try again to get hold of it merely because you have written me a letter which it rejoiced me much to receive. Otherwise " I have much to say for that same Falstaff," and there are some of your accusations which I think I could convince you of the groundless-ness of. I cannot imagine for instance why you should charge mathematics with being useful and with strengthening the memory, when you may easily know that all of the science which we learn here is perfectly devoid of all practical use; and I can give you plenty of testimony that it may produce the effect of very thoroughly spoiling memories naturally good, besides giving you psychological reasons why it should do so if you wish for them. Nor do I think that you quite fairly represent the nature of our progress in scientific knowledge when you talk of its consisting in the rejection of present belief in favour of novelty ; at any rate if the novelty be true one does not see what else is to be done. But, to tell the truth, I am persuaded that there is not in the nature of science anything unfavourable to religious feelings, and if I were not so persuaded I should be much puzzled to account for our being invested, as we so amply are, with the faculties that lead us to the discovery of scientific truth. It would be strange if our Creator should be found to be urging us on in a career which tended to a forgetfulness of Him. But of this perhaps another time. I have undertaken to preach at St Mary's next February, and may possibly take that opportunity of introducing some of my own views on this subject. If I do, you will believe me when I tell you that they were formed before your sermon was preached, and without any reference to other people's opinions. I say this not that I am at all afraid of your taking ill my differing from you—I do not think we are likely to quarrel on such matters—but because I would not be suspected of making the pulpit a place of controversy, or even of the wanton exhibition of a difference of opinion. I am exceedingly obliged by your invitation for Christmas. If I can, it will give me great pleasure to visit you, but I am at present exceedingly uncertain if I shall be in England. I have a project, at present undecided upon, of visiting Paris at that time in company with Sedgwick. I do not wonder that you like our Master so much, for he always strikes me as most admirable in respect of principles, affections and temper. I hope his Vice-Chancellorship will prevent his being so hard a student as to hurt himself, which I suspect he sometimes is. If they will make him a bishop I think they will do a good work, but I fear that time will not yet come. I am, I confess, heartily grieved at the chance of seeing French Lucasian Professor. It will be making the office contemptible, and will besides be a clear proof that there is no greater disposition here to select people for their fitness to offices than there has been in previous times; that we do not feel the responsibility of our situation. I wish Babbage had any chance. He would be an admirable person, and so would Airy who is also a candidate. If your friend Kaye votes for French, which I suspect he will, it will unsettle my notions of his uprightness a good deal. As for the Mineralogical Professorship the lawyers have it in their hands, and nobody knows anything more; and as I am tired of thinking about it I suppose other people are too. I shall see that all is right about the money which you have paid; but on looking at your letter I see that it is paid at Child's, where I have no account. Esdaile is my banker, and if you have an opportunity it will be better to transfer it there—to the account of the Cambridge bank, for me. I do not see why you should suppose that I am so thoroughly tutorified as to make that a condition of writing to me, but I suppose that in point of fact you did not. I hope you are well, which you do not say anything about. I have nothing of Germany to tell you.

Yours ever affectionately W. WHEWELL.

TRIN. COLL., Dec. 12, 1826.

MY DEAR ROSE,

I hope my procrastination in answering your letter will convince you that I am not very fiercely disposed to fight you, whatever other inference you may draw from it. The fact is that I wanted to be able to say whether I had much chance of seeing you at Christmas, well knowing that two people may understand each other and make out whether they agree or differ by a *vivâ voce* communication, which it is not always easy to do by any finite quantity of epistles. I am sorry to say that there does not appear to be any great probability of our knocking our bodily heads together for the present, as Sedgwick and I set out for Paris in a week, and probably stay there all the vacation, or as much of it as can be turned to any purpose of absenteeism. As to your attack on experimental science, I dare say that when you had explained all the angry language that you thought it fair to use, when you found her poaching beyond her own domains, we should not be very wide apart. For I will agree with you that she may poach. It is possible to make that study too exclusively predominant, as it is any other. Only one thing I protest against, and that is the being called upon to shew what is the *best* and *highest* and *most fitting* of human studies. For my own part I altogether dislike the moods comparative and superlative, and I have often thought that men would be much improved by being debarred all use of them except in cases of mensuration. What I do hold is that inductive science is a *good* thing, and, as all truth is consistent with itself, I hold that if inductive science be true it must harmonize with all the great truths of religion ; nor do I see how any one can persuade one's self to believe that all this tempting system of discoverable truths is placed within our reach, as it were on purpose, while it is at the same time tainted with the poison of irreligion— a sort of tree of knowledge and of death, both in one, without the merciful prohibition attached to it. It appears to me that our faculties for discovering and enjoying truth, and our faculties for making champagne and catching turtle and then making beasts of ourselves by a too intense perception of their beauties, are altogether different things. I do not know what thinking is if it be not fixing our minds on what is or appears to be true. Nor do I think we can reasonably expect to be able to lay down maxims as to what kind of truth is most to be pursued ; only this I think we may presume, that the right way of pursuing any of the kinds will not make us forgetful of the greatest and most comprehensive of truths. And this I think can be shewn ; and I should perhaps venture to say that a person who is unacquainted with science is blind to many and wonderful views which, properly considered, it gives him of the relations of ourselves and the world to the Deity. And it appears to me so far from reasonable to consider all acquisitions of such knowledge useless here, because they will be superseded by the full perfection of another state of existence, that I should almost say the reasoning would be equally good if used to shew the uselessness of cultivating our virtues here. At least this appears in the highest degree probable—that our enjoyment of that kind will be proportioned to the degree in which the previous advance of our intellectual cultivation has prepared us for it. Without something of this kind it would not be enjoyment. And we are taught to expect exactly an analogous rule with respect to our piety. Even here this enlargement and exaltation of our views is, or should be, perpetually taking place. For it is by no means correct to consider advances in science as rejection of what was known for something new. The novelty, if the philosophy have been duly inductive, *includes* old truths and shews them from a new point of view. Nor does it appear to me that experimental philosophy is much more an exercise of the memory alone than is mathematics. As for the contemplative devotion you speak of, I am not at all disposed to depreciate it, but I do not see how it can be made the object of education or communication.

But I must conclude. You see that there is no end of this, and there must be one to my letter immediately. Do you know that Airy is our new Lucasian professor, the best they could have chosen? Hare says that you may manage the copyright in any way you choose; he does not see however what Murray can have to do with it. Smith declares he long ago sent Boeckh to the Library.

I am writing in haste, and I dare say not after the clearest fashion. I am very sorry I cannot see you. Can I do anything for you in Paris?

Yours affectionately, W. WHEWELL.

TRIN. COLL., *Dec.* 10, 1826.

MY DEAR JONES,

If I do not write and tell you, how will you ever know that I have engaged to preach the course of sermons you wot of in February, taking the afternoon turns at St Mary's? I have not yet written the greater portion of what I am to say and have finished none of it in the form in which it is to be said, so that I am somewhat behind. And what stands still more in the way, I am just going to set off with Sedgwick on a long-projected expedition to Paris. If when I get there I find that I cannot work at this matter, I must come back again, for I shall otherwise get somewhat frightened with my undertaking. With time enough I should not fear the greater part of the work—all the argument about the activity and omnipresence of the Deity—but when I come to the indications of benevolent design in the moral frame of society, I have not such an habitual familiarity with the view of the subject in its details as to write with the confidence and vehemence which would be becoming. I have no doubt I should get on better if I had you at my elbow. It might have been better to put off the matter a little longer, but I had no choice if it was to be this year, and my idea might have faded before another. I believe you will think me rash, and I almost think myself so, but some parts I am persuaded I can make striking[1].

Do you know that we have settled the Lucasian Professorship by giving it to Airy? Babbage came here to canvass, and is now here. He has effected a complete conquest of your Master's good graces and is staying at Caius Lodge. Herschel has also been here to support him, but all in vain. In fact I think they have done well in electing Airy who will reside and give lectures— practical and painstaking ones—who is *par éminence* a mathematician, and whose reputation will all go to the account of the University. French withdrew, finding the office untenable with his present one—a sign of grace, partly, I believe, brought about by our Master.

How are you and your lucubrations? I hope well. I shall be glad to hear of any advances they have made. I am afraid I shall not see you this vacation. I go to town next week and dine with Herschel on Tuesday. I shall get to Paris as soon as I can, and if I leave that place shall come here immediately. We are just at the end of this term, which like all other terms went on very well after one got fairly into it.

Can I find or do anything for you at Paris, or can you suggest any person or thing more advisable to see than another? I have not yet determined what hotel to go to, but of course I shall be found *Poste Restante*. In London I am at the Old Hummums.

Remember me to Mrs Jones. Be well and strong and wise and industrious as ever.

Yours always W. WHEWELL.

---

[1] For an account of this course of sermons see the first volume, page 323.

MY DEAR JONES,

I am sorry my letter gave you even a moment's uneasiness—I think unfounded—but I see that I ought to have explained myself more. I never intended to publish, and I do not think that I shall preach, any thing which will even brush the most delicate bloom of novelty off your plums. The only part of my plan where I shall take moral considerations at all will be my last sermon, the other three being mainly or altogether on physical sciences. And then, though in my sketch I have talked of taking more than one example of the false philosophy of the irreligious school, I shall not in fact have by any possibility room for more than a very short attack on the principle of population. And what I have to say about that is in sum this—that the false induction consists in generalizing the impulse to increase, and not generalizing, to a coordinate extent and with corresponding views of their bearing, the moderating and controlling influences which the nature of society and of man contain. I am aware quite that I see all this far more clearly than I should otherwise have done, in consequence of our discussions on such matters, but I do not think that such a general view will at all stand in the way of having due credit given to any exposition of the argument in a more detailed and didactic manner. You will see, if you recollect at all the time in which that part was written which I read to you, that with me detail and even very formal reasoning is out of the question. I must keep people a long way off my arguments and let them see them at the edge of the horizon by their reflexion in the skies. The principal advantage I shall derive from knowing your views—and indeed it is one of the greatest possible—will be that I shall talk confidently of that which I do not prove, and assert roundly that a great deal more may be proved and will be so when any body comes with the right kind of wisdom to the task. I may in this way declare that rank and inequality of ranks and progress of agriculture and such matters are good things; but if I do, it will hardly occupy more than a single sentence.

Do you dislike this? If you do, I can manage to alter my plan and will readily do so, for I know that at any rate my sermon will smell of Brasted, and I do not want you to think that I am going wantonly to spill the odorous spirit. I can either make my three sermons into four and leave out the moral part of the subject for another time, when I can set my back against your book and fight out the battle; or I can take something which does not exactly come in your way, as the selfish system of morals, or something of that kind. I think I could contrive to make that example answer my purpose, though my present leaning is towards population. However we can talk of this when I see you.

I cannot anyhow get off till next year, and I think we shall easily devise so that there will be no necessity for it. I suppose I misled you about the day of dining with Herschel, but I am glad to think I shall see you next week.

Airy is going on very straightforward with his professorship, as he does with everything. He has already given notice of experimental lectures next term. Adieu.

Yours always W. WHEWELL.

TRIN. COLL. *Feb.* 26, 1827.

MY DEAR JONES,

I have just preached my last sermon, and have delayed writing to you till I could say so. I should have sent you some indication of my being on English ground long ago, but I found that talking about my sermons disturbed the serenity of my thoughts as to the subject of them, so I gave up the attempt. I have got through them without getting quite up to the moral part of my subject, so that I had no opportunity of giving any of the views which we have talked about, if I had been tempted. No population, and in short nothing but one or two analogies from the natural world to illustrate the probability of our being very fairly ignorant of the more general laws of the moral world. The rest was an attempt to make science fall in with a contemplative devotion, which I don't think was difficult, though people seem, from the notion they had of scientific men, to have thought it must be impossible. I believe I have succeeded pretty well on the whole, but I have not time to ascertain what people think in general.

*   *   *   *

I forgot to say that I doubt much about publishing. I wrote at last in haste, and I believe I had better wait.

Yours ever W. WHEWELL.

You will see that my plan altered its shape much in arriving towards its execution. I still look forward to the rest, but it grows big to look at. Are your laws extending under your hands still ?

TRINITY COLLEGE, *June* 1, 1827.

MY DEAR JONES,

*   *   *   When I told you of Herschel's intention of going to Madeira I might as well have added what was told me at the same time, that he had some half intention of going farther, probably to Teneriffe, to add to his stock of volcanic observations. I was told too that he did not wish these plans to be talked of, as his lady mother will most likely set her face against them. I am quite willing that you should like Lunn. We had him here for a few days, and I found him a very pleasant companion. I should like to see you during the present month, but I fear it is not easy. I have got another task of sermon-writing on my hands. I have engaged at our Master's request to preach the Commencement Sermon on the 1st of July, and I must set about it very soon. It would satisfy me to let you see it before it is preached, but I hardly see how that can be, as I have promised to go into Lancashire this summer, and intended, if it were possible, to put this in execution before the Commencement. This will make it necessary for me to set out in about a week, and if I can get on with my composition so as to be secure on that head, I shall take that plan; if not, I must go to Lancaster in July, which will much abbreviate my foreign travels. Indeed I sometimes doubt whether I might not better stay at home and read than go abroad and ramble; for I have a very grievous weight of ignorance which I should like to diminish, and I do little to that effect in the term. Moreover, I have some speculations which I should like to try to evolve. And it would be but decent that I should make myself somewhat acquainted with mineralogy, if I really am to be professor. The decision approaches respecting the mode of election, though still slowly. To confess to you a secret, which I hardly confess to myself, I much doubt whether this same professorship is a business I shall make any thing of. It will take a long time to turn me into a good mineralogist and lead me into paths out of the way of my favourite pursuits. At any rate, as I have put my hand to the hammer I ought to be practising with it a little. So you see, I am at present all afloat. I shall perhaps come after Commencement and try to persuade you to go to the Netherlands, which will be a better reason for going there than any I have at present.

Yours ever, W. WHEWELL.

TRIN. COLL. CAMBRIDGE, *Nov.* 11, 1827.

MY DEAR JONES,

* * * I shall probably soon have to decide on my professorship. Send me hint of the way in which it appears to you, to help my decision. I will tell you the case. It seems to be conjectured that Richardson's judgment will be adverse to us. I can hardly think he will determine for *nomination* by the Heads, for there is no single precedent and but weak analogies. He may very easily and I think will determine against *open election*. But if he do declare the election to be by *nomination*, I am doubtful whether I shall not throw the matter up. I do not intend to devote my life to mineralogy, and should probably not keep the professorship many years (except &c.). At the same time I do intend to make myself master of the subject at any rate, and I have got reforms, as appears to me important and irresistible, to introduce into it. These the situation of Prof. would enable me to promulgate more influentially and better every way. I should also (why should I not tell you so?) somewhat like the kind of rank which it gives here, but this is a thing which has two sides, and has at any rate not much weight either way. I have a wish to pursue several other studies and objects by and bye, but these plans are yet too indistinct and round-about to affect me much for a year or two. On the other hand I have got work enough if not too much already, and might, by being ignorant of what I had taken upon myself to know, get discredit, at least at first. And again a consideration which weighs with me more than I suppose it will with you. I think the election by *nomination* so manifest a usurpation of the Heads, so unjust, and also so prejudicial to the University, that I should be very glad to back up my testimony against it by refusing the professorship in which it is first attempted to be exercised. I may just hint to you also that I cannot make up my mind to Divinity. All such plans must be given up. Now with these materials, when the time comes, I shall manage somehow to decide ; but if you will give me your view, which will be quite dispassionate and clear, it will be of great service to me. I look with great pleasure on my mineralogical speculations, whether I am to be amateur or professor, and am sanguine about their success. This is better and wiser than desponding about one's fortune as you sometimes do. I suppose, like the people in the epigram, we shall both be mistaken.

*   *   *   *   W. W.

TRIN. COLL. *Nov.* 23, 1827.

DEAR HERSCHEL,

I am much obliged by your letter and glad to find that in my reformation of chemical notation I have stumbled upon the same views which you had adopted. I ought to have recollected your formulæ, for I recognize them now that you mention them. A part of my system was to introduce, where it is useful, the oxygen O as an algebraical quantity ; but in mineralogy it is just as you say, that for brevity's sake the use of dots is better. I have got some other reforms in mineralogy which I am quite confident about as to their desirableness, and only grieve at the small chance there is of my ever making those original discoveries

and advances in the science which give a man the right and
power to regulate its external clothing. I shall stick to mineralogy (at least so far as I stick to any matter of speculation) for
the present. As to our Professorships here, Turton had not one
when he took that of Divinity, and if Woodhouse's were anyhow
vacated, Airy would be a candidate, and I most fervently hope
a successful one. So far as income is concerned, the Lucasian
is rather a starving matter, for they compelled him in taking it to
give up all his College employments and all emoluments except
his fellowship. Peacock would not take the Plumian, I believe,
and, independently of all other considerations, I could not have the
face to set up for an observer. I do not know what would come
of the Lucasian in the case I have supposed, for, to tell the truth,
I have very little thought of the matter. One good thing is, that
Airy would as Plumian Professor continue to give the lectures
which he gives in his present capacity. I am very glad of what
you say respecting Jones. I wrote to him a week ago and have
just heard from him. If the University do not print his book it
shall be no fault of mine. I am fully convinced it will be on
every account a thing they will afterwards desire to have done;
and I should not much fear bringing it about, if it were not that
the ominous phrase of "Political Economy" may frighten them
beyond all control. I am quite satisfied both of the truth of
Jones's general views, and also that they possess the great property of true doctrines, that they are capable of being proved
at any step of their induction between the most general and the
most particular. You will laugh at this criterion, but it is capable of being exemplified nevertheless. And one beauty of it is,
that it brings in moral considerations exactly in their right place.
I am sorry you have not heard any thing respecting my recommendation of Ritchie to the Virginia University, because it seemed
to me possible that the arrangement might be advantageous to
both parties. I mentioned him again two days ago to Mr
Lawrence, an American gentleman, who was here on the same
errand. I do not at all like what I hear of the projects respecting
the R. S. What possible good of any kind can result from placing
Peel at the head of it? I am by no means disposed to take the

matter so quietly as you seem to be, and I do not know any thing more likely to be fatal than the counsel which recommends those who are conscious of meaning right to be silent and inactive, while others are forward and busy. I doubt whether the chance of having one's motives misinterpreted is enough to justify this quietism, and I am by no means certain, in my own case, that my insignificance as to such things and my ignorance of the details of the case will prevent my coming up at the election to vote, with the hope of being then able to see whether one way of acting is not more right than another. I think the day is the 30th. Do not fancy that I shall come resolved to do something because it is an opportunity which does not often occur. But it does appear to me that with all their evils such societies as the R. S. must have much influence, may have much influence for good, and are most likely to have this, if those who are really intelligent and disinterested will do what they think best, and leave others to make out that it is so when the clamour of times and parties has subsided.

I am very glad to hear the account you give of Babbage. If any epistolizable matter occurs to me I will make a shot at him.

I will be of what use I can to Griesbach *even* though (what an even!) he is not my pupil.

Ever truly yours W. WHEWELL.

TRINITY COLL. *Dec.* 9, 1827.

MY DEAR ROSE,

I can hardly believe it to be so long as I fear an exact chronology would prove it since I received your last kind letter, for undoubtedly my determination to answer it forthwith was so strong that I can hardly imagine it has not been executed. But I have to attribute to the life I lead here not only this delay, but the wrong estimate I had formed of the length of it. For one consequence of having one's time occupied with a perpetual succession of small and similar matters of business is, that all portions of it appear of undistinguishable duration: like travelling in a perfectly level plain on a perfectly straight road, where you are unable to judge of distances, and find the short eight miles which you have before you at the beginning of your stage, and the long two miles when you are near the end of it, present much the same appearance. However the reason of the thing may be, I fear the fact is undeniable that it is near a month since I heard from you, and that it is still uncertain whether I shall produce a letter which is post-worthy. I did not however neglect your request with respect to the MS. in the Library, though in truth my assistance was in no way necessary. Your brother may in his simple capacity of M.A. copy the notes you mention or any others; and if it be thought right, as it perhaps is, to ask some permission to publish them, I imagine that Lodge's, which he is quite willing to give, will be amply sufficient. This I believe your brother already knows, and has probably told you. Getting the MS. out of the Library might perhaps be more difficult, though possible, if it could not be dispensed with.

I do not think I can tell you anything very distinct with respect to the counsels which you are to have in common with Hare. I told him of your desiring and wondering by means of his not writing, and his account of the matter seemed to be that he had got nothing to say, that is nothing to tell you which you did not know on the matters in question. He may possibly have written to you since, for I have not heard him say anything recently thereon. I am sure I wish you the best success in your undertaking, and this at any rate can do you no harm, though it will neither much enlighten nor fortify you.

I am in a state of the grossest ignorance with respect to all literary and scientific matters, and must therefore perforce hold my tongue on such subjects. I have got such an infinitude of that trifling which men call business upon my hands, that writing and reading are out of the question; a little arithmetic is all that I can occasionally indulge in. I did however read Bretschneider's book against you, and would have sent it you if I had thought it likely to fail of reaching you. I did not think it very formidable for you, but I think one may feel some indulgence for those persons who never having looked on Christianity in the light in which we look upon it, do not feel, and cannot comprehend, the horror with which their views are received. When I begin to

learn my alphabet again, and to resume the practice of reading, I
shall lose no time in looking at the articles which you mention.
I have some hope that this return to ancient habits may take
place after the 16th, but at any rate during the vacation I expect
some such remarkable event. I should indeed be much rejoiced
if any thing in the way you mention were to bring you to Cambridge as an annual or perennial. And though we can here too
sometimes be ignorant and dogmatical, I shall be quite content
that you shall have the fear of such attributes before your eyes, if
you will think this apprehension makes it necessary to seek our
pure streams as the only place to wash off such stains. When
you come here we can easily consider further whether and where
such things are. I got your letter written for Bonomi which, if I
recollect right, he sent in a parcel from London. I did not know
he had been here, and am sorry I saw him not. Though nothing
but an author's vanity could revive the subject at this distance of
time, I must thank you for what you say of my sermons. I seem
to myself to have got a perception of one or two truths, which, if I
can make other people also perceive, I am sure it will do them good.
Mais cela sera pour une autre fois. At present I can neither read,
mark, &c. Among other dainty devices I am moderator, which
will keep me here much of the holidays. Adieu for the present.
Give my regards to Mrs Rose, and believe me

Affectionately yours W. WHEWELL.

TRIN. COLL. Jan. 6, 1828.

MY DEAR JONES,

*　　*　　*　　*

During the last term I have been almost too busy either to
write or read. I took upon myself a number of employments
which ate up almost every moment of the day. This I might have
foreseen and did; but, nevertheless, the inability to follow at all
any studies of my own, and to read anything except for my daily
needs, made me feel as it always does intolerably ignorant and
stupid. I am now just going to enter upon the Senate-house
examination as moderator, which is one of my functions, and to

carry into effect some reforms that we have been devising. My professorship of mineralogy is still in uncertainty. We have got a decision which is smack against us as to the mode of election, and which also declares the professorship to be terminated by a vacancy except it be re-established, and it is not certain that the Heads will consent to revive it[1]. We have another professorship vacant by Woodhouse's death. Airy will probably succeed him, and the vacant one will be the Lucasian for which Babbage was a candidate. Some people here hope that Herschel will take it, which I much doubt. I should rejoice to have Babbage, but I am not so sure that he (one of the πολλοί) would succeed, and not at all certain that he would now offer himself. He is at Naples and therefore long to get at.

*　　*　　*　　*

Yours ever, W. WHEWELL.

TRIN. COLL. *March* 6, 1828.

DEAR HERSCHEL,

I hope you will be as glad to hear as I am to tell you that Babbage was to-day elected our Lucasian professor. It was entirely in consequence of your letter, which I got yesterday, that I bestirred myself in the matter, for before that I did not consider myself sufficiently authorized to press his claims upon the electors. Peacock, Higman, and I, wrote a letter to each of them which was sent last night about eleven, and to-day at 2 the election took place. Nothing could be more distinguished than the mode of giving him the chair, for though at such a period one might have naturally supposed a large proportion of the votes to be engaged, out of 11 electors only 2 voted for one other candidate, and 1 for a third. I hope Babbage will take the matter in the same spirit in which I am sure it was received here: for I am quite clear that the Heads elected him in a very cordial veneration for his talents and confidence in his goodwill to the office and its duties. I

---

[1] See Vol. I. page 32; also Dean Peacock's *Observations on the Statutes of the University of Cambridge*, page 134, and the *Report* of the *Cambridge University Commission*, 1852, page 67.

suppose you or some of his friends will write to him immediately and inform him of what has been done. It was at first suggested that a difficulty would arise in consequence of the impossibility of his being *admitted* immediately, but it appears that the words of the deed require him to be admitted "proximo opportuno tempore," which allow any reasonable latitude of interpretation. In what I had to say to the electors about the matter I referred them to their own interviews with him on the former occasion for information as to his intentions with regard to the manner of discharging his Lucasian functions.

I rejoice very much at the event of this business, for it is very honourable to him, and, as appears to me, to the electors. Nothing but the wish to obtain the best professor for the University and to do justice to Babbage can be supposed to have influenced them; and having been obliged to act in some haste, and to take a responsibility which might have been disagreeable, if matters had turned out otherwise, I am very glad to find that, judging favourably of our Heads, and not despairing of the result, we have been led to that which will I hope be a source of satisfaction both to our new professor and to us.

Yours very truly, W. Whewell.

Trin. Coll. May 21, 1828.

My dear Jones,

*   *   *   *

Among other ties I want to get my mineralogical sketch printed, and it will hardly be finished much before that time. I avoid all your anxieties about authorship by playing for lower stakes of labour and reputation. While you work for years in the elaboration of slowly growing ideas, I take the first buds of thought and make a nosegay of them without trying what patience and labour might do in ripening and perfecting them. To tell the truth I say this with no conviction of the wisdom of my proceeding; but so the matter is, and I shall circulate my system as I now have it, and let all that choose to read decide as they like, while I, so long as I work at minerals, shall try to develope its details. I believe this to be very indifferent economy of reputation and thought, but I do not see at present how I can do better.

    *   *   *   *

Ever yours, W. WHEWELL.

TRIN. COLL. *May* 30, 1828.

DEAR JONES,

    *   *   *   *

My mineralogical *aperçus* are printed, and it is to my eyes a very amiable-looking book. I have put some little speculation into it of the kind you desiderated, but I could not consistently do much that way, as my ostensible purpose was to compare what had been already done, so as to make the first step towards a catholic system of mineralogy. I must know much more about the matter before I can pretend to go alone.

I do not see anything to discourage any one in his hopes of reasonable fame from any science, and most especially yours. Take care to be first right in your principles, then bold and persevering in your assertion of them, and you cannot fail.

By way of exemplifying my doctrine on a small scale I am going to send my pamphlet to all the mineralogists I know in London, Edinburgh and elsewhere. I feel, however, far more curiosity to know what will come of the method when I have time to follow it farther, than to know what is thought of my present guesses, which is fortunate.

    *   *   *   *

Yours ever, W. WHEWELL.

TRINITY COLL. CAMBRIDGE, *Sept.* 9, 1828.

MY DEAR JONES,

How do you do this fine September? I am just returned hither—viz. to-day—from the uttermost ends of the kingdom, and I feel a desire to put myself in communication, as military men say, with its more central parts. I cannot pretend to give any very satisfactory account of my sojourn in Cornwall, which is just ended. To tell the truth, I consider that we have failed in our object nearly or altogether. Mainly in consequence of a rascally piece of steel deviating $\frac{1}{10000}$th of an inch from a straight line, by the fault of Thomas Jones, of Charing Cross, who is as great a reprobate as his illustrious namesake was, and whose sins are written in brass to our confusion and his. I conceive that this defect of our oscillating apparatus will prevent our drawing any sound conclusions at all from all the labour which has been bestowed upon them, but Airy, who opposes a face of adamant to a fist of iron, thinks he can still make something of their vagaries. And I must needs say that our labour has not been small nor our perseverance easily exhausted: but what I most regret, is the entire loss of the summer, in which, if I had used it better, I might have melted and measured the whole mineral kingdom. However this has not been without its compensations, and I have been hammering the coasts of Cornwall very vigorously at intervals. But you—what have you been doing?—Have you been living upon the stock of health and leanness acquired during our journey in Wales? Have you been cultivating Rent, Profits and Wages, and getting them ready for undying types? You must have been doing this; for all things call upon you. I have been reading a pamphlet, which you very likely know, as I read the third edition, concerning the "True Theory of Rent," by a certain Mr Thompson of Queens'. If you have not read it, read it forthwith on various accounts. Now one inference to be made from this same pamphlet is, how ripe the world is for your speculations, and how they will become less striking and original by all delay. Here you have the fallacy of rent being the excess of richer soils, the case (no doubt most imperfectly) indicated of rent in countries where this does not apply, the bearing of taxes in the various cases, the influence of moral causes and national habits. All these topics, no doubt, very slightly touched and with no consciousness of their extent and general principles, but still shewing how the opinions of clear-headed and inquiring men tend. On this account especially it is that you, who are in possession of the general views which connect and systematize these *apercus*, and of the collections of instances which illustrate them, should linger no longer. In the same way, so far as I can understand concerning Mr Sadler from the *Quarterly Review,* he has got hold, probably combined with much folly, of some of the true circumstances of the progress of population, and of the preventive checks. All these fermenting principles must converge to system and unity before long; the political economists are not all the war;—if they will not understand common sense because their heads are full of extravagant theory, they will be trampled down and passed over; and it will be the height of indolence and bad management if you allow other heels to take the *pas* of yours in this most meritorious procession. You will think that finding myself unable to make any positive assertions about my physics, which I made my business this summer, I am remunerating myself by very resolute operations with regard to matters that I know still less about. But as far as I have spoken I am quite clear. Nobody can doubt, for instance, that as to the matter of Rent Mr Perronnet Thompson is right, and Messrs the Westminster Reviewers wrong. And it may be noticed that his method of fishing out the very phrase and word in which a fallacy resides is what I have often mentioned to you as a useful exercise. I suppose many people are not assisted by it, but I am sure there are many others that are. I say nothing in defence of the mathematics with which Mr Thompson has filthified his subject, and not much for his very bold colloquialisms, but I think that I recognize in him several views which I learnt from you[1].

I shall probably come to London the first week in October. If I do, perhaps you can come at the same time, or else come here like a reasonable Caius man (a rare animal now), and let us talk our way to the bottom of these matters. At any rate, let me hear from you in what state of preparation you are. Give my kind remembrances to Mrs Jones. I hope she has long ceased to be an invalid.

Adieu, yours, W. WHEWELL.

[1] See Vol. i. p. 309.

TRIN. COLL. CAMBRIDGE, *Oct.* 14, 1828.

DEAR HERSCHEL,

I received this morning a letter from Lardner, in which he talks of a certain Cyclopædia which he is engaged to superintend, and to which you have promised your aid. He tells me that you have spoken of undertaking something with respect to acoustics and crystals in conjunction with me. And I should be glad to know whether any project of this kind was a serious speculation of yours, and what sort of work you think might and would be produced by any such collaboration. He says you have engaged to furnish an article on light, and I suppose therefore that you approve of the plan and artists of the book in general.

As to sound, I know nothing about the matter, and intend to put off my education till your treatise in the *Encyclopædia Metropolitana* comes out. You will judge, therefore, how likely I am to assist you on that subject. With regard to crystallography, I should on many accounts be much delighted to be engaged in any joint speculations with you on any such topics. But my views as to the best and simplest ways of treating such matters are hardly yet fixed, and even if we agreed about the principles to be followed, I think it would take us some trouble to select among the systems published or to work out the details of a system worthy to supersede them. Do you think this a thing to be attempted?

*　　*　　*　　*

TRIN. COLL. CAMBRIDGE, *Nov.* 18, 1828.

DEAR JONES,

Don't you recollect promising that if ever a proposition was made to admit Bachelors to the Public Library you would come to Cambridge and vote for it? Such a proposition comes before the Senate on Wednesday, the 26th. The Bachelors are, according to the law which we propose, to have the privilege of taking out five books. A few restrictions are added to make the matter more palatable to some, but none which will render it at all inconvenient to the youths. So come and exercise your functions as a wise and beneficent legislator without fail.

I am puzzled what to think, that you will not send me your MS. or bring it. Have you resolved not to look to the University to help you? If you are still doubting and delaying from doubt, I can add nothing to what I have already said. You perhaps imagine my hopes to be smaller than they are. I have good hopes; but I will not talk to you of them. What is the use of speculating about the future except so far as we can influence it? To act is our business now. We are men, and shall not lose our good humour and our resources whatever the event may be. Otherwise, what is the good of our philosophy and self-dependence? It will not do at the same time to despise men, and to depend upon their opinion for the comfort of our lives. For my own part, I have long been teaching myself to do neither—not unsuccessfully I hope. But at any rate let us not delay, when delay is at the same time a discomfort and a disadvantage. No doubt it is impossible not to be anxious and curious about the future. It is no small matter, but what of that? It is exactly the reason for doing now quietly and carefully what is manifestly to be done.

I should not wonder if my wise apophthegms make you laugh; but when is a man to utter moral sentences if not in the case of his friend's doubts and difficulties? So in case they happen not to apply, reserve them and return them to me on some similar occasion.

If you will come here *now*, you shall hear both Sedgwick lecture and Smyth speculate on the French Revolution. These are things you ought to do. Moreover, I daresay I shall find a number of my own troubles and humours, for which, if you like the office, you may be the depositary to my no small advantage.

At any rate, say or do something.

Adieu, yours ever, W. WHEWELL.

[The proposition respecting the University Library was not carried on this occasion, for a member of the *Caput* interposed his *veto*.]

TRIN. COLL., *Feb.* 4, 1829.

MY DEAR JONES,

*   *   *   *

Babbage has been here in his capacity of professor, and I have really enjoyed his society much, having seen him more closely than I had done before. But his anxiety about the success and fame of his machine is quite devouring and unhappy. It seems to me that you authors who embark the thought and labour of several years in one bottom cannot sometimes muster fortitude and philosophy enough to look tranquilly to the end of the voyage; and when I see how oppressive your anxiety sometimes is, I am tempted to be less angry with my own fiddle faddle way of swimming cork boats in a gutter. However if your anxiety be great you must recollect that it is because you look with no unfounded trust for a golden argosy of fame and influence; so push off your boat like a man. I am just setting to work with lectures and a dozen other things, but I will not let you forget that I am expecting to hear from you.

*   *   *   *

Ever yours, W. WHEWELL.

TRINITY COLLEGE, CAMBRIDGE, *Feb.* 15, 1829.

MY DEAR HERSCHEL,

It was with very great pleasure that I learnt from your letter that your marriage is to take place so soon. After the way in which you express your wish that I should be present on the occasion, it would be no ordinary difficulty which would keep me away, and I am happy to say that I shall be easily able to manage it. I am really pleased and grateful for your desire that I should be with you at the period which is to mark your entrance on a happier life, for I do not think you have any friend who wishes your happiness more zealously, or who anticipates it more hopefully from your character and prospects.

All this I say, you know, only so far as you will allow me to say it without a personal knowledge of her who is to be the great element of your happiness. And this is a deficiency in my fitness to congratulate you adequately which I shall think myself fortunate in being able to remove speedily. It would give me great pleasure to be introduced to Mrs Stewart on the 27th, but that I fear is impossible. It is the middle of term and I dare not play the truant twice together at times so near as that and March 3rd. I shall be in London some time this week, probably on Tuesday evening and on Wednesday, and shall then try to find you out. I speak somewhat doubtingly of my hope of success in this, for you have, I believe in conformity with the most authentic precedents in similar cases, omitted to give any date to your letter; so that except I can find Babbage or some one who knows of your whereabouts, I shall be at a loss where to look for you. I shall be to be heard of at the Athenæum and probably at Ibbotson's, Vere St., Oxford St. Pray give my best thanks to Mrs Stewart, whose willingness to consider me as an acquaintance I hope I shall soon have an opportunity to shew my sense of.

It seems unreasonable to disturb the current of your feelings at present by any of the concerns of every day life and therefore you are not to imagine that I attach any importance to the small matters which I am going to mention. Do you recollect if you ever returned to me one or two numbers of Berzelius's *Jahres Bericht*, I think the 4th and 6th, which I left with you once upon a time ? The one which I want is that containing B.'s views on the new chemical system and new chemical nomenclature which modern discoveries had suggested. I am more and more convinced that chemistry and mineralogy, which have been coquetting together so long, should be indissolubly married in order to ensure the happiness and dignity of both parties, but this is a union about which I do not expect to interest you much just at present. However I suppose you will give me joy if I can do anything to forward the match. The other matter concerns myself personally, and I beg you not to take a moment's trouble about it any further than occasion may throw it in your way. I think you are a member of the University Club. Lodge and I are balloted for on the 26th. If you can put any body in the mind to give us a white ball apiece you will do us a service.

*   *   *   *

Yours affectionately, W. WHEWELL.

RASTADT, *July* 31, 1829.

MY DEAR JONES,

I have got so far on my road towards Switzerland and should have been further, but that the churches have turned out so excessively entertaining that I could not possibly go on without looking after them. I was not so much pleased as I expected to be with the buildings in the Netherlands. The best churches are of late date, and the civil buildings though of curious architecture have little harmony in their style. They seem to belong to the period just preceding the revolt of the provinces. But when I got to Cologne I was quite captivated by the odd ways that the churches have; and there are in that city such a lot of them that I had very nearly never got away from them. They have just resemblance enough with our churches of the eleventh century to enable us to class them, and at the same time such a number of novel features as to shew that one's former notion of the manner of building at that time was very confined. As I went on I found the peculiarities of one group illustrating each other, and at the same time leading me on to another set, again a little different, in such a way that it made the most amusing progress that can be imagined, and I made my other objects give way to this. The whole of the middle Rhine from Cologne to Spires is full of interest in this way; and at Cologne better than anywhere else the introduction of the pointed arch as an artifice of vaulting, and *afterwards* into the details, might be abundantly exemplified. And then you have large and splendid buildings before the change took place, going back up to very early times, and some of the earliest and most gorgeous specimens of the completely formed Gothic, as the cathedral at Cologne; so that the series, though not clearly perfect, is very full and almost satisfactory. I wonder how it is that this quantity of materials never set the German heads to work to classify and theorize, a task they are so fond of when their materials are more scanty. I suppose they are like that kind of spider which does not like to have its thread attached to too many fixed objects, but lets one end of it float loose to hoist itself up in the air. So many facts would no doubt cramp a bad theory very materially. These Germans are undoubtedly strange hands at system-making. I met with a fellow the other day who had made a system of what he called Biotomy, by which he explained how the elements of time in a man's life and space on the surface of the earth had a marvellous analogy with one another. You would not think it worth while to attend to the details of this wonderful theory, but the man (a very grave professor at Bonn) declared to me confidentially that he considered it as great a discovery as the Copernican system— provided, he said, it be true, of which I cannot doubt. Another man has got a project to determine whether the moon is inhabited by rational creatures; which is of this kind. You must build a huge wall on Salisbury Plain in the shape of the 47th proposition of Euclid. If the lunarians are rational they must by this time have made out a system of geometry, which must be the same wherever reason is. They will see your diagram. They will answer you by building something else, I suppose the 48th proposition. I think I shall go and see this learned Theban who lives at Munich.

I have been so much occupied with my churches that I have not run much after anything else. Mineral collections take much time to see them to any purpose and are not so amusing as old monasteries; and Professors, except they turn out very good fellows, are not worth making acquaintance with; so I shall trouble myself with science no more than is convenient. I found one or two little books about the population of the Netherlands which you may like to look at. They seem to have paid a good deal of attention to the subject in that country. The region I have come through to-day and yesterday would, I think, have interested you. You have long villages, in which the houses seem to be all of the same rank. No poor, no rich. The houses large, wood frames and plaster, white-washed, separated by orchards between each, and generally with a trellised porch of vines in front. The peasants, tall fellows in large three-cornered hats, white linen trowsers and gaiters. There is not in all the villages I have passed through any house at all resembling in rank either the peasant's or the surgeon's in an English village. For anything I know you can give a reason for this. The country is a good wine country on the slopes, corn and potatoes in the plains. I am now writing in the room next to the one in which a German *table d'hôte* is feasting. Their meal is supper and I suppose some 30 of them have been feeding. They make a horrid din, and have indeed attempted three cheers, but in consequence of calling out Oo, Oo, Oo, instead of any more congruous sound they failed signally in this undertaking. However, they still go on gabbling forth gutturals and vowels as broad as the Rhine, in such a way as to put the partition of the room in a state of vibration at which Herschel would be puzzled. I am established on my side of the wall with a flask of Nierstein and a crust of bread and shall be able to stand their siege some time longer, though I hope they will let me go to bed in reasonable time, as I must be off early to-morrow. I go to Baden, (there is a fellow laughing in the next room till he crows like a cock and a woman giggling an accompaniment). I go to Baden mainly because *they* were there last summer, for I cannot find time to stay there; then to Schwarzach (an old monastery) and Strasburg. I return by Munich, where if you can write you may write to me. I epistolize you in preference to eating a greasy supper with my good friends in the next room; but also because I often think about you as I grind along with my *voiturier*. I bestow every now and then an especial act of recollection and good hope upon your *Political Economy*, which from this time is, I expect, destined to have no stop to its progress except such as Kennedy may insert in proper places. Shall I find it " quick and hasting into birth " in October. I hope so.

I travel alone, as I cannot otherwise stop when and where I wish, and have of course abundant time for reflexion. The sum of my reflexions about myself is much what I have told you, that I have undertaken more than I can do and that I shall have in return no satisfaction except that of proving myself a goose.

The good folks in the next room become more and more noisy. It is it seems an annual party of the landlord's. I have stipulated that it shall not last more than half an hour longer. Remember me kindly to Mrs Jones.

Adieu, yours ever, W. W.

MY DEAR JONES,

I dare say you cannot see me here in a little hospice about half as high again as your Snowdon, but I assure you I look down with great self-complacency upon you mortals who are condemned to dwell on the ground storey of the earth. It is rather cloudy but through a little hole in the clouds I can see you very tolerably. You are looking with great satisfaction at half-sheet D of your Political Economy and just beginning to discover the merit and the difficulty of a proper division into paragraphs. You have got some shockingly ill written heaps of paper lying beside you, which you are going to make a little more seemly to look at as soon as you have done correcting your press. Mrs Jones is asking in vain for the meaning of various ejaculations which escape you from time to time. Be a good boy and take pains with all the base and mechanical parts of your task, and do not, as I did, execute it so imperfectly that you are impatient till a second edition enables you to correct your blunders. I have been rambling for some days among the sources of rivers and am now very near the place where your friend Chevalier Boufflers must have been when he described himself as being able to spit into the Mediterranean on one side and the North sea on the other. I can at this moment hear the roar of some of the fountains of the Rhine, and early to-morrow I shall see the Rhone issuing from his icy urn. I have been leading an odd vagabond life but one of great enjoyment. Two days ago I came along a road which was as wild and savage as anything I hoped to see. I can seldom find occasion to apply these words, but in this case they were appropriate enough. The path was a pass from Glarus to the Grisons, where the way is varied by a huge glacier up the slope and over the back of which it is necessary to climb. The glacier is surrounded by enormous precipices down the chasms of which other glaciers stretch their ragged skirts. The lower parts of this prospect were occasionally shewn by the gambols of huge volumes of mist and cloud which at other times completely enveloped us. The ice was intersected by large chasms which we were obliged to coast in order to turn their ends, and often to turn back because they only led to others. One fellow went first with a rope tied to him to pull him out of a chasm if he should tumble through. The inside of a chasm in the ice is very curious. The lower part of the walls is blue, as if you were looking into the caverns of the ocean, and there are hung about and across, spotless shroud-like sheets and festoons of snow, which make it look as if it were prepared for a tenant. By dint of $2\frac{1}{2}$ hours of climbing we got to the upper edge of this icy district, after which the danger and most of the difficulty were over. The mountain over whose shoulder the pass goes is a very considerable fellow, being above 11,000 feet high, but the clouds never allowed us to see more than the skirts of the surrounding rocks. I suppose our path might be about 8700 feet above the sea. Besides my regular guide I was obliged to take another fellow from the last village, and to give him a sovereign for the job. However this man was worth his money, for besides that I could not have got over without him, he turned out to be an old chamois-hunter who had got a lameness from exposure to cold, but who was a hearty lively companion and told me stories about chamois and chamois-hunters in abundance. But what was not the least entertaining part of the adventure was my domestication for a day and a night among the herdsmen of one of the highest Alpine pastures. I do not know whether you have any notion of their situation and arrangements. Two men and four assistants inhabit a wretched hovel built loosely of fragments of slate rock. I ate curds and whey, hot milk, cold milk and various combinations of these with butter, &c. But my sleeping is not so easily described. Over a shippen[1] filled with pigs the floor is entirely covered by two coarse heavy cloths under which I, my guides and, so far as I know, all the cowmen slept. The place is dark except so far as the crevices let in light, and is not high enough to sit upright in. These cowmen have a hard life of it. They are called Saennen (San in the singular) and inhabit the Saennenhütte or *chalets* in summer for the purpose of making butter and cheeses. They have various

---

[1] *Shippen* is a Lancashire word for a cow-house.

alps or pastures at different degrees of elevation. The winter drives them from the Higher Sand-alp, where I was, in about three weeks from this time. They then take their cattle to the Lower Sand-alp, which is about 1500 feet lower, and about 5500 feet above the sea. They *hire* the cattle from various districts, generally distant ones, for the summer half-year, from St James to Martinmas, and pay 20 gulden a piece for them, (a gulden is about 20*d*.). They also pay a rent of 700 gulden for the Alp. With 70 cows, (I think) they make three large cheeses a day, (a cheese sells for 10 or 11 francs), which are the next day carried down to the lower station. They have also a flock of goats on similar terms, and a flock of sheep which they buy in spring and sell fattened in autumn. They have also a herd of swine at board, which live on the whey—a dark-brown well-made breed. The men, of course, go out in all weathers, (and the weather is generally bad enough,) morning and evening to milk the cows, with great hoods of sackcloth about their heads and shoulders. The cowherds have 48 francs for the summer. The rest of the day they are employed in making cheese and butter. The hut just holds two large cheesepans, a stone for a press, and a revolving churn. They have all their wood to carry up to the hut, for they are far above the region of pines. The old man who was my host complained that it was hard work to make a living out of the employment. I have this instant had a visit from my present host who is no other than a Capuchin friar with a long brown beard, a brown frock and cowl, and a rope round his waist, with whom I have been talking about the condition of this place, which is curious enough. He declares that 10 months of snow make a very healthy climate, and that his people live very well with no bread, meat, or firing. I have been lucky in my companions; the other day I had for my extra guide a crystal-hunter from whom I learnt many of the localities of minerals which are usually described erroneously as coming from St Gothard. I have picked up some minerals but no very great quantity. I must in a few days begin to turn the visor of my travelling cap northward. I shall do it with regret, for I have a great desire to follow out my architectural researches in Italy; but this must be for another time. I shall return by Munich, and, if possible, both by Weimar and Heidelberg ; but I believe this is too wide a project.

They are making a new road up the valley of the Reuss and the wages are from 20*d*. to 30*d*. a day; but this it appears is not enough to tempt the Swiss, and all the labourers are Italians from the other side, which my guide says is because they can live upon *polenta* which the Swiss cannot. Adieu. I shall drop this letter down to you as soon as I have a convenient opportunity. My kind remembrances to Mrs Jones.

Yours alway, W. WHEWELL.

7, SUFFOLK STREET, *Jan.* 10, 1830.

MY DEAR ROSE,

I had rather have talked with you than written about the business you suggest. To tell the truth, I do not feel any strong disposition to engage in the project. If it is set a going I shall wish it well, and shall very likely put in my word now and then : but I see some objections to it which perhaps you and the friends you mention have considered more thoroughly than I have done. It seems to me that a provincial paper, even from a University, could not make its way very rapidly, except it were to season its articles somewhat in the John Bull style, which would not be well. I do not think that fairness and good writing, even if you could get a constant supply of them, would do the business. You might circulate very well among clergymen and university men who are already convinced, but among the liberal revolutionists you would probably get no hearing though much abuse; and even the common vacillating popular mind is, I think, out of the reach of a voice issuing from such a quarter. You would moreover be obliged to make a *system* of politics, not only on the controverted points, but on all; which in these mutable times it is difficult to do till one is forced to it. I say nothing of the chances of writing feeble, extravagant, or any way bad, which may happen to unpractised people; but there would be a good deal of difficulty in getting your active writers to agree on a system of opinions neither too lax nor too rigid; and some chance, perhaps not a great one, of splits and collisions arising from differences of opinion. I think too that your arguments and even your assertions and proofs would come with less weight when issuing from a professional source.

I have for my own part a profound horror of a personal share in the responsibility of providing food for any periodical creature—even if it were to be like a household dog to protect you from thieves and robbers. But this I suppose other people do not feel to the same extent. In the case of a newspaper, I do not think I could surmount this dislike. No good could be done in such a case without throwing into what you wrote your whole strength, and soul, and *temper*, as well as a considerable slice of your time (considering that all writing requires materials and preparation). And these are no small sacrifices, especially for people who have enough and more than enough already of occupations forced and voluntary.

I have told you all that comes into my head against your project, which I hope you will consider as a good-natured reply, inasmuch as all that can be said for it you probably have thought of.

I go to Cambridge on Wednesday. I have not seen the poet and cannot guess the news you hint at. But it seems to be so disagreeable that I hope his poetical imagination has had some share in bringing it into existence. It would be worth something to get him to write politics, which I think he is now quite ripe for. Peacock's war cry is uttered, but he is not so fierce as he was *vivâ voce*, and I do not despair of our coming to some reasonable understanding. Adieu. Give my regards to Mrs Rose and all the rest of your party.

Yours affectionately, W. WHEWELL.

TRIN. COLL., CAMBRIDGE, *Feb*. 28, 1830.

MY DEAR JONES,

I have put off writing to you because I thought I might send a letter along with your next sheet, but that cannot stir till you send us some more MS. I am glad that you have satisfied yourself as to your arrangements with Murray, and glad too that Jacob[1] is getting a little light as to your merits. I do not think you need be surprized at his criticism: it means apparently that he understands the latter part of your book much better than the former, which, according to my conception of his character, is just what one should expect. Take care you do not set too much your heart upon immediate brilliant success now that you have thrown off all shame in the declaration of your passion, and I will guarantee you against failure.

I will get somebody to look over your sheets, but it will be more to the purpose if you and I are more careful, for nobody will give half the pains that we can. I can work pretty well in that way if matters go on with spirit, but not otherwise. I see your bookseller has already advertized you as printed at the University Press (in the *Literary Gazette*), and I do not conceive that any one will see any harm in it.

As to my own matters, I think your notion about the title page is sound and good, but unfortunately it did not reach me till I was already a volume. You will see when I send you the book (which will not be till you send some MS.) that I have not put my name in the title page. I have much intended to do it; and the mighty matter about which I tried to get your advice with a pertinacity worthy of the importance of the occasion was whether I should call myself "a summer tourist[2]." Some of our people here thought it not suitable to the dignity of my character—God save the mark—of which they, as well as you, seem to have a wonderful idea. To tell the truth, when I compare your notions about the management requisite with regard to my reputation, with anything which I am doing, or am likely to do, I feel a sort of disposition to abuse myself as if I had somehow unwittingly imposed upon you.

*   *   *   *

Adieu. Yours ever, W. WHEWELL.

---

[1] Probably Mr William Jacob; see Vol. I. page 7.

[2] See Vol. I. page 42.

TRIN. COLL., *May* 21, 1830.

MY DEAR JONES,

I cannot but be vexed that you persist in tormenting yourself in such a variety of ways, and when other methods fail that
you should set seriously about being frightened of my own worshipful self. I think I must, some comfortable evening when we
are together, try to shew you that idleness and procrastination, in
spite of the most luminous conviction of their folly, are not faults
towards which I can afford to be very intolerant; and that remorse
and shame for all sorts of intellectual and prudential absurdities
are feelings with which I can offer a very hearty sympathy. When
I have done myself justice in this way, which will not be very easy,
I shall expect that you will consider me a very reasonable and
comfortable sort of confessor when you are a little prickt by the
consciousness of transgression in such matters; and that you will
not let me be the instrument of converting the peevish self-dis-
satisfaction which is the natural retribution of such failings into a
really bloody-minded terror and despondency. I had no notion
that your birth-throes would have been so painful, but I have no
doubt the bantling will be worth them.

I do not think it is wise in you to force yourself to write in
haste if it worries and tears you so much. Why should you not
lay aside the continuation of your task till all the goblins have
disappeared which you have contrived to summon about it. One
obvious mode of proceeding, and one which offers some advantages,
is to finish and publish *Rent*, and to leave the rest for another
season. I have no difficulty in conceiving the annoyance of
travelling on with a pack of printer's devils in full cry after you.

*   *   *   *

Yours ever, W. WHEWELL.

[The next three letters relate to Geology. Dr Whewell reviewed the first volume of Lyell's *Principles of Geology*[1]. The
author wrote to the reviewer a letter beginning thus, "I cannot
refrain from thanking you for the great service you have done all
geologists, and me in particular, by your splendid article in the

---

[1] See Vol. I. page 51.

British Critic, which I have read with the greatest pleasure."
The letter then proceeds to ask for advice with respect to nomenclature, and the subject is continued in two other letters. In a
note dated Oct. 27, 1873, Sir Charles Lyell said, with respect to
Dr Whewell's three letters in reply to his own: "It strikes me
that there is much spirit, vigour, and originality in the three
letters on nomenclature of the dates of Jan. 31, Feb. 19, and
Feb. 22, 1831. As the terms then suggested were used throughout the eleven editions of my *Principles of Geology*, and in my
Elements and *Antiquity of Man*, and adopted in France, Germany,
and Italy, the points discussed in these letters on which I had
consulted Dr Whewell may be said to form part of the history
of Geology, and will be interesting and quite new to many students
of science."]

TRIN. COLL., Jan. 31, 1831.

DEAR LYELL,

Your letter was a source of great delight to me, first because
your praise is well worth having, and next because I was rejoiced
to find that you did not think my review too presumptuous, considering that all my knowledge on the subject is derived from
persons like yourself. I was afraid after I had finished, that in
my speculations upon the trains of thought which your book had
raised, I might have omitted to express the great respect I felt for
the work and the strong pleasure with which I had read and re-
read it. I shall be much interested to see your next volume, and
not at all disposed to quarrel with any case you can make out on
account of a myriad of years more or less. So far as you can
prove an order of things interwoven with the present order by
organic evidence, and not separated from it by dislocations differing either in extent, position, connexion, or other circumstances
from supposable recent dislocations, you have my good will in
going back if you like ten times as far as the Chinese creation.
But it seems to me to be made out that in this retrogression we
do come at various intervals to certain epochs where you have
evidence of a transition of another kind. However this will do

for another time; and I shall be glad to see your musketry brought to bear upon the species of cold-blooded animals, which, though it be death to them, will be, as it was in Esop's time, sport to spectators. But you must recollect that you have hitherto been firing with blank powder, and have not, so far as I have seen, a single species to produce which you have fairly destroyed. My engineering consists rather of mines to overturn your houses than of shot to kill your men.

I am glad you are trying to concinnate your nomenclature. I will tell you the result of my speculations thereupon. The termination *synchronous* seems to me to be long, harsh, and inappropriate. For the fact to be described is not that the species are contemporary with US, the wretched materials for future anthropites; but that they are *identical* with the *recent* species which we take for our type of comparison. I would therefore use a term expressing either *identical* or *recent;* perhaps better the last. Then your terms would be

1 aneous, 2 eoneous, 3 meioneous, 4 pleioneous.

Do you like this? They are shorter than yours. The misfortune is that the termination *neous* would rather look like a mere termination, as in *erroneous,* than a derivative from νέος. I had once thought of these from ταὐτό

1 ataatic, 2 eotautic, 3 meiotautic, 4 pleiotautic;

better in form, but neither so short nor so significant. Either, I think, is preferable to pleiosynchronous, &c. The words meioneous and pleioneous (or whatever terminations you take), might be spelt mioneous, plioneous, &c., and would be so according to the Latin and old English spelling, in which the Greek ει become *i;* but this has often been violated lately. I like your introduction of ἠώς, though it is somewhat poetical, for it will be well remembered: but I wonder you never thought on the same principle of calling the early tertiaries 'nyctosynchronous', which is a very pretty looking word and quite new. I have been more puzzled with the task of grouping these classes into two general divisions, and cannot suggest anything better than your own terms ;

I Preliminial *or* Tertiary;

II Liminial *or* Penultimate.

By putting them thus you might in referring to the latter division use Penultimate, whenever it was spoken of absolutely, and Liminial, whenever it was mentioned in correlation with Preliminial.

I thought at one time of trying to make use of the analogy of the tenses of verbs, the imperfect, perfect, &c., but I could make nothing of it. In that method of description the next period, when our organic remains are the materials of speculation, would be the *paulo post futurum*.

I shall look with great interest for your classification, for it is, I conceive, the key to all geological reasoning. I shall try hard to be at the geological dinner, though I believe I can only manage it by playing truant. Sedgwick is here at present constructing his annual speech. I do not know whether you have heard that at a late meeting of the council we voted the first Wollaston medal to Wm. Smith, I think most wisely and properly.

I shall be very glad if your Oxonians are excited by good said of them: I think the prospect is so far promising, that this mode of treatment has not been much tried. I wish Brewster had not been so thoughtless and prejudiced as to say what he did, for such things are much easier to repeat, than to weigh, for most people, and I do not think it will be easy to neutralize the effect of such an article as he wrote. Adieu. I shall be happy to see you *crop out* again.

Yours truly, W. WHEWELL.

It has occurred to me that καινός is a better word than νέος, and I propose for your four terms, 1 acene, 2 eocene, 3 miocene, 4 pliocene. The termination *cene* is right, as in *epicene; αι* becoming *e* as well as *οι*. For eocene you might say spaniocene, but I like your *eo* better. Is not this shortest and best?

ATHENÆUM, *Feb.* 19, 1831.

DEAR LYELL,

I am delighted that you find anything which you can use in my suggestions, and will try if I can hit on anything else to make your nomenclature more tidy. *Cenary* or *Cœnary*, is a bad word; the termination is a Latin one, and it will not be done scholarly and wisely to stick it to a Greek root. *Cenogenous* is perhaps better than your *cenophorous*. But if you want to distinguish the *cenes* from the preceding formations I think *paracene* and *procene* would be two good words for them and their predecessors. As to the diphthong I do not think the ambiguity of κενός very important. Nobody will know the origin of pliocene, &c., till you tell them, and nobody will mistake it when you have told them. In other cases we brave this ambiguity. Nobody would care if there were a possibility of *economy* coming from ἥκω. Still, if you like the diphthong, use it. It will certainly mark more distinctly that your termination has something in it, like Lord Burleigh's head. My objection to it is that it is not English; we have no such diphthong, and people do not now write *Cæsar*, *sæcular*, &c. You have been very meritorious in making geological words English in their form; I delight in your plesio*saurs*, and pterodac*tyls* ; and if it falls to my lot to review your second volume I believe I shall retrench your superfluous vowels.

If I were you, I think I would take the whole word *antepenultimate*, and leave the next generation to dock it if they find it too long. You will never get people to alter their accentuation— it is harder than learning a new word. If you choose to have *antepenúlt* and *penúlt* I see no mighty harm in them.

As to your *luviality* I hope you will not adopt such a beastly compound as *prot*alluvi*on*, which is wrong at both ends. I cannot imagine how alluvi*on* ever got its termination. *Primalluvium* would be tolerable. But if you are content with an arbitrary distinction, I do not see why *il*luvium may not rank with *di*luvium and *al*luvium. Perhaps, however, a significant word like proluvium, praeluvium, or primalluvium is better.

I like your scruples about *acene*, which I think ought to exclude it, besides other reasons. You can say 'the earliest paracene' or 'the latest procene' formations, which will answer the purpose. I should hope that this opposition of paracene and procene might, in the end, exclude the word tertiary, which no doubt is desirable and will ultimately be done.

I think, between you and Elie de Beaumont, we have a good chance at present of getting a chain constructed which will hold the world together pretty well. You must allow some links to be longer and stronger than others, and he must allow that they all hook into one another. I have no doubt that the present order of things is but as a watch chain compared with the chains of Baron Trenck, if you set it by the side of the Alp-making times. But of this another time.

Yours ever, W. WHEWELL.

TRINITY COLL., Feb. 22, 1831.

DEAR LYELL,

For *thalweg* and *riggin'*[1] I do not think you can do better than take *daleway* and *ridgeway*, which seem to me to be good-looking English words, and likely to become both familiar and distinct by usage. I am mightily delighted with the appearance of your language now that you have completed it. Your objection to *procene* and *paracene* is in the spirit of the true philosophy of terminology. These words indicating classes of a *higher order* than pliocene, &c., ought not to have the same form. I am reconciled to *cenogene* from your economical plan of saving substantives by it; though I am afraid homogene and heterogene are hardly respectable enough to countenance you. However, it would be better if they were so, for it is really a scandal upon the luxury of the times to see a word, of which the radical and essential part is overlaid with mere termination longer than itself. If I had not Brogniart's book, I would take it of you as well as thank you, which I do. I got it a little while back precisely as a specimen of wanton and worthless terminology. I feel a strong confidence that your phraseology will obtain permanence and prevalence—for this reason mainly, that it refers the reader directly to facts, and does not connect itself with them by any theory or influence. This inestimable advantage, and the importance and quantity of the information which I foresee in your second volume and which will be clothed in this garb, cannot help, I augur, making it the language of all geological mouths in the course of ten years. I am

---

[1] *Rigging* or *riggin'*, according to Sir C. Lyell, is a Scotch word expressing the summit line of the ridge between two parallel valleys.

not quite clear about *hypogene*, not that I object to the word for being not an exact definition in all cases. In natural history sciences it is, I believe always, certainly generally, impossible to have words which are such definitions. What you want words for is to *classify:* technical classes are not limited by definition, but bound together by the aggregate of their relations; and instead of drawing a boundary line, people should take a central and normal group and range cognate objects about this. But I do not quite see whether the leading fact in your hypogenes is their *being under* the paleogenes, or their containing evidence of some mode of origin. In either case, however, I see no objection to hypogene; but I am disturbed by D'Hallog's trilobite in hornblende slate, and by your understanding hypogene in such a way as to include this. However, when I have time to look your letter more carefully over, I dare say I shall see the matter more clearly.

I will stand by you in cutting off the tails of the anoplotheres and paleotheres. I am confident it will be doing a good action. We have long been in the habit of borrowing for the uses of science words resembling the exotics of a *hortus siccus*, being dried and lifeless, with no power of germination even to the extent of a final *s*. With a clear perception of what you have to say and of the idiom of the language in which you have to say it, I do not think you need despair of infusing a little more character and vitality into our scientific phraseology than it has lately had.

Do not fancy I meant more than I said when I spoke of reviewing. But really there is a great satisfaction in speaking one's mind about you geologists, for you seem to be so confident that the main substance of your lucubrations must be valuable and permanent, that you allow one to pull about its extraneous covering without any sort of ill-humour. This it is to have to do with pachydermatous animals, instead of the usual tribe of authors who are like molluscs deprived of their shells.

Adieu. I have not time to ponder very exactly what I have written, but I hope I have given you *good words* as far as they are intelligible.

Yours truly, W. WHEWELL.

TRIN. COLL. *Feb*. 10, 1831.

MY DEAR HERSCHEL,

I lose no time in doing that which is a great delight to me, sending you Jones's book—long looked for—sometimes despaired of—always desired—published at last. I think you will find it improved by every page which you have not seen. I do not see how it can fail to make its way *at once*—in the long run its failure is not possible.

I hold myself to be one of the most fortunate men of the age in having before me at the same moment, just published by two intimate friends, two such books as yours and his. I am also bold enough to think that I can point out the meaning and merits of these same books better than most other people. But of this more anon.

I beg my regards to Mrs Herschel. I conceive this to be a case for mutual congratulation among all right-thinking people, and I beg you will represent my sentiments accordingly.

Ever, dear Herschel, truly yours, W. WHEWELL.

TRIN. COLL., *Feb*. 1831.

MY DEAR JONES,

Senior's notions about the action of science are most admirable, and I quite rejoice to have such specimens of what is to be avoided. I will refer to the passages and revel in their absurdity as soon as I have a moment to spare. I have a most firm confidence that by and bye the whole world will think them as nonsensical as we do. I shall be glad to see your speculations about induction when they are finished, for among other questions it is certainly an important one how the true faith can best be propagated. I have done what I could in my review of Herschel. I do not know whether you looked at it in that view, but I intended it to be as good an attempt as I could make to get *the people* into a right way of thinking about induction. What you would make of a popular exposition of the matter applied mainly to moral, political and other notional sciences is what I do not so well see. I do not believe the principles of induction can be

8—2

either taught or learnt without many examples. Of induction
applied to subjects other than Nat. Phil. I hardly know one fair
example. Your book is one. A good deal of Malthus's population is a beginning of such a process, excluding of course his
anticipatory thesis, the only thing usually talked of. What else
can you produce? or how can you expect to lay down rules and
describe an extensive method with no examples to guide and
substantiate your speculations? You may say a number of fine
things and give rules that look wise and arguments that look
pretty, but you will have no security that these devices are at all
accurate or applicable. Even the Master himself has not escaped
error when engaged in such an employment. It is not that I do
not think induction applicable and the examination of its
rules important in the notional sciences, but I do not yet see my
way to anything which I could propound concerning generals.
There are various subjects which are well worth an examination
for this purpose:—*language*, a good example but requiring much
labour; *antiquities*, especially architecture, of which something
may be made; Montesquieu's *Esprit des Lois*, which is I fear too
imperfect to do any good with; the enquiries concerning *taste* and
beauty, which are hardly begun inductively. In what you call
intellectual philosophy I see scarcely a possibility of exemplifying
induction; so if you can make anything of the matter, I shall be
very glad to see it. I cannot imagine, however, that you will do
your pupil any great good by treating him to all our wise aphorisms. You may enable him to vapour about the successive steps
of generalization, and to call his own opinions inductive, and those
he rejects anticipations; but I am persuaded he will have no
profitable perception of any valuable truth. If there be any
practical inference to be drawn from the nature of true philosophy,
it is this, that general propositions can no otherwise be understood than by understanding the instances they include, and how
the poor boy is to get any good by your declaiming to him about
generalization, when he has not learnt any portion of any science,
is past my comprehension. Let him read Euclid and Algebra,
and when he has done that, Mechanics and Newton, and then
there is some chance of his knowing in his third year what

induction is. By your way of proceeding I cannot but think you run a good chance of making the aphorisms just as useful to him as the moods and figures of syllogisms, and of giving him into the bargain a very groundless contempt for those who have learnt one verbal science, because he has learnt another.

Sedgwick is really reading your book and seems much pleased. He has not read enough of your predecessors in the science to see all you have done. But I am by no means certain that I shall do your bidding and tell you of all the sayings I may happen to hear about it. I have as yet heard none, but I think you would not be wise to care about *such* judgments at any rate. The sentences that people utter about a new book at first have generally little meaning, often said merely for the sake of saying something, always expressing unformed and worthless opinions; and you must be greedy of disquiet to thirst after any such praises or to attach any sort of importance to either praise or blame in such a shape. If you cannot disregard such matters, I shall think you are hardly worthy to have written so good a book.

As to the review, I had thought of trying to get Lockhart to put it in the next, but really and truly was obliged to reject the intention as impossible. I do not know what authority you have for saying the next will not be out till the end of April or May. L. says the delay will not be more than a week, and I could not write it in less than a month if I were to devote myself to it day and night, with my other employments. Herschel's took me a good fortnight on a subject quite familiar to me, and when I had nothing to read. It would do you little good as well as me, if I were to write it in such a hurry as to make it contemptible. Indeed it is with great difficulty I can now get through my daily work. You *must* let your book have time to make its way. There is no fear of its being forgotten, and if you had written with the pen of an angel it cannot have the kind of notoriety which a scandalous novel has. I suppose it is now published, as our Cambridge booksellers have got copies. My recollection of Lockhart's note is very distinct, that he expected " the Herschel " in the March and "the Jones" in the June number, so I cannot comprehend your interpretation.

I am afraid I have written you a cross letter, but I dare say the next will be better. In the mean time get your induction into a readable shape, and give my remembrances to Mrs Jones.

Yours always, W. WHEWELL.

TRIN. COLL. *April* 24, 1831.

MY DEAR JONES,

*   *   *   *

I like much your aspirations after a reform, or at any rate a trial, in the way of reviewing for ourselves. I have always had and have great misgivings about engaging systematically in anything of the kind, because partly from business, partly from meddling too widely, and partly from other causes, I find great tracts of ignorance and shallowness in the middle of my knowledge, which might be very inconvenient to a reviewer: but still I think we could do something. And I have a very strong conviction that taking such a line of moral philosophy, political economy, and science, as I suppose we should, we might partly find and partly form a school which would be considerable in influence of the best kind. In the lighter kind of articles I should not fear our doing as well as our neighbours. Criticism would be somewhat of a puzzle, for I do not well know how the people would agree whom we should probably get together: but I suppose we could avoid all public and indecent quarrels like a discreet couple. However I leave you to finish your ruminations upon the hopeful project which you seem to have in your inside.

*   *   *   *

Adieu. God bless you. Yours ever, W. WHEWELL.

[The next two letters were written to J. D. Forbes, afterwards Professor of Natural Philosophy at Edinburgh.]

TRIN. COLL. *June* 9, 1831.

MY DEAR SIR,

I should sooner have answered your letter, but that I have been much engaged and was desirous to have first a few minutes of leisure that I might consider how far I could offer you any suggestions likely to be of service. I am truly glad that your visit here left an agreeable impression upon you. We have reason to be gratified with the good opinion of persons who are, like yourself, zealous and disinterested admirers and cultivators of science, and we are happy to find that we obtain it. I rejoice, for my own part, that you came to make acquaintance with us personally, and I shall have a pleasure in continuing to cultivate our intercourse on all future occasions. And if I can be of any use in any of your scientific pursuits I shall hold myself very fortunate. The education of an English university has no doubt its advantages which are great, but I am not sure that in the points to which you refer, the acquisition of mathematics as a preparation for physical research, there is anything communicated to the student which may not be amply supplied by such a course of study as you may easily go through for yourself: and I will describe as exactly as I can what appears to me the most advisable course of reading for such a purpose.

I by no means think it necessary to go through Lacroix's *Traité*. You would find a great deal there which would be of no use and would occupy much time. At the same time in the present state of science and its prospects there is no part of pure mathematics which can be considered as *beyond* the wants of physics, and what I would propose goes on the supposition that you have acquired so much as to be able to go on with any part of Lacroix, or any similar work, *which is wanted*. The requisite preparation for this appears to me to be some elementary treatise on Differential and Integral Calculus (Boucharlat's as good as any) and something on Finite Differences (Herschel's Treatise in the translation of Lacroix for instance).

I should then recommend you to read Poisson's Mechanics and Hydrostatics *carefully*, to see both the mode of generalising Physical principles and of applying pure Mathematics to them.

After this I should say the best thing to be done is to read some good treatises on some branches of physics which include the higher parts of mathematics; and, when it is necessary, to turn back to your treatises of pure mathematics to supply whatever you find you want in that department. I would mention in particular Airy's Tracts (of which he is just preparing a new edition to include the vibratory theory of Optics). If you can read this I do not think you need fear anything, and yet it does not appear difficult. I would also mention a treatise on the Mathematical Theory of Electricity in the Encyclopædia Metropolitana which I compiled from Poisson's papers, and which I particularly refer to, because the mathematics is of the most general and abstruse kind which has yet been introduced into Physics, and is also in a great measure of the same sort as that which occurs in Poisson's researches on heat which appear to come into your line of investigation. I do not think you will find any insurmountable difficulty in this Article, but if you do and will let me know, I will do what I can to elucidate it. Also if you read Poisson's papers on the laws of the distribution of heat and will inform me of what appears difficult, I will gladly look at it. This will be the less trouble to me as I shall have in a short time to read these memoirs carefully. I may also mention among such applications of mathematics Herschel's Treatise on Light. I think the study of such books will do more to fix the useful parts of mathematics in your mind than the express reading of pure mathematics.

There is a book by Pouillet on Physics and Meteorology which has already reached three volumes. I have not read it, but have heard it well spoken of. The author is Gay-Lussac's colleague. It contains but little mathematics.

I do not know whether my recommendations will fall in with your plans at all. If I can improve them by any further explanation, I shall be glad to do so. I hope I have not delayed writing so as to put you to any inconvenience.

We shall probably meet somewhere during the summer. In the mean time believe me, my dear Sir,

Yours very truly, W. WHEWELL.

TRINITY COLLEGE, CAMBRIDGE, *July* 14, 1831.
MY DEAR SIR,

I am sorry to hear you have been reminded of Sedgwick by a touch of his familiar ailments, and even he was not so desirous of your sympathy in his dyspeptic lamentations as to hear without regret of your being a fellow-sufferer. I still think you cannot do better than go on with Poisson's *Mécanique*, which seems to me much better adapted than mine for your purpose. I refer particularly to the mode of getting very general problems into equations, which is no small portion of the battle, and of which the *Hydrostatique* in Poisson is a good specimen. I believe I spoke of Poisson as a writer on the theory of heat, but the writer whom I intended to mention is Fourier, and I should recommend you his *Théorie de la Chaleur*, as soon as you have got Lacroix and Poisson under the dominion of your gastric juices. Fourier's book is a moderate quarto, not very hard when you once get hold of his reasoning, and a very instructive example of the application of analysis. I think it would every way be a better study than either electricity or optics. If you have any difficulty in procuring the book, and think it worth while to have my copy sent you, I will pack it off to you with great pleasure for as long as you want it. Connected with the subject of heat, you probably know Dulong and Petit's researches about gases. I do not think there are any analytical investigations connected with them, but I believe a little mathematics comes in. Herschel's Calculus of Finite Differences is not to be had separately that I know of, but if you cannot easily get hold of it, read the treatise on the same subject in the *Traité Élémentaire* of Lacroix, which will set you up sufficiently for the present.

I have been much interested with your speculations on climate in Brewster's Journal; an important and as yet infant subject. I shall be very glad to hear of anything further you do in that way. Airy goes on here pursuing Optics among other matters—predicting the most curious phenomena from calculation, and finding his predictions verified by experiment. He is printing his Optical Tract, which will, I think, be a very valuable present to the student of that subject.

I am afraid I shall not meet you at York. Even if other circumstances allowed me, I should feel no great wish to rally round Dr Brewster's standard after he has thought it necessary to promulgate so bad an opinion of us, who happen to be professors in Universities. He seems, with respect to such people, to have the power of imagining the most extraordinary things without a vestige of foundation: and just as he chose to fancy before that we each had a thousand a year (which notion he persists in referring to), he has now chosen to fancy that we are all banded together to oppose his favourite doctrine of the decline of science; though the only professor who has written at all on the subject is Babbage, the leader of the Declinists. It requires all one's respect for Dr Brewster's merits to tolerate such bigotry and folly.

*   *   *   *

I am much gratified by your taking the trouble to read my paper on Political Economy. Pray recollect that I profess only to trace the consequences of Ricardo's principles—principles which I am sure are insufficient and believe to be entirely false. I was much obliged by your list of errata. I had detected most but not all—and am ashamed of the length of the muster.

Sedgwick is still here expecting to have done with his pen and to have to take to his hammer in about a week. Few others of your friends are here. If you come and see us in October or November you will hear our summer stories.

Yours very faithfully, W. WHEWELL.

TRIN. COLL. *July* 15, 1831.

MY DEAR JONES,

You will probably have seen by this time that Lockhart has published 'the Herschel' and suppressed 'the Jones,' at least for the present. I am very wroth with him for the whole of his proceeding, but my anger has long ceased to be of a character which is likely to blaze out in letter writing, and indeed has received small accession by this last step of his. He shall go on as he likes and hear nothing from me, and I am not sure that I shall take the trouble of disliking him for what he has done.

I regret his not publishing the review of your book, but I trust it has grown out of any want of such help, and, as I told you, I had much restrained my laudatory disposition. I cannot say that I much like the review of Herschel now that I look at it in cold blood, and I have a strong persuasion that all the philosophical part will repel most and puzzle the rest, except it be seized upon by somebody who resolves to prove it wrong.

I am amused to hear of your passages with Whately; I am quite ready to fight, and very confident of getting the better. But people care so little about all matters of method abstractedly considered, that I should be better satisfied of obtaining notice if we had more to shew in the way of example of the right method —in short if you would get your Wages published. The feeling on which Whately founds his opinions, that principles of action are known by consciousness and do not require detailed observation, is not true as to masses, but it is plausible and generally assumed, and it will take some trouble to eradicate. The analogy between physical and political or economical science is yet to be shewn. There is no want of abundant means of shewing the actual folly and essential barrenness of their method, but if this be made the prominent feature of the dispute, people will ask you to reckon your fruits: so vindemiate as fast as you can. Though I talk so wisely I enjoy the prospect of a row, and can hardly help setting to work with the Logic at once. But I must see or hear from you more and learn what your plan of operations is. It is a filthy business having to do with these reviews, except one could have one's own way as to time and so forth, which editors will not, and I suppose cannot, allow. Apropos of him of the British Critic— I hope you have heard of him. He has the reputation of managing his business very negligently; and I find it was in consequence of this part of his character that I got no coin from him. It is soon to come.

You may perhaps recollect when you see the Quarterly Review that a passage about the Family Library near the beginning, and one about Sedgwick and others near the end, are interpolations of the worthy editor's. With the latter, however, I by no means quarrel, except at being made to praise Leslie's style.

I have been busy with vile business—abhorred of all philosophers, and most of all of me—ever since I saw you. Yesterday was my first holiday, so I bought a copy book and began writing astronomy. I think I see my way and like my work, but I have a good deal to read and hardly know how I shall get through it, for I cannot keep free from disturbance here. If I can reduce my books within reasonable compass, I will elope with them. My Induction is a perpetual snare to me, for everything belongs to it; but I shall keep my nose steady towards final causes for some time, and then look about me and see what can be collected on my path as materials for the Temple you talk of. Do you know I think it would be by no means beneath the dignity of our office to speculate somewhat about what the Master calls the Georgics of the subject. I think the practical rules and cautions for making experiments and collecting laws from these, as far as modern lights can help us, may be a very proper chapter. I wish I could get Herschel to write it.

*   *   *   *

Yours always, W. WHEWELL.

July 23, 1831.

MY DEAR JONES,

I do not think I shall get away from this place immediately; indeed I cannot very well foresee when such a thing may happen. My subject or subjects grow upon me faster than I can dispose of them, and I do not like to leave unrecorded and unemployed a number of trains of thought which, if I lose at present, I shall never be able to resume with equal advantage. Among other things I have got hold of a new science, which is altogether admirable both for my theology and for my induction—one of which, I confess, I augured no such matters, and for which I had somewhat of a contempt. What do you think my new pet is?—Meteorology. The people have been collecting facts for a very long time (ever since Noah), and are now just beginning to get a notion of the general laws and properties into which the mass is to be resolved. I do not know any subject which is at present in so instructive a con lition. Moreover, those who pursue it talk excellent *philosophia prima*, as is always the case among people so employed; the wisdom which people utter under the unconscious tutoring of practice is very noticeable in the business of the intellect, as well as of the hands and the heart. There is in especial one man whom I have always despised because he is certainly not clear-headed as to his mathematical conceptions, and who generally passes for a stupid man: but he is an excellent fellow for induction, and even talks much to the purpose about methods.... I should not wonder if *you* glory in finding that mathematics is not essential to philosophy. But this want makes the man a most troublesome fellow of a writer. Moreover, I have got out of this matter a most admirable set of Bridgewaterisms, and I am indeed mainly employed in getting these upon paper. I am hugely delighted with what I am doing, but I begin to have less trust in this feeling, having found that it may delude me; however, you shall see what I have done, and tell me how far I am right.

I think still I can get through the most of what most presses on me now in another week or so, and then I will run away. But I think my time will then be almost come to go northward, which I must do before September. However, when I see my way better through the haze of works and days I will tell you, and will try to get my valuable copy book shoved under your critical nose. Only pray use your eyes as well as your nose; for this is to be the best part of my book; and it much behoves me to know whether I have, or can obtain, any chance of getting hold of people's attention and approbation. In this matter, where I am to receive a thousand pounds, it is by no means enough that I should know that I am right; though that will be the main concern when I write for my own satisfaction and the instruction of the world, like yourself and other great philosophers.

How are you disposed at present towards the Logician? Is it peace or war? If you will give us illustrations and examples of the ascending method applied to moral sciences, we shall have no difficulty in fighting the 'downward road' people. I have got some additional views, but no removal yet of my doubts as to the

identity of the scientific method (that is, the method of making a
science) in physical and moral sciences. I may perhaps be able to
tell you more of my notions of this by and bye, when I get rid of
the clouds and climates.

*   *   *   *

Yours ever, W. WHEWELL.

[The next two letters were written to the Rev. W. Vernon
Harcourt of York; they relate to the origin of the British Association for the Advancement of Science.]

LANCASTER, Sep. 1, 1831.

SIR,

Without pretending to decide immediately on the probable
success of such a project as you suggest of an Association among
the men of science of this country, I conceive that such a proposal must be looked upon with interest by all lovers of science,
and I shall be happy if I can contribute any hints which may be
of service to you in developing it. If such a plan could be carried
into effect, it would certainly give to the meeting at York a higher
utility than that which would be likely to arise from the personal
intercourse of those whom such a motive might bring together,
though even this may no doubt have very great advantages.

The objects of such an Association as you mention would apparently be, to place before its members a view of the present state
of science, to select future subjects of examination, and to encourage in some way or other researches on such subjects.

I conceive that such a mode of action might be of very great
service to the scientific character of this country. The researches
of our men of science have been too much insulated from each
other and from what is doing in other countries; and the bearing
of what they have done upon the present state of science has not
been often clearly placed before the public. This inconvenience
might be remedied by the publication of reports, drawn up by
well qualified persons, of the recent progress and present condition
of the different departments of our knowledge. The annual
speeches of the President of the Geological Society, and in some

degree those on similar occasions in the Astronomical Society, have been good examples of what such reports may be: and these have been of great use. In most other branches of science we have had no such views presented to the world in this country. In Paris reports of this kind upon science in general are drawn up by the Secretaries of the Institute, and excite much interest. Berzelius every year presents a similar report to the Royal Academy of Sweden.

I offer you the following suggestion as one course of action among others for your consideration. The meeting at York might, in any way that was thought best, select one or two of the most eminent men in Britain in each department of science, and might request them to draw up respectively a report upon their own subject; stating what had recently been done both abroad and here, what is the present state of the science, and what appear to be the points most to be recommended for investigation at present, either in consequence of their importance or their promise of discovery. Such reports might be presented and read at a future meeting either a year hence or at some shorter period. The extent and degree of the interest excited by the prospect of such reports would enable the founders of the Association to judge what chance of success the temper of the country gave them; and these reports thus collected would be very valuable guides and materials in all future proceedings. I conceive that any man of science applied to to compile such a report, would feel himself distinguished by the selection, and would feel also the obligation of research and impartiality which such an office would involve. The reports would of course be printed, and the reading and the consideration of them would be a prominent part of the business of the next general meeting. I conceive that, independently of any ulterior steps, the publication of such a view of science would be both very interesting and very instructive. I will mention a few names which occur in connection with different subjects as examples merely, and nothing more, of the persons who might probably be engaged to give you such sketches of the aspect and prospect of their studies. In *Geology* you know how rich in such persons you are. Sedgwick, Conybeare, Buckland, Lyell, &c., &c., would any of them do the work well, and indeed as I have already said it is almost done already. *Astronomy* might be separated into *Physical Astronomy*, of which the principal cultivators in our own country are, I think, Ivory, Airy and Lubbock: and *Observing Astronomy*, where you have many good names, Herschel, Airy, Brinkley, &c. In the recent researches about *Light* you have Brewster, Herschel, Airy, who have occupied themselves with experiments, and the two latter more especially with the theory. The properties of *Heat* are another subject, in which much has been done and much remains to be done. Dalton and Leslie have been our principal discoverers. B. Powell, of Oxford, has been pursuing a portion of the subject; and I believe several of our best chemists have attended to it. Connected with this is the subject of *Meteorology*, which appears to be making considerable progress. Dalton, Luke Howard, and Daniell are the principal persons who have prosecuted it, and the two latter would be able to tell you in what condition they conceive it to be. You probably know much better than I do who are most likely among our *chemists* to take an impartial and enlarged view of the present state of their science: Turner appears to be eminently well informed and candid. I hardly know who at present can be considered as peculiarly well acquainted with *Magnetism:* Kater and Barlow may be mentioned. *Electricity* I suppose would be got from some of the chemists: as might *Electro-magnetics* and *Thermo-electrics*. Professor Cumming of Cambridge, has pursued these latter subjects with considerable care. In *Botany* you have the names of Lindley, Hooker, Wallick, &c., but with this and other branches of *Natural History* I will not presume to meddle. You will, I suppose, find no lack of students of them. *Comparative Anatomy* is another division of the subject which ought not to be omitted. I should suppose that one of the most difficult of the sciences to speak distinctly and impartially about must be *Physiology*, but it is a very important and apparently a very progressive one. I will only add that Mr Willis, of Cambridge, and Mr Wheatstone, appear to be the persons who principally attend to the subject of *Sound:* and that it would be proper, I should think, to make *Geography* one branch of your report, which would lead you to Greenough, Beaufort, Basil Hall, and the other London geographers. I have only mentioned these subjects, and these gentlemen, as exemplifications of what you have before you; and I think if you could induce a number of these to join in preparing, by next year, a joint representation of the present condition of British Science, and of the views, concerning the points now to be attended to, which its most eminent cultivators entertain, you would do something to give it a fillip.

This refers to a part only of your object: but after such a survey of the place where we are, you would be better able to judge of the mode of advancing. There would be no difficulty in finding subjects to recommend to the notice of scientific men either by means of prizes or in any other way. It would however be desirable to select subjects where we appear to be approaching towards discovery, or where asserted discoveries appear to require confirmation. You know how easy it would be to suggest such in *Geology*, if we could send persons provided with knowledge and money to other countries. In *Chemistry*, no one in England appears to pursue the subject of *Isomorphism*, the most important and promising step, I should conceive, which has been made in the science for many years. The connexion between the chemical composition of crystals and their form is another subject where we cannot but hope that something will shortly be known. The condition of the upper regions of the atmosphere as to heat, moisture, and wind, and the physical differences between different kinds of clouds, is another promising subject. Every one probably could mention such in his own department of science.

I can hardly pretend to suggest any thing at present as to the arrangements concerning prizes, funds, &c. As to the materials of the Society I would ask, for the sake of consideration, whether it might not be better to make your Association consist of all persons who have *written papers* in the memoirs of any learned society. It would be desirable I think in some way to avoid the crowd of *lay* members whose names stand on the lists of the Royal Society.

All Committees on memoirs presented or on any subject concerned with science ought to give *public* reports of their views. The neglect of this practice appears to me a serious deficiency in the arrangements of the Royal Society.

I have written very hastily the first ideas which offer themselves, and I must beg you to consider them rather as thrown out for your consideration than as asserted in preference to anything else. I shall be extremely glad to hear that your plan or any modification of it is likely to be carried into effect. I am really sorry that it is out of my power to attend the meeting in person. I am one of the examiners for College Fellowships this year, and our examinations take place at the same time as your assemblage. I think if you can get such a set of reports as I have described it will be the best beginning for any good plan, and another year will shew what can be done.

If I can offer any further suggestions likely to be of service I shall be happy to do so. I shall be at Harrogate probably on Tuesday or Wednesday next, and at Cambridge by the end of the week. At either place your letters will find me.

I am, Sir, yours very faithfully, W. WHEWELL.

[It may be interesting to record as an illustration of the expense of postage in former times, that the charge for conveying the preceding letter from Cambridge to York was two shillings and sixpence.]

Sept. 22, 1831.

DEAR SIR,

I send you along with this a letter containing some of the suggestions which I mentioned in my letter to you, expressed in a form in which they may be communicated to any of your scientific visitors if you think it useful to do so. I am far from wishing you to notice them except you think they may be serviceable; and I think it very likely that those who are present at the meeting will be much better judges of the advisable steps to take. I shall therefore be very well satisfied if I find you have not thought it necessary to make any use of them. I was much obliged by your considerateness in accepting my former letter as in some degree confidential, which indeed I intended it to be. The suggestions and names were intended to be, possibly, some assistance to you in developing the plan you mentioned; and

I would on no account offer my opinions to the meeting on such a subject.

In answer to your enquiries I beg to say, that if I were invited by such a meeting as I conceive yours will be, to prepare a report on the state of the science of which I am Professor, I should do so with great readiness and to the best of my ability. With respect to holding any office in your Association I should wish at present to be excused. It is very possible that I may be absent from England the whole of next summer, and it may perhaps be best at first to select your officers from those who are present at the meeting, and are of course the most obvious persons.

If you were to determine to hold the meeting hereafter in succession at different places, I am sure that a large portion of our members would be gratified with your selecting Cambridge. The Philosophical Society, whose members would naturally feel a strong interest in such an arrangement, cannot now be consulted. There might be some doubt about the best time of our receiving scientific visitors, for in the vacation (from June to October) the university is deserted by most of its usual inhabitants, and I should be sorry, for the sakes of many of them, that they should be absent on such an occasion. The following is a suggestion which I beg leave to offer you. The Visitation of our Observatory is always held in May; and on this occasion we have generally had some strangers here, who took an interest in science, and have been happy to see all such. If the party becomes next year more numerous and more distinguished, we shall rejoice the more. I cannot at present tell you the day in May, nor offer any invitation beyond my own; but upon the strength of this you may inform your visitors, if you think it desirable to do so, that we are always glad to see men of science here at the period of our Visitation, and will do what we can to entertain those who may come here next May. Any further arrangements which may be necessary can be made hereafter.

I wish you the best success with your Association, and shall be extremely desirous to hear of its proceedings. Believe me, dear Sir,

Your very faithful and obedient servant, W. WHEWELL.

9—2

TRINITY COLL., *Sept.* 18, 1831.

MY DEAR HERSCHEL,

I was with Airy in Derbyshire when he received a letter from you in which, as I understood, you told him that you had some intention of writing a book about *Photonomy*. You know perhaps how solicitous I am that my friends should not use bad language; and accordingly I have been minded, ever since I knew this, to write to you to beg you to think twice before you give such a name to the part of science which treats of light.

I conceive that you form *Photonomy* by analogy with *Astronomy*, and suppose the latter word to refer to the *Laws* of the stars. If you suppose this, I believe the supposition is wrong. The termination ονομος has nothing to do with *law*, but comes from νέμω, the verb, which means, I suppose, *to have to do with*, or some such matter. Αἰγονόμος, βουνόμος are fellows who keep goats and cows, οἰκονόμος manages a house; in like manner ἀστρονόμος is a star-driver or star-manager or star-monger, and φωτονόμος is a light-monger, that is, a *wax-chandler*. If you doubt this, look either in the lexicons, or in the beginning of Delambre's Astronomy.

It is very true that we very much want a name for the part of science which treats of *Light*. We also want names for some other branches of science, as that which treats of *Heat*, which is going on almost as fast as *Light*. I have been in the habit (in my MSS.) of using names which seem to me to have as close an analogy as one can establish with the names of recognised sciences (Mechanics, Hydrostatics, &c.). I have called one *Photistics* and the other *Thermotics*: (φωτιστικόν and θερμοτικόν are as good Greek as ἀκουστικόν). I do not care whether you take these or any other words, but I should be sorry that a large family of *onomies* should come in under false pretences to reinforce the brood of *ologies*. Jones has got his own peculiar *onomy*. *Illæ habeat suam*...Don't you go and make him fancy that we are obliged to put on his uniform. I want to have coats with no tails—not to terminate either in λόγος or νόμος—each of which is a horrid "black fish." I would stick to the root of the matter.

Light is light, heat is heat—there is an end of the essentials, and why should you be so superfluous as to stuff in anything else ?

So much for this weighty business, which I should not wonder if you think hardly worth a letter. But there is another affair which I should have been glad to consult you about, and if I were not inextricably engaged for this week, I would have come over to Slough on the chance of finding you at home, and willing to "trust the Ruler with His skies" for half an hour. The managers of the York meeting have applied to me, as I believe they have to you, for hints about the possibility of managing it so as to make it useful. I do not think there is any chance of exciting the kind and degree of interest about such occasions which they produce in Germany; but if there be any obvious prospect of stimulating the zeal of men of science and giving a useful direction to their labours, I should be very unwilling to refuse a share in the task of raising the requisite shout : albeit being much more willing to follow my own devices, and very doubtful about the chance of doing any great good by this machinery. I shall not be able to go to York, but the advice I gave was to this effect; that the meeting should select eminent persons in each department of science, and beg them to make, by next annual meeting, reports as to the present condition of their respective provinces, and the points where research will apparently be most useful; that the purport of these reports, and the degree of interest which they may excite, should be the guide and basis of future operations of this Association if it continue ; and that at any rate such a collection of reports, if it can be procured, be printed, by which means the *witenagemot* will not have met in vain.

I want much to know whether you see any chance of any good arising from such a proposition. I put it forth rather as a conjecture concerning a possibility than as anything else. One of the suggestions which my correspondent makes is, that the meeting may be held at Cambridge next year, if we think it good. One would not repel people by any apathy or backwardness when they ask your sympathy in such a case; but at the same time I should be sorry to urge men of science to come here and make a huge congregation, as if we intended to say we would make it worth their while to do so. If it really comes to a question whether we shall encourage the *sçavans* to visit us *en masse*, it would be very desirable that we should be able to guess whether the best of them would accept the invitation. Suppose that it were proposed to hold such a meeting in Cambridge next year, would you be likely to countenance us by taking an interest in it, and joining the party? If you do not see any good likely to come of such a proceeding, I shall think there is no great hope to be entertained of it; and if you keep aloof from the meeting, I shall not expect it to assume the character which would make it attractive and useful. My enquiries are very hypothetical, for I think it rather unlikely that the proposition to meet at Cambridge next year will be made on either side. My intention at present is to tell my correspondent that the Visitation of our Observatory is always held in May, that we shall then be glad to see as many *sçavans* as possible, and will do our best to entertain them; and I will tell him the precise time as soon as I can. I wish I could have seen you, for it is hardly possible to expound to you how little of substantiality there is in these speculations, but, at the same time, how desirous I am of knowing something of your views about this august proceeding.

I do not know whether you have looked at what I have said of you in the Quarterly Review[1]. I am very little satisfied with it myself; and, least of all, with the good I have said of you, which is far less than I thought. Jones is also turning reviewer. I have just been with him to Haileybury, where Malthus and he had divers palavers of no common length, I hope to their mutual instruction and comfort. Pray give my regards to Mrs Herschel.

Yours always, W. WHEWELL.

TRIN. COLL., Nov. 12, 1831.

MY DEAR LUBBOCK,

I think you are one of the Committee for revising the laws of the R. S. I have been applied to to join this Committee, but have been obliged to decline for want of time. But there are

---

[1] See Vol. I. page 54.

one or two suggestions which I will mention to you and which you may use or not as you like. I am strongly persuaded it would tend much to give both spirit and value to the proceedings of the Society if papers were referred to Committees, who should give written reports upon them and abstracts of them ; and also if annually or at some proper interval, the president, secretary, or some other person, should be called upon to give a survey of the progress of science during the intervening interval, and of the share in it which the Society's proceedings might occupy. I would refer to Cuvier's and Fourier's annual reports of this kind as admirable instances of the beauty and instructiveness of which such compositions are capable. In a more limited Society I should think the practice of *Eloges* on departed members also a good institution, but in the R. S., constituted as it now is, there would be difficulties about this. If you or any body concerned wish to have reasons pointed out for the opinions I have stated above I think I could give strong ones; but I will not detain you with them now.

I send you some memoirs which I have received for you in a *pacquet* from Quetelet. By the way there is a problem in *Biometry* (if you choose to call your calculations on lives by a Greek name) which may perhaps be included in something you have done, and if so I should be glad if you would point out the solution. It occurs in some of Jones's speculations. It is this:

"It is said to be ascertained that to put off to a later period of life the average age of marriage, does not diminish the average number of children to a marriage. This being assumed, to find the effect on the increase of the population produced by a given retardation of the average age of marriage."

I may mention, on another of your subjects, that you will find some American *tides* in the American Almanack, which you may get at a bookseller's who lives, I think, in Henrietta Street, Covent Garden.

Yours always, W. Whewell.

[The next letter was written to Professor J. D. Forbes.]

TRINITY COLLEGE, CAMBRIDGE *Dec.* 16, 1831.

MY DEAR SIR,

I was much interested by all that your letter told me about the York meeting, and have conceived the best hopes of the future fortunes of the Association from all I have since learnt. I am particularly glad to find that you are to give us a Report on the science of Meteorology. The subject is so important and so interesting that I am rejoiced it has been picked out from the crowd of sciences, and is to be treated of by a person who has given his attention to it. I have myself read a good deal on this and related subjects, and shall be glad if you will let me make a few enquiries as to the extent of research which your Report will include. Meteorology is connected with so many other subjects that it will require an exercise of discretion to fix its limits; and all the subjects on which it most depends are in that rapid state of change and progress which makes them very attractive topics. You will of course tell us what has been done with the barometer in this and other countries. But I should be desirous of knowing how far you intend to go in the very important subject of Hygrometry. If I am not mistaken this will turn out to be the branch of Meteorology which most requires cultivation. The theories of vapour, Saussure's, Deluc's, and especially Dalton's, must be decided upon in the progress of the study. The different Hygrometers, especially Daniell's, and Jones's improvements on it, are of great consequence. I believe the greater part of meteorological phenomena will be found to depend on the combination of Dewpoint and Temperature in different parts of the atmosphere. Humboldt has recently published two volumes on the *Géologie et Climatologie Asiatiques* (Paris, 1831). You will there find some speculations on isothermal lines and various other matters which I suppose will belong to your domain; and it appears from what he says that in his experiments he used an instrument which a M. August has invented and called a Psychometer, to answer the purpose of Daniell's and Leslie's Hygrometer (Tom. 2, p. 375). It appears that August has written an Essay, *Sur les progrès de l'Hygrométrie dans les temps modernes.*

The formation of tables of Meteorological observations is a labour which up to the present time does not appear to have been attended with any very valuable results. Cotte (*Traité de Météorologie*, 1774) gives an account of what had been done up to his time. I believe Toaldo is one of the best authors of this kind. Howard "On the Climate of London" is a very interesting book; and his classification and nomenclature of clouds is a very important step in the modern history of the subject. One must allow this, even if we do not assent to his system; for an arrangement of clouds according to their formation, their appearance, and the circumstances of weather which precede and follow them, is highly important and must have much to do with any sound knowledge of the phenomena of weather. Howard attaches great importance to the electricity of the atmosphere, and considers that agency as very operative in producing clouds, rain, &c. Whether this be so or not, the subject is a very curious one. I think it was a man of the name of Reid who published in the *Philosophical Transactions* and elsewhere on atmospherical electricity. Dalton's papers on clouds, rain, &c., of course you know.

One point on which I want you to inform me is how far you intend to introduce into your Report the recent researches on the relations of heat to vapour and gases; a subject which I have been in the habit of calling *Thermotics*. Much has been done in this department of late, but perhaps you will think it too extensive to make it an appendage to Meteorology, though it has the closest bearing upon that study. I refer particularly to the laws of the specific heat of gases, the heat developed by compression, the rate of evaporation, &c. The part of Thermotics which refers to the expansion of fluids and solids, to latent heat, &c., is more obviously out of your way. I am glad to hear that Professor Powell is to give us a report on the subject of the researches on radiant heat.

It has been usual in England to place what I have called Thermotics in treatises of Chemistry. Perhaps you can tell me whether Mr Johnstone, who has undertaken that subject, intends to treat of these enquiries. They are eminently important because recently so much has been done in them in a very masterly and decisive manner, and much remains to do and may be done by your countrymen if they will be zealous and will take the trouble to understand the present state of the science.

I shall be very glad to hear from you on these matters if I am not giving you too much trouble. I think there is a hope of something being done in Meteorology, and certainly there is much to be done. I was at a council of the R. S. the other day when we were on the point of ceasing to require Meteorological observations, because no one could shew how they could be made so as to be of any value. I hope this will soon be altered. I think the Hygrometer will be one of the most important agents in the change, but of this we can talk hereafter.

Believe me, dear Sir, yours very faithfully, W. WHEWELL.

I have neither time nor paper to talk of your review of Macculloch. I think in many things you deal with him full gently; but the strongest point of attack seems to me to be his want and contempt of the knowledge of organic remains. The geologists will not allow that they can afford as yet to quit details for theories.

TRIN. COLL., *Feb.* 3, 1832.

MY DEAR ROSE,

"For aught I see" your friend's sounding board should be as good, or better than if it were parabolic. But I recommend you rather to trust to experiments than to any reasonings of us theorists; for the doctrine of sound is by no means as yet in such a state that you can deduce complex conclusions (such as the audibility of the same speaker in all parts of a large room,) from simple principles, with any great confidence. One may say, however, that no good theoretical reason can be given why the parabola should be better than the sphere, and that there are some very plausible reasons looking the other way. But by all means attribute most weight to the trial.

I can easily imagine that the spherical segment is much the cheapest. Observe, moreover, that you need not have the form circular; it is enough if the surface be spherical, and it would not be difficult to give instructions for framing one of any other form, as these, merely for example[1]:

One of the most essential points would be that the inside of the sounding board should be very smooth. Probably if it were of metal, as tinned iron, the effect would be stronger, but this would, I suppose, make it more expensive; and then it must be painted, or the preacher would look as if he were roasting in a Dutch oven. With a spherical board a little adjustment as to position and distance would be requisite, but this would be easy to manage either by trial or by calculation. Some ornament also might be desired for the edge of the board, and this might be invented if the form were fixed on.

If you can get such a magazine as you speak of set a-going it may be a very good thing, especially if you can drive out of the heads of laymen the notion that the church consists of the clergy only, and that they have nothing to do with it. I have had one or two letters lately from Jones about the St Simonians. He took from Cambridge some books about them, and is struck be-beyond measure with their mixture of meaning and extravagance. In the mean time he complains much of his health.

I am glad to hear from your brother that you are to be here soon. Give my kind regards to Mrs Rose.

Yours affectionately, W. WHEWELL.

Feb. 19, 1832.

MY DEAR JONES,

Your letter which I got this morning made me glad and sorry. I was grieved to hear that you had been worse in health than I supposed. The Roses who are just come here seemed to consider you as by no means a sick man, though Mrs Rose did tell me of your weight, which did not much surprise me. But I rejoice that you have set seriously to work to make yourself well, for I have long had a very strong conviction that you were leading a life by no means wholesome, and that it was far more

---

[1] The letter contained three sketches of sounding boards.

necessary than you seemed to be aware to exercise a great
deal of self-denial: a very disagreeable prescription no doubt,
especially when it has to be taken habitually, but the condition
of well-being and well-doing in most things. I hope you do not
expect that you can make yourself well in three months, and
then take what libations you like, because I imagine such an
expectation could only lead you to disappointment and discomfort:
but if you will consider self-denial and bodily exercise as para-
mount obligations for a long time to come, I do not see why
you should not have an old age as prolonged and as healthful as
that of Cornaro. This will in the long run be all the better for
the book, though I confess I want much to get Volume two
a-going, and do not see why this should not happen immediately,
if you have not deceived yourself about your state of preparation
with your MS. As to your polemical quiver, I will try to fit
arrows to the string as soon as you send me them. And when
you have time you shall write your fill about the economical
conditions of political institutions, but pray do not set off on
this cross road at present. I can easily imagine how tempting it
must be, but your self-denial must be exercised with regard
to speculative as well as bodily appetites. This is a doctrine which
you have preached to me of yore, not needlessly, nor I hope fruit-
lessly. When you do set to work with this speculation I hope
you will not be so heretical as to "start" with any "principle",
however liberal or plausible. If any truth is to be got at, pluck
it when it has grown ripe, and do not, like the deductive savages,
cut down the tree to get at it. I could tell you more about
this, but another time will do. But I too have been reading
the St Simonian; who is the man that writes the *Exposition?*
He must be a fine fellow; I am entirely charmed with the
beauty and coherence of great part of his theory. It is full of
the most wide and striking notions, admirably combined and with
a marvellous air of truth about many of them. Of course I need
say nothing of the gratuitous and nonsensical blasphemy in which
it pleases him from time to time to talk of the man St Simon;
but his theory of organic and critical periods is constructed and
followed into its various developments with consummate per-

ception of the present state and tendency of men's thoughts and the cravings of their nature. I do not think the doctrine of the perpetual diminution of antagonism quite so well made out; and the assumption of a complete difference in kind between the next organic period and all preceding ones is, as appears to me, quite forced and illogical. I can hardly help imagining that this theory must have been invented by some German. Lessing began such a one, but I think in modern times some countryman of his must have run on further in the same line. But then a German would have left it merely a speculation. It is only for the practical men of the English and French democratic states to urge such principles into the field of action, to make the first glimmerings of Political Economy a ground of legislation, and to preach a theory of the progress of man as a new religion. But I have really been interested and stirred by these speculations beyond most things that I have read. There are, as you say, several right notions about the character of science, one in which they have hit on the same word which I have used for nearly the same thing—the *conceptions* which must exist in the mind in order to get by induction a law from a collection of facts; and the impossibility of inducting or even of collecting without this. However, I do not think they have done anything to diminish the necessity of developing and verifying this notion, and I am very sure that when this is done we shall have a much clearer view of the nature of general truth and the way of getting at it—including political and perhaps something of moral truth. But I dare hardly turn my thoughts a moment towards these regions at present, for I can barely get on from day to day with other matters.

I saw Herschel at the Geological anniversary on Friday and learnt from him that he was going to visit you. I should like much to come to Brasted that I might see how you take to abstinence and exercise, but I think it will be impossible till the term is over, even for a day, and I have no chance of seeing you, except you will come to London for half a day, in which case I will manage to meet you. I think exactly as you do about Babbage's book. He told me also of his project of publishing a book "like Herschel's," at which I could not quite suppress an internal smile. But still there is a great deal of ingenuity in his speculations, and the one you mention about skilled labour is, I think, the brightest of them. Moreover the book is of a kind which will receive its full meed of praise in these days. Give my kind remembrances to Mrs Jones, who I hope looks with approbation and comfort on your system of self-tormenting.

Adieu. Yours ever, W. WHEWELL.

TRINITY COLLEGE, July 1, 1832.

MY DEAR JONES,

I have been on the point of writing to you several times, and should have done so but that I could not make out any good prospect of coming your way soon. I am now got back here after a *giro* to Oxford and other places, and my present intention is to sit still and mind my book, as boys are told to do, for I do not know when I shall have such another opportunity. I am very sorry that you are ill and out of sorts, and cannot imagine why the rheumatism should be so insufficient for your annoyance that you must call in the pictures of coming evils to complete your discomfort. For my own part I see nothing for me to do with regard to public matters, and therefore I leave them to take their course, hoping the best. "The last dread curse of angry heaven, The gift the coming ill to know," I leave to you, or rather I would leave it to those to whom I wished ill, for when we cannot act I do not see the good of it even supposing the knowledge unerring. The only moral I can extract from such anticipations is the importance of getting our speculations into such a form that no calamity or adversity shall have the power, by putting an end to us or to the command of our faculties, to destroy the chance of our beautiful theories coming before the world, in our time or afterwards. As for smaller matters I find my interest in them decay quite as much as is desirable. I wish you would follow my example, and why cannot you? You know as well as I do that those who theorise rightly are in the end the lords of the earth.

Our meeting at Oxford seemed to all of us that were there to go off as well as was possible. Every body of note was there. Even Babbage, Brown and Fitton took an active share in the proceedings. A deal of work was done of which you will see nothing in the newspapers; and the show proceedings went off very well indeed. What more would or could you have? We meet at Cambridge next year, and shall have to make our arrangements a little more orderly in the mean time, but I think we shall prosper.

I hoped I might have heard from you something to print by this time. I am here now for a month, probably longer, and shall be a beautiful critic. I wish you would try me; and do not wait till I become blue-devilled with hard work, which perhaps may happen in a few weeks. If it do I shall migrate, but I do not know whither.

*　　*　　*　　*

Adieu. Yours ever, W. WHEWELL.

COUTANCES, *Aug.* 23, 1832.

MY DEAR JONES,

For fear I should forget anything which I may learn that is in your way, I will put down what I have collected hitherto, though my information appears to be mighty scanty. Indeed, when I begin to interrogate about such matters, I am oppressed by the consciousness of my own ignorance, which prevents my understanding or recollecting the details which would be most decisive. At Caen my principal acquaintance was a M. Le Prévôt, a man of very considerable station and of great intelligence. I could not induce him to give me or direct me to any estimate on the subject of your problem as to the time in which the number of proprietors is doubled by the law of equal partition; but he told me that many estimates which had been given on this subject were to be received with caution, being made by persons who are hostile to this law. It does not operate, he says, in any great degree to break up the large properties, inasmuch as the possessors of these have very seldom

more than two or three children, but among the small proprietors
the subdivision goes on.   In the meantime the people get more
and more into the way of having comforts beyond mere necessaries.   They eat meat, and as another instance they have all
got *umbrellas.*   The small proprietors have almost all a trade
besides their property to depend on.   This I found confirmed at
other places.   The man who opened the door of the church of
Haute Allemagne for me told me that he had a property of about
200 francs, but was besides by trade a carpenter.   He grew corn
for himself and his workmen on his ground, but bought his cider
from Basse Allemagne, a neighbouring village, where apples grow
better.   The people are much attached to their property, and,
as I understood, seldom sell it.   This is an open country covered
with corn in small strips; and when a plot is to be divided among
heirs, this is done by means of certain marks (*devises*), with small
trouble and no inevitable legal expenses.   There exists a very
great jealousy on the part of the small proprietors towards the
large ones, and the only chance the latter have is by making
common cause with the former.   I saw something of the expression
of this attachment to property at a place a little way out of the
road from Coutances to Bayeux, when Rickman and I were
attracted by the grand appearance of a fine open tower of Early
English work, with tall lancet-windows supported on bundles of
shafts and the light of the sky shining through.   While we were
drawing, the population of the *commune* all turned out into the
streets and got more and more troubled (you are to observe it is
a mile or two off the road), and at last one of the National Guard
put on his uniform and came and told us to follow him to the
maire.   Wishing to see what would come of this we made no
resistance, and were taken to a dirty farm-house, where the
servants were eating a dinner of herb-*potage*, bread and cider,
in a room of which the main furniture was a large bed.   The
maire was gone to Caen, and the *adjoint* was not to be found;
so after being kept prisoners for an hour and a half we were
marched to Bretteville where our carriage was, under the guard
of our first friend the sergeant-major of the National Guard,
who seemed especially indignant at our intrusion, two other

fellows with white belts and sabres, and two others with fowling pieces. On the way I talked with our convoy, and they declared that if a man had but a single *sillon*, he would keep it, and that it would be found that all attempts to change such a state of things would be fruitless. You will easily suppose that their patriotism was somewhat excited by the laudable service in which they supposed themselves to be engaged. When we got to Bretteville, the *maire* of that place dismissed us, observing to our guard how important it was that nothing should occur to interrupt the good understanding of England and France; but I think he would have some trouble in pacifying our sergeant-major. The country through which we have passed to-day (south-west from Bayeux) has been quite different, being divided by hedges as clearly as your Kent, and covered with apple-trees, so that from a hill it looks like a forest. It is described as "agricole; propriétés très-divisées; système de fermes où le fermier fournit ses instrumens," &c. When I have inquired once or twice how the division would be executed if *one* of these small fields were to be divided, the people all seem to think some of the inheritors would sell to the others. In some places also I was told of the small proprietors letting their land to large farmers. But here my host speaks of the small proprietors, who have no cattle, being assisted by their richer neighbours in ploughing "gratis," and then in return they help to reap their neighbour's harvest. I think the cottages here are not ill furnished. They have in general a large *armoire* and a clock. The former is a piece of furniture of some splendour. A house (farm and hedge cider house) at which we were to-day contained in the sitting-room two of these *armoires* (besides a sideboard, or table and cupboard, and an enormously high bed) They were of oak, carved with some taste,—one of them almost new, which our hostess told us cost 300 francs a few years ago, of very good workmanship—the other, about thirty years old, was her mother's. The one which she opened was quite full of cloth and linen furniture. We met this afternoon a great number, I should think 200, carts, coming from the sea-shore laden with sea-sand for manure. Some of them must have had to go and return thirty miles on a hilly road with a very heavy load. Team generally a horse, then oxen, and another horse, all in one line; very good carts made expressly for the purpose.

I will send off this letter for fear I should lose it, though it will do you little good. We have got on well in the matter of architecture, having seen much and speculated wisely. I shall be in England again in the first week of September. I hope you and Mrs Jones are well. My best regards,

Yours ever, W. WHEWELL.

BAYEUX, *Aug.* 25, 1832.

TRIN. COLL., CAMBRIDGE, *Oct.* 2, 1832.

MY DEAR WILKINSON,

Since I received your last letter I have been for the greater part of the time in France, where I went with Rickman the architect to look at the churches which they have got in that country.

We travelled over great part of Picardy and Normandy with great satisfaction, and, I think, with some profit so far as our knowledge of their architecture was concerned. We have classified to our own content the styles of church-building which we found there, and I think Rickman will probably illuminate the world upon the subject as soon as he can find time to put his notions upon paper. The most remarkable circumstance which happened to us, besides seeing on the average five churches a day for a month, was that at a certain village called Norrey, between Caen and Bayeux, we excited so much alarm by looking and making memoranda that the National Guard of the place put on their uniform, turned out and took us prisoners.

We were marched across the country under a guard of three sabres and two fowling pieces to the next magistrate, who luckily turned out to be a reasonable man. When Rickman produced his printed card, he observed very sagaciously, that Monsieur " n'aurait jamais improvisé un timbre comme celui-là," and moreover, that it was very important to preserve a good understanding between France and England, and so dismissed us, much to the discontent of the National Guard of Norrey. In other respects we got on extremely well, and found some very intelligent architectural antiquarians at Rouen and Caen. On my return I went to Herschel, who is, among other employments, grinding specula to take with him to the Cape of Good Hope, where he is going next year to observe the Austral nebulæ and double stars. He has a charming wife with two nice little girls, who I dare say will turn out as intelligent as your eldest daughter, who I hope has not forgotten me.

If I recollect, you were somewhat disturbed at the thought of my having written a book with the view of superseding the reading of Newton's text. I do not think that, if you were to see it, you would apprehend much mischief of that kind from it, and I would send it you if I thought it were worth the carriage so far. I believe those who do not read the text of Newton, having this book, would not have done so a bit the more, if they had it not. The fact is, as I think was even in *our* time, that no persons do read the text except men of mathematical heads, and lovers of original books, and those will do so still. The rest have been in the habit of having recourse to manuscripts generally illogically constructed, and substituting something else for Newton's reasoning (which I have not done), and in this way "Newton" has been "a subject" which interfered mightily with any consistent study of Dynamics and Physical Astronomy. However, whether I have done any good or no remains to be seen, for the book was only published just before the long vacation, and has not yet taken its place in our reading. I am on the point (in a week) of publishing another *Mechanics*, which is to be an easy book explaining the matter about as fully as I know how, and containing as much mathematics as Wood does. It is an employment at which I sometimes grumble, to have to write so many elementary books,—a very difficult and ungrateful office; but when you have to lecture and instruct about these things, you have the defects (or what appear such) of existing books so strongly and repeatedly forced upon you that it is difficult to abstain from trying to remedy them. However, I hope I have almost done with this kind of work. Gwatkin has left

Cambridge, as I presume you know. The Trinity men of our time still stick to the college in a wonderful manner, but your Johnians begin to grow out of my knowledge with a few exceptions.

Dear Wilkinson, yours always, W. WHEWELL.

Dec. 2, 1832.

MY DEAR JONES,

I have been in London and have heard that the business of your Professorship is not yet settled[1], so I am less surprised at not having heard from you. I am vexed at this procrastination, because it wastes your time, and will, I suppose, more or less annoy and occupy you. So far as your prospects of being Professor in the end are concerned, I suppose they are good. I had to call on the Bishop of London, and I asked him if the office was filled up. He told me there was a difference of opinion whether it should be continued, but that he wished it to go on, *because* you were to be the Professor. I thought him extremely sensible, and was very glad to find he had got such a notion: but the delay is wearisome and troublesome. I *calculate* you are writing or thinking the finest course of lectures that ever was delivered, but I wish they had not half uncorked the bottle so long beforehand.

How goes on your state of *physical* suffering? I lament much that there should be such need to dose you, but I hope to find you all the better for it when I next see you—not a greater but a happier man. Is there any chance of our meeting soon, and will you come here? I want to see you about my own book, though I am afraid I shall not get you to mind it much. I have now put above half of it into, or nearly into, its final shape, and think I have given it a coherence and obviousness of object to which it had no pretensions when you saw it; being then only in fact an attempt to get the arguments into words—an attempt which in several instances only served to show that they must

---

[1] This refers to the Professorship of Political Economy in King's College London, to which Mr Jones was appointed shortly after the date of this letter.

be quite recast. I am coming now to the speculations about the effect of the study of science upon the mind; where I think I have a good point or two, but hard to get into unexceptionable shape. If I can succeed reasonably well in making my two or three views clear, I shall hasten to the end without more delay, and then go to press immediately. We have all engaged to publish between March 1 and July 1, and I feel a wish at present to be among the earlier ones. You will have heard that we have an election for the University coming, and by the time you receive this you will hear that Peel has retired[1]. I think this must puzzle the Tories hugely, for they had got up a strong declaration here in favour of the *sitting* members, and will now have to look out for a new man, with adverse skies, and the recollection of the difficulty they had in finding the last. I shall vote for Lubbock, as my own pupil and particular friend and as *the* mathematician of London. But I meddle not with Whig or Tory. Do you intend to come and see the humours of this election ?

I wait to hear how your stomach and veins bear treatment so different from what they have been accustomed to. Write or come. If you have been kept so low as you say, I shall venture to dispute with you on your strongest points. Kind remembrance to Mrs Jones.

Yours always, W. WHEWELL.

TRINITY COLL., Dec. 23, 1832.

MY DEAR SEDGWICK,

When you had scribbled down the last sentence of your sermon after the bell had stopt, and had succeeded by a sort of miracle in reading your pothooks without spectacles, omitting, however, half the sentences, and a quarter of the syllables of those which remained, I dare say you thought you had done marvellously well, and had completed, or more properly had ended, your task. In this, however, you were mistaken, as I hope soon to

---

[1] William Yates Peel. The persons elected were Henry Goulburn and Charles Manners Sutton.

make you acknowledge. The rising generation, who cannot err,
inasmuch as they will discourse most wise and true sentences
when you and I are laid in the alluvial soil, declare that their
intellectual culture requires that you should print and publish
your sermon. I will give you a list on the other side of the
names of the persons who have joined in expressing this wish[1]
I undertook very willingly to communicate this their desire to
your reverence, inasmuch as I thought your sermon full of notions,
as the Americans speak, which it will be very useful and beneficial
to put in their heads ; or rather to call them out, for a great number of those good thoughts are already ensconced in the excellent
noddles of our youngsters, like flies in a book-case in winter, and
require only the sunshine of your seniorial countenance to call
them into life and volatility. I do not know anything which will
more tend to fix in their minds all the good they get here, than to
have such feelings as you expressed, at the same time the gravest
and the most animating which belong to our position, stamped,
upon a solemn and official occasion, as the common property of
them and us. And I also think it of consequence that, when they
on their side proffer their sympathy in such reflexions, we, on
ours, that is, in the present case, your dignified self, should not be
backward in meeting them, by giving to all parties the means of
returning to and dwelling upon these reflexions. Such is my
thinking about this matter, and therefore I have undertaken to
urge their request; and I hope you will be able to extract from
some abysmal recess your manuscript, and to place it before the
astonished eyes of the compositor. It is probable that he will
look, as Dante says the ghosts looked when they peered at him,
like an old cobbler threading his needle,—but never mind that.
The fronts of compositors were made to be corrugated by good
sentences written in most vile hands ; so let him fulfil his destiny
without loss of time.

I send this to Askrigg, where I suppose by this time you are
established in Lodge Castle. I should be very glad to be with
you, but at present 'I have got some fish to fry.' I also am
bound to the press, and hope to be tormenting the men of ink

[1] The letter contains a list of names, in number twenty-five.

with my Bridgewater hieroglyphics before the new year is many days old. I have also one or two visits to make : and among the rest I think I shall go to Hastings and see Lady Malcolm, from whom I have had another very amusing letter. She would fain see you there too, but I must explain to her how impossible that is.

I am blown in pieces in all directions with a vile cold, which produces explosions worse than anything which General Haxo could contrive. However, I still hold out, and make a sortie now and then.

*　　*　　*　　*

Adieu. Yours ever, W. WHEWELL.

TRIN. COLL., *Dec.* 27, 1832.

MY DEAR JONES,

Thirlwall says he can be with you on the 9th, so I will come to you on that day also. I suppose we shall not come together, for I leave Cambridge to-morrow, and he stays here a few days longer. I am going to London for a couple of days, and shall then come back to Lord Braybrooke's, and to Cambridge again in about a week. I do not know what Thirlwall's limits may be For my own part I do not think I can undertake to stay more than three days with you. But when I see you I will expound to you my visiting projects.

I hope to find you quite got over your indignation about the King's College business and ready to set to work, or rather forward in completing what that absurdity interrupted. If you are resolved to maintain your first volume to be a failure, I will not fight against you now : but at the same time I am bound to say, in virtue of my philosophy, that it is no bad success for a book not deductive like Ricardo's, but inductive, and in its inductions, as you must allow, far from complete, to obtain so much notice as yours has done : and I dare say that half the impression is a great deal more than the Wealth of Nations or Malthus's Population or any book of similar novelty and importance sold in the same time. However we may convince ourselves of the truth

and importance of your doctrines, we cannot alter the laws of the
progress of knowledge; and to the common herd of readers, time
and your labour, and that of the persons who understand you,
must be applied before your doctrines can be brought home.
I have always told you you shall have fame enough, if you will
wait a little. I tell you so still. I always meant to tell you that
your fame would not come immediately; so I am no whit surprised or disappointed by anything that has happened; indeed,
I should say that your book has succeeded better, rather than
worse, than my anticipations. M‘Culloch's review gave you or
your friends an opportunity of urging your opinions in a polemical
form : that opportunity was lost by your procrastination : I think
the loss was an advantage; and shall think so the more if it
urges you on to publish what remains. I still tell you that you
shall have plenty of fame in your life-time if you will work for it;
certainly the mere consciousness of deserving it will not bring it,
and in truth, from all we can see, making it one's object is not the
best way to get it. I had rather, for my own part, consider the
discovery and promulgation of truth to be a sufficient employment
and reward in itself. I do not know what is the good of knowing
and admiring such people as Herschel, if one cannot learn this
from them. But I know my fancies about this matter are likely
to appear extravagant to you, as they do to other people.

   You once talked of writing an article for the Philological
Museum. Can't you do it now ?—for Thirlwall is rather in want
of grist for his mill. I have written a paper which I think will
amuse you. I will try to get a proof-sheet to bring with me.

                         Adieu. Yours ever, W. WHEWELL.

                                   BRASTED, *Jan.* 14, 1833.

MY DEAR HERSCHEL,

   I did not answer your last letter immediately, thinking
that what I had to tell was hardly of consequence enough;
but Jones informs me that he is going to send a missive to you
under the shelter of which this may travel. In the experiments
with Mr Malthus's eyes I used the prism with refracting edge

downwards, so that the upper image was the green one, as I think you conjectured[1]. It was quite clear to me in the experiment with the chromate of potass that the idiopt perceived scarcely any, if any, steady distinction between the two images. I would however have gone again to Haileybury, having provided myself with a fluid prism, but the Malthi have left it for the vacation.

You will be very glad to hear that the people at King's College have at last ascertained their own minds and appointed Jones their Professor of Political Economy. This will be an excellent thing both for them and him with common good fortune. They will have the credit of first exhibiting this science in its right form with its noble views and really good lessons, and he will work more and be more known than has been the case hitherto. He is now writing an Introductory Lecture and a Syllabus which are to be published before long.

I am going to do something about tides. Pray think for a moment on the propagation of a tide-wave as a hydrodynamical question, the undulation travelling along a channel of variable section, and tell me whether you think that the solution on the *common suppositions* can be accepted even as an approximation to a real case, such as that of the British Channel.

Give my best regards to Lady Herschel. I wish you would contrive to let her see Cambridge before you go to the other hemisphere for change of stars. If you do not like to go to the Observatory (which would be inconveniently distant no doubt), you can tarry at some of our inns at the worst of times. If you will come in vacation time, as for instance at Easter, we can house you in College, as we did Mr and Mrs Murchison. Airy has had a little girl. I want him to call her Urania to correspond to the boy who is the Georgium Sidus. Many happy new years to you and yours in this or the other hemisphere.

Yours always, W. WHEWELL.

[1] Mr Malthus was unable to perceive colours properly. The colours in the rainbow and in the prism always appeared to him to divide themselves into two mainly distinguishable halos, blue and yellow. Red heat he called yellow.

Trin. Coll., *Jan.* 30, 1833.

My dear Herschel,

I think you have got our Transactions up to the present part which we have just published, so I send you that. I send you too an article[1] which is by and bye to appear in a Journal published here, where you will see that your *Leucocyclite* is criticised, perhaps too unceremoniously, but I am not afraid of your taking offence at what is said. You will have heard from Jones that he is preparing his introductory lecture for the 27th. If he gets an audience of any extent and intelligence, then I conceive that his popularity for the future is out of the reach of doubt; but I have a notion that the fate of all such undertakings in London is a matter of entire accident and incalculable caprice, so I will not indulge in any anticipations.

I rejoice to hear that Lady Herschel has presented you with a young experimental philosopher. I will hold you any wager that, young as he is, he will learn more than you in the next twelve months.

Yours ever, W. Whewell.

Trin. Coll., *Feb.* 2, 1833.

My dear Jones,

*   *   *   *

Another of the inclosed papers is a sonnet by Hamilton, the Dublin Astronomer, about which I want your advice. It takes my fancy extremely, and from the time when he first shewed it me (it has never been published), I thought that I should like to print it at the beginning of my Bridgewater book, either on the reverse of the title-page, or at the end of the preface. In doing so, I should say in the preface that the statements in one of my chapters concerning the tendency of mathematics to lead men's minds from religious views must be held to apply to some cases only, as was clear by such an example as the author of these lines, one of the first analysts of the age.

---

[1] The paper *On the Use of Definitions;* see Vol. I. page 65.

This would be no more than justice, for he is a superb analyst and a noble fellow. Tell me whether you think this would produce a good or bad effect; whether it would be looked upon generally as exaggerated, affected, or the like; or whether it would be a graceful and reasonable thing to do, supposing it done in good earnest. And tell me this soon, for I must write to him about it if I do publish the lines[1].

*   *   *   *

Yours always, W. W.

Feb. 11, 1833.

MY DEAR JONES,

I certainly am somewhat amused at seeing you so soon relapsing into the human frailty of definitions, after having just finished a most eloquent and convincing sermon against them; but I am also somewhat alarmed at seeing you in so dangerous a path. It appears to me a striking and edifying lesson on the dangers of looking with scorn at other people and thinking of oneself as exempt from the weaknesses and necessities of human nature. However, as you are overtaken in this way, I will give you such help as I can. I certainly should not have supposed that you could need a formal definition, exprest in studied terms, not afterwards to be modified. If you use "wealth," or any other term of which the signification is too lax or too wide in common language to convey your meaning properly, by all means give a *description* of what you intend it to include, and refer to and recall this description by such artifices of language as you best can when you want it. But as for your laying your hands upon an English word and conspiring with another man to make it mean some part only or modification of its usual meaning—if you do this, I as an Englishman protest against it, and will do what I can to raise a hue and cry against you if you make the attempt. A proper business indeed it would have been if you had written to Malthus! He I daresay would have given

---

[1] This sonnet has not been identified by the representatives of Sir W. R. Hamilton, who have kindly searched among his papers for it.

in to your folly, and at this very time you would have been
arguing the matter with him and fancying yourselves most
usefully employed. I will tell you however what Malthus *ought*
to have answered to your solicitation. " My dear Professor Jones,
I do not know what are the doctrines which your lectures are
to assert and explain. Language, my excellent young friend, is a
very good thing when it is used for the purpose of asserting
truth; I recommend you to employ such words and in such a
sense, that your propositions shall be true, and, if you are so
fortunate, shall be important and general truths. But I cannot
at all conjecture whether your circumscribing the meaning of
the word *wealth* in the way you propose will help you in doing
this; I should suppose that you, having, I presume, some notion
what your doctrines are, can tell whether the wealth to which
they apply is in all cases such as your phrases point out;—if so,
you may either use the word or the phrases wherever there is
occasion for one or the other;—but if it seems to you a matter
of importance to avoid using the phrases by previously giving
a formal definition of the word, this circumstance would make
me suspect that you have somewhat wavered in your notion
during the progress of your reasoning. As for my own part,
I cannot possibly agree to your proposal that we should endeavour
to give circulation to your proposed definition: for I do not
know what my next speculations may be, and it is very possible
that they may relate to something for which *wealth* is the best
word, but to which your definition does not apply. At any
rate, I will not put myself in the way of having any verbal
disputes if this should be so; and for the same reason I should
wish to use the privilege of greater age and experience and to
recommend you to keep out of all such rash engagements." After
receiving this long lecture from your economical pastor and
master, you would, I suppose, strike out all the formalities of
your definition and content yourself with saying, that when
wealth is spoken of in the following reasonings it is not intended
to include any but material possessions, &c. &c. I have another
special objection to all your definitions whether conveyed by
means of one part of speech or another. I cannot see that they

come to anything more than this—"By wealth I mean *property:*" and if you mean *property,* I cannot see why you should not say *property.* If *property* requires any explanation, certainly every possible mood, tense and participle of the verb *appropriate* requires it quite as much. If *property* is not a satisfactory word in the case where you use *wealth,* then your definition does not include all that your notion does, and will get you into scrapes and deserved ones.

All this I believe to be sound inductive logic, and so I send it you with my blessing.

I was going to write to you independently of this case of conscience of yours. I shall go to London on Thursday, and attend the Geological anniversary on Friday. Can you come then? Though you will not have finished, we can talk, and I want to show you my dedication. I shall write to Mould's for a bed for two nights. At any rate, let me find a letter at the Athenæum. Adieu. Yours always, W. WHEWELL.

I am glad Mrs Jones is better. Give my remembrances to her, and tell her that I have no fear of her making my sermon into *wealth,* that is something which is appropriated.

ATHENÆUM, *Thursday, Feb.* 14, 1833.

MY DEAR JONES,

I will stay in town on Saturday, so come here on Friday and we will still have our talk. Notwithstanding all your virgin tremors I have no fears for you, and no doubt that you and your public will have a pleasant honeymoon of it. The rule of prudence here, as in other cases, is to think well and kindly of your partner that is to be, but not to invest him with any supernatural powers or perfections. Any danger which there is of your not doing yourself justice arises from the chance of your thinking too highly of your public, or lecturing as if you did—I mean thinking too highly of their philosophical capacities. The highest flights of your generalisations, the most magnificent anticipations of future progress, the things which strike you most will not touch them much; some will be stupid and unable to understand; others prejudiced and unwilling to assent. But views, which are not too wide to be presented with complete clearness and illustrated by pointed detail, will be sure to win you applause. So you must clip your eagle's wing to prevent his soaring too high and make a smart flycatcher of him like a swallow.

I cannot persuade myself to entertain a moment's apprehension about the facility and propriety of your delivery. Of course you will not lecture in the *bow-wow* style in which you sometimes preach, and the new position will prevent you falling to the tone of familiar conversation. And then as for your wanting words, I suppose that even you never dream of such a preternatural occurrence. You must allow that you have no experience to give any colour to such fears.

I am playing truant so long this week that I cannot leave Cambridge again next week; so I hope you will come to-morrow. I am at Mould's. I will come and hear your lecture if I possibly can, but am quite willing to make an engagement not to bother you about it when there is no good to be done.

In the meantime are you printing—*printing* your lecture and syllabus? You ought to be. Why should you not give yourself a reasonable chance of full success? Your procrastination has prevented your doing this in every instance hitherto—which is often enough. If you come, I hope you will bring your lecture and syllabus—so goodbye for to-day.

Yours always, W. WHEWELL.

TRINITY COLLEGE, Feb. 17, 1833.

MY DEAR HARE,

I should at any rate have been delighted to receive a letter from you, but I was more especially so because I had long been nourishing an intention of writing you an epistle, but might easily have kept it still in the condition of an intention till you returned to England, if there had been no special cause to determine its transformation into an act at one time rather than another. It

was not that I had got much to say to you (you recollect perhaps
a person who could not conceive that men ever wrote to one another except when they had some definite *what* to communicate),
but I did not like so long an interruption of our intercourse after
it had been so frequent: and, moreover, I felt as if I had never
acknowledged or responded to all the kind thoughts and words
which there were in your last letter before this; which, considering
the great solace and joy they were to me, would have been by no
means a reasonable proceeding. I include in this agreeable recollection all the surplus beyond what I deserved; for you are not to
suppose that I was so fastidious as not to be pleased that there
was such a surplus. But to talk of other matters. I shall finish
my letter to-night, and send it off forthwith as the best chance of
not delaying it much longer, for I foresee a week full of engagements rising up before me. I have to-morrow coming to dine
with me Clot Bey, the Frenchman, who has learned Egyptian and
taught the Egyptians to get over their reverence for the human
body, which they have entertained for five thousand years, I suppose, and has induced them to dissect. You will not feel any
strong admiration for the man on this account: but I have also
another stranger coming whom you would much more probably
like. This is also a Frenchman, named Rio, a friend of all the
Malcolms, who made his acquaintance at Rome and are quite enchanted with him—Lady Malcolm especially—and he is, as might
be imagined, enraptured with them, especially Lady Campbell.
He is, moreover, a philosopher of the school of Bonald, an intimate friend of De Maistre and of Schelling, and a most earnest
Catholic, enthusiastic about painting, music, and languages—now
should you not like to know *him?* as girls sometimes say to one
another. But his errand to England appears to me the most
amusing I have heard of. He is a Breton—a Celt—he holds that
the Celts are the only sound part of the French population, the
only part which has any religion, or any social vitality. From
the French Celts the regeneration of France must come, if it come
at all. But the French Celts are poor, and have been oppressed,
and have let some of their Celtic spirit and culture slip away from
them. This is to be restored by a reinfusion of Celtic poetry and
history. So M. Rio is come to cultivate the Welsh, and to see

what he can find in their literature worthy of being employed to fill the minds of the Bretons with worthy thoughts and hopes. Now is not that a pretty project? Poor Rio has been grievously disappointed hitherto in his attempts both to find Welshmen and Catholics in London; I hunted with him a whole morning, but we could not catch any creature of any value of either description. Since that time he has detected Digby, and has of course struck up a very close friendship with him. But I am very sorry to learn from him that Digby has been very ill, and is still ill of a rheumatic fever—his not unusual visitant. Digby has been publishing a Vol. III. of the Mores Catholici, as beautiful, I think, as ever, and even more remarkable for his assimilating power of extracting from everything nutriment for his own opinions and feelings. I have left myself but little room for any account of our other friends. Thirlwall's lectures are admirable, and the men take to them with great earnestness; especially the second and third year. He has got the third volume of Niebuhr, so you shall not need to be idle when you return. Sedgwick is well and prosperous; two days ago I heard him talking at the anniversary dinner of the Geological with his usual eloquence and whim. He preacht (I hope you are sensible of the politeness of my spelling) our Commemoration Sermon, which, though most unskilfully written and delivered, had in it so much of good, that the undergraduates begged him to print it; perhaps in half a year longer he may do so. The Marchesa is well, and is, as usual, one of the greatest of one's comforts here: also my Bridgewater Treatise is going to appear in a fortnight; and this transition is not so violent as you might think, for the Marchesa has just been criticising my Dedication (to Blomfield). Romilly has a cold, and Thorp has a lame leg by a fall from a horse; but they will be well long before you get this, even if it travel quicker than yours, which seems to have been six weeks on the road.

There is no use in writing politics: but as I came from London this morning, a man begged me to buy a newspaper containing Lord Grey's Bill for the *Suppression of Ireland.*

My kindest regards to Worsley; I am glad he is still with you.

Yours affectionately, W. WHEWELL.

TRIN. COLL. *March* 24, 1833.

MY DEAR JONES,

Why have I not written to you? Why, I was expecting every day to receive your lecture all properly finished. I cannot imagine what you have been about all this while, seeing that you had it ready for the press 8 days after it was delivered. However, I shall look to see it soon. I think my Bridgewater makes a very pretty-looking book. All the people here say they like it much. I have not yet talked with any one who has read it through. I want most to know what impression is made by the part containing the contrast of induction and deduction. I hope that will take hold, and then I do not care how soon people forget that there was a time when they were blind to the difference. Pray set about dispersing your lecture as fast as you can. I have no fears at all about your other lectures: they ought not to be so high-pitched as the first; they ought to be what they will be, a clear exposition of principles, with instructing views of realities illustrating them. In short I cannot conceive any danger or even serious doubt in your task now.

I think I shall go to London on Wednesday and stay for a few days—will you come? I am going to get more materials for my paper on Tides, which is now assuming the form almost of a treatise of respectable size. Also I want to talk with you about getting statistical information, if the British Association is to be made subservient to that, and which I think would be well.

I hope you will succeed in your attempt on the Archbishop. I do not know on what principles they act in such cases, and can only hope that there may be some rule or habit, which I am not aware of, which may give you a living of reasonable fatness. I hope at any rate that you will not have the hardship, which to your temper I do not think would be a slight one, " in suing long to bide."

I write having no time and nothing to say and must end. My regards to Mrs Jones.

Yours always, W. WHEWELL.

[The next two letters were written to Professor W. R. Hamilton of Dublin, afterwards Sir W. R Hamilton.]

CAMBRIDGE, *March* 18, 1833.

MY DEAR SIR,

I hope we shall have the pleasure of seeing you here at the meeting of the British Association in June. I shall reckon upon it, and shall provide rooms for you in College. I shall be especially glad of any occasion which gives me the opportunity of seeing you, and I hope, if you and people like you will help us, the Cambridge meeting may turn out as agreeable and instructive as I found the Oxford one.

I do not know where it is likely that the following meeting will be held, nor have I had any talk with any person on the subject. For my own part I see many good reasons in favour of Dublin, if Dublin thinks so, far more clearly indeed than 1 saw them when we were at Oxford. I do not know, however, what is to be urged for other places.

I have published two books since I saw you, one only within a day or two, which I wish to send you and will tell my bookseller to convey. The last is my "Bridgewater Treatise on Astronomy and Physics in connexion with Natural Theology." Your sonnet, which you shewed me, expressed, much better than I could express it, the feeling with which I tried to write this book; and I once intended to ask your permission to prefix the sonnet to my book, but my friends persuaded me that I ought to tell my story in my own prose, however much better your verse might be.

The other book is partly historical and partly speculative on the laws of motion, which I much want you to see.

In the hope of seeing you in summer believe me, my dear Sir,

Yours very truly, W. WHEWELL.

I thank you much for your Lecture on Astronomy.

TRIN. COLL. CAMBRIDGE, *May* 17, 1833.

MY DEAR SIR,

I am extremely glad to hear that we shall have the pleasure of seeing you here at the meeting of the British Association. I reckon upon you after the enquiry which you make, for I am not satisfied with myself that you should have such a question to ask. Whenever you arrive, you will find a room in College ready for you, and you will have such "provants" as our college ways afford as long as you can stay—the longer the better. It would be hard indeed if you, a professor and a Trinity College man, to say nothing of other claims, could not be housed in *our* Trinity College for a few nights in vacation time.

I beg leave to offer you my most hearty good wishes and congratulations on the change in your position, which left any uncertainty on the subject of the place of your abode during your visit here. In truth I somewhat share in the disappointment of Prof. and Mrs Airy, that you do not bring Mrs Hamilton in your hand, that she may see what English Savans are like, and that we may have the pleasure of forming her acquaintance. But we shall be very glad to have you by yourself, since better may not be. We are to see Herschel and his lady, Buckland and Murchison bring their wives, Davies Gilbert his daughter, and, I believe, many others.

I received and read your sonnets with great pleasure. I think you are to be envied for being able to keep your mind in such a frame, but then you have aids in the task which all of us have not.

I shall be very glad to have a more detailed account of your optical views laid before our meeting.

*   *   *   *

I directed my London publisher to send you my little book on the Laws of Motion. I have, however, said little on the logic of Statics as distinct from Dynamics, for I wrote for our "non-reading men" who have a very limited taste for metaphysics. However, I think my notions of the distinction are more clear than when we talked on the subject before, and I shall be glad to discuss it further.

In the hope of seeing you on or before June 24th, I remain,

Yours very truly, W. WHEWELL.

TRIN. COLL. *Ap.* 28, 1833.

MY DEAR MURCHISON,

You have probably seen Sedgwick, as he was to be at the D. of Sussex's, and is to come here to-morrow However, I proceeded upon your authority to take possession of my own copy of your speech, which I wanted much. Sedgwick's ailments, I hope, are quite dissipated by the air of the Hastings sand and sea.

I am glad of your opinion of my book. After reading the whole I do not fear that you should think I have done any injustice to the mathematical *developers*. As to the Quarterly Review of it, if I felt much anxiety about that matter, I should not at all depend on Lockhart's giving it to Jones: I think he has rather Brewster in his eye. I do not expect a good review from Brewster, for though an admirable discoverer he is a bad critic : nor do I expect a favourable one, for he hates all talk about Bacon and Inductive Philosophy, which I hold to be the best part of my book. After the impression the book has made on those whose opinions I most care for, I do not feel very much about the Review, except it were by some better hand— some one likely to take a more commanding view of philosophy than Brewster can. Sedgwick had for a time a strong propensity to review me; but I think the time and trouble would repel him when he came to the job. In many ways he would do it admirably : but we *freshmen* reviewers are too serious for Lockhart : if I ever review again, I think I shall know my trade better than before.

TRIN. COLL., CAMBRIDGE, *April* 27, 1833.

MY DEAR FORBES,

I send you herewith notices which are intended as invitations to some of your Scottish societies to send delegates to the B. A. I leave it to you to decide whether those to Glasgow and Inverness are to be sent, as I do not know the character and merit of the institutions. I send you also a few blank copies, which you can fill up and dispatch, if you think that there are other bodies which it is proper to invite. I find from Phillips that almost every little town in Yorkshire has its "Philosophical Society," and that it has been thought proper to invite them; so you need not be very exclusive. I send along with this invitations to some of the persons you mention; such of them as I know. It does not appear to be thought proper that we should officially invite any persons who have not already given us their names; but individuals may make such applications and offer such inducements as they think proper. If you can induce the noblemen you mention to come, we shall be most happy to see them. Sir John Sinclair is a member, and has been written to I send you some of our blank circulars.

When we meet here we can talk about Mechanical books and the like. It was, I believe, from running too hastily over your *red cross* lines that I missed what you said of my last little book. I acknowledge the want of such a book as you describe, but I do not think it could be made without a considerable sacrifice of time, an article which I find it more and more difficult to command. I become also more fastidious about style and manner, and I have moreover still to print (and write) the second part of the new edition of my Dynamics; so I can promise nothing. I shall be very glad to advise and suggest, if any one else will work; *you* for instance, why should not you write or select such a book? I dare say another person would choose better than I should. I directed my Bridgewater book to be sent to you; there are some parts of which I should like to know your opinion.

Pray seal the accompanying letters, and send them to their destination.

Believe me, dear Forbes, yours truly, W. WHEWELL.

*　　*　　*　　*

TRINITY COLLEGE, CAMBRIDGE, *June* 10, 1833.

MY DEAR FORBES,

I hope this letter will find you still at Paris, as I want you to deliver a message there. I am told that there is a hope of seeing Arago at the meeting of the British Association. This information comes from Mr Pentland : if it is well founded, pray let Arago know that we shall be delighted with his joining the meeting, and invite him to take up his residence in Trinity College as long as he can stay with us, and to share such hospitality as our College can supply. If Mr Pentland comes with Arago, we will establish him in the neighbourhood of his illustrious friend. It would on all accounts give me the greatest satisfaction to see Arago here, and the feeling would be equally strong on the part of the President, and all the members of the Association. If you find they are likely to come, or any other eminent Parisians, let me know as soon as you can.

I suppose you are looking about with an eye to anything which may further the business of our meetings. It would be well, I think, to have some good subjects ready, either for discussions or reports, on which some light is likely to be struck out. If the meeting is held at Edinburgh next year, this will very materially affect its success, and will therefore interest you. I do not think there is any subject on which we are so likely to do something, as yours, Meteorology. I should think it likely that we may get such cloud maps, as you speak of, drawn for a given time. Another point which appears to me to require consideration is the theory of clouds and rain. If I am not mistaken, Gay Lussac a little while ago countenanced the theory of clouds being composed of *vesicles* of water, which appears to me quite fantastic and incredible. But how are they supported ? Again, what is the immediate cause of the conversion of clouds into rain ? You still speak well of Hutton's theory, but that theory appears to rest entirely on the doctrine of the solution of water in air, which you, rightly I think, reject. Then as to other questions. Is there any chance of obtaining any information of the condition of the higher regions of the atmosphere ? How high does the direct current of the common winds extend ? Can any facts be collected to verify or test Herschel's theory of hurricanes ? (p. 132 of his Astronomy just published). Again, will any analyst solve for us the problem of the pressure of an elastic fluid *in motion,* so as to apply to the barometer ?

Of course there is no lack of such questions, but it would be well if you would select a few which may be likely to receive answers, either by co-operation or otherwise.

I should be glad if you could procure me the *Annuaire* of 1825, which you quote as containing Arago's speculations on climate; and indeed I should like to have all which contain scientific articles of his.

I am much obliged by what you say of my Treatise and by your corrections, which I have adopted, if that part of the second edition is not already printed. I imagine such works are not much to the taste of the French public at present. If you hear of any intention of translating it, I should be glad to send corrections.

You will bring all the scientific novelties which are afloat at Paris, and we shall be glad to see you here before the meeting begins, if it suits you, for we shall have preparations to make.

My best regards to Mrs Somerville if you see her.

Yours very truly, W. WHEWELL.

TRIN. COLL, *July* 27, 1833.

MY DEAR LUBBOCK,

Finding nothing which made my presence in London necessary for more than a few hours, I came to Cambridge by the mail last night. I found here an answer to my application to Mr Yates about the Liverpool tide observations. I send you the paper that you may see how our affair stands. You will find that the tide observations which they possess refer to an earlier period than any which Dessiou has already calculated, and to a period when, I suppose, there were none made at Brest. As it appears that the original papers will not be allowed to leave Liverpool, it is worth considering whether it is best that Dessiou should go there, if it can be allowed. I suppose the discussion of a year's observations would not take him long, and, if he could set about it soon, he might possibly be at work when I go into the north, in which case I could look at the result of his labours and enable myself to form an opinion about the desirableness of proceeding. Indeed I think it is well worth your considering whether you might not take Mrs Lubbock to Liverpool. It is quite as good a sight as Birmingham, besides which you might see the railroad and travel along it to Manchester, and then cross over into Yorkshire by the English Apennine chain.

In the mean time I shall be very glad if you will look at the question and consult Dessiou and any one else, so as to further the business as much as possible. It would at any rate be desirable to get things *en train* before your wandering begins. I suppose either you or I ought to communicate with Baily and Peacock when any decisive step is taken, though I think we have their authority for doing all that I have suggested. By way of shortening Dessiou's absence he might pick out his transits beforehand, and indeed all the part of the process which would require his presence at Liverpool would be his transcription of the tide-times, which, when his skeletons were prepared, would be soon done.

I believe my geological ramble in company with Sedgwick and Airy will begin in the middle of next week : but if you write to me here, I will direct your letters to be forwarded in case I should be gone.

I send a copy of my address of which I wish to beg Lady Lubbock's acceptance, as well as one for yourself, which I had not when I was with you. Believe me,

Yours very truly, W. WHEWELL.

I should wish to have the enclosed paper returned to me when you have an opportunity.

TRINITY COLLEGE, CAMBRIDGE, *Oct.* 31, 1833.

MY DEAR LUBBOCK,

I send you two letters which I have written; one to the Managers of the Lyceum and one to Mr Yates. If you approve of them, pray seal them up and send them off—enclosing the one to the Managers in the other. You had better probably get an Admiralty, or some other, frank, so as not to impose any charge of postage. If these fail we must try some other way.

I have been going on with my comparison of your Tides with

the theory, because I really could not help it. The conclusions I obtain interest me so much that I am quite incapable of waiting for your new observations. Moreover I can get very good formulæ from the old observations; and I believe these will agree with the theory *with a little modification,* which I will explain when I see my way more clearly. At any rate I shall get formulæ which will represent your tables very well, and I am persuaded that I can calculate tide tables from my formulæ, which will agree with observation as well as any extant tables or better. I should like to do this, and when Dessiou is calculating London tide tables, I should wish to send him my tables of the corrections for parallax and declination, that he may use them at the same time with yours (which will give him very little additional trouble) and thus get out a set of tide tables of my own. I hope you see no objection to this. I have got my formulæ and tables for the times, and am going to see about the heights.

As you appeared to doubt the possibility of obtaining the laws of the inequalities from the observations alone, I had no hesitation in setting to work to try, and I have now not only done this, but also got nearer the theory than I expected to do. My lucubrations on these subjects I have drawn up as a paper which will be of some size, and which I intended for the R. S. If it is likely to interfere with what you are doing, I will look elsewhere, but I think I can make something of the subject, and shall not be content till I have worked out my views. They make me more desirous than ever to have the Liverpool and Brest observations discussed, for it will be infinitely interesting to see whether the inequalities due to parallax and declination assume the same form there as at London.

I know so little about the methods of calculating the orbits of comets that I can hardly assume a right to interpose my judgment on the subject of Halley's Comet. If I see Stratford I will mention the subject to him, and give him my opinion of the superior value of your elements so far as I am able to form one : but not having the comparison of the other orbits with observation before me, I cannot judge relatively. Yours in your first note appears to agree very well upon the whole with the observations. Airy shewed me to-day a diagram, which, I understand, is to be inserted in the Nautical Almanac, containing your orbits, Pontécoulant's and Damoiseau's. I should think this would answer most purposes, as the comet itself, if it comes at all near the time and place of its appointment, must decide which is the best orbit.

I shall look with great interest to your new London observations; the more so, as I have now got the laws which I *expect* them to follow. But the observations of some other place not discussed are now the subject of my most intense curiosity.

Believe me, yours truly, W. Whewell.

Trin. Coll. *Oct.* 21, 1833.

My dear Jones,

*   *   *   *

I went to London, as I told you I should, and saw Herschel and his wife—a great pleasure to me, as it always is; and the unhappiness of parting with him is almost done away with by looking at the views and feelings with which he goes, and the temper in which she accompanies him. I went with them to the ship in which they are to make the voyage—a very respectable structure to a landsman's eye, and a very safe-looking "craft" to a sailor's, as I collected from what I heard there. The Captain talked of leaving Portsmouth about Nov. 1, and I have not since heard of any more recent determination.

I should like to see you much if it were at all feasible. I suppose it is not, for I am now fairly fastened here, and cannot well go even to London for the next month or two. I want to know something of the prospects of your second volume, and, if possible, to expedite its progress. My conviction of the importance of doing this by no means diminishes. There are other matters concerning myself about which I should be glad to talk to you— not, I think, from any feeling of indecision; but because in any considerable change it is pleasant to speculate on probabilities with a friend, and there is no one here now with whom I can discuss this matter—I mean my project of resigning the tuition, of which

I have spoken to you before. My wish to do so grows upon me : the "business" remains as tiresome as ever—I do not say that it becomes more so : and my Induction "invites my steps" every half hour that I am left to my own thoughts. If I am ever to do any good, I must set about it soon (I shall be forty in half a year)— I shall never be able to give my mind to my tutor-work, though the work might well employ any body, and is by no means ungrateful, either as to the chance of doing good or of making money. As to the latter point, I have made possibly enough to make me easy for the rest of my College life, and it would take ten years more to make a fortune, with the chance of losing credit and money in the mean time. I think I have now a good opportunity of resigning with credit, and therefore in two or three days, when the Master returns into College, I shall propose the matter to him, and try to make some arrangement by which I may, in a year or so, unwind myself from my disagreeable web of small filaments, and become master of my own motions.

I have as yet not mentioned this project to any living soul, except one person whose personal concerns rendered the communication necessary. If you see anything very absurd or insane in it, write and say so, for I have no disposition to rush on headlong. But the scheme has grown up in my mind so gradually and steadily, that I think it little likely any further consideration will alter it.

I was much afflicted to hear of the fortunes of your face, which have certainly been most disastrous. I hope matters are now in a more comfortable and prosperous condition.

Pray propose the following problem to Mrs Jones—Why is the lady of a certain great astronomer like a hodmadod ? I suppose she knows that hodmadods are the snails with green and yellow striped shells, which are found in gardens. This problem is of Cambridge produce, and was propounded during the meeting of the British Association in June. Adieu. This is the first day of lectures, so that it is wonderful my being able to write a letter to you—however it is between 11 and 12 at night.

Yours as always, W. W.

TRIN. COLL., CAMBRIDGE, *Nov.* 13, 1833.

MY DEAR JONES,

Why have I not had the end of your last letter which you promised me ? What is come of your winter plans, and where am I to seek you, supposing that I am able and willing to do so, in the Christmas vacation ? Has St Leonards or Weymouth carried the palm with you ? I should be glad to hear any news of Hare and of his new abode which you can give me. Worsley is just come here and tells wonders of his collection of pictures, bought during his tour in Italy. All this you must write and discuss to me without any extravagant delay.

As to my project of giving up the tuition, it was an admirable device, but I have perhaps done better still. I have made an arrangement with Perry and Thirlwall, by which the former is to take all the bookkeeping, which of itself is more than half the plague of the tuition and necessarily carries with it half the remainder, and I am to remain as head tutor with considerable emoluments, and with easy work compared with what I have had. This is excellent good in itself, but better still in this, that it makes it a very easy and short affair to give up the tuition altogether, if I find that my remaining official employments are burthensome. I can slip my cables at very short warning when I once resolve to make a run.

This being so, I am meditating the returning forthwith and in earnest to my beloved Induction. I have been employed all the term hitherto upon a thumping paper on the Tides, which I intend to be a step of some consequence in the theory. I wish I could explain to you how useful my philosophy is in shewing me how to set about a matter like this, and how good a subject this one of the Tides is to exemplify it.

If I ask about your book, I suppose you will tell me about your face. I am very sorry for their interference on all accounts, and wish heartily that your face may diminish, and your book increase. I wish this not on your account only, but because it is clear that, after a little while, books of theory will assume a very important sway over men's minds, and it will be bitterly provoking that there should be no sound and philosophical one published on your subject, when there is one almost written.

Between my Tides and my Induction I have given myself a holiday of three days to write letters, of which this is one. I look forwards with infinite satisfaction to the prospect of setting about the book in earnest. I have a week or two's work to bestow upon the history of Astronomy as subsidiary to it, and then I shall try my hand at expounding myself. But it is best talking of such things when they are done.

Do you know when Herschel sailed (I suppose it is now a matter to be spoken of as past), and do you know anything of him for the last week or two of his stay in England? If you do, pray impart your information. Probably Mrs Jones has had some adieux from Lady Herschel, for I find Mrs Airy has.

Ever truly yours, W. WHEWELL.

TRINITY COLL., CAMB., *Dec.* 10, 1833.

MY DEAR JONES,

Your letter has delighted me. I was deeply grieved by your former one. Not by any cowardice that I found in it, but by your courage and your manifest conviction that a great effort of courage was needed. The amendment which you describe is beyond anything which I had hoped : pray go on getting well as fast as you can, and leave poison to rats and arsenic to mineralogists. As to my visiting you, I think the best thing you can do is to make your movements independently of any regard to mine. I hardly know how to get away from Cambridge, for, besides all other entanglements, I have begun to write my book, and have got about 40 pages in my head, which I have been trying in vain to find time to jot down upon paper : and I think I shall be loth to go hence before I have done this, which may not be soon, as bills and pupils are at present likely to take more time than usual in consequence of the process, which is going on, (pleasant in other respects) of transferring a large share of the trouble to Perry. I assure you my book is assuming a very orderly and plan-like shape, and what I write now is expected to stand; so that, if you do not take care, I shall get the start of you with my next volume. I think a year and a half or two years will help me through a good deal of the part I contemplate publishing first; however more of this hereafter. When I do leave Cambridge I go to London, to Milton, to Audley End, and perhaps elsewhere: and when I can take a jump to you, it will make little difference whether I find you at Worthing or at Brighton. Either place will do well to enjoy one or two days of Christmas holiday.

I give you a very vague account of my probable motions, but I cannot do more. I will write again when I know better what I am going to do. My best remembrances to Mrs Jones. I hope the air is as good for her as for you.

Yours always, W. WHEWELL.

TRINITY COLLEGE, CHRISTMAS DAY, 1833.

MY DEAR HARE,

Your letter came to me very opportunely, on the morning of our Commemoration day, as I have no doubt you intended that it should. I am glad to see that you are, as of yore, an observer of times and seasons, for Worsley, who dined with us to-day, tells me that he received a letter from you this morning, besides your recollection of Kate's birthday, which is an occasion too peculiar to be disposed of by any mere process of classification. But as to the Commemoration day, which I must not dismiss so immediately, I was very glad to have it distinguished by such a mark in addition to others. It was not without its appropriate glories, for Sedgwick's Sermon had been publish*ed* [1] (if you will allow me to say so) a day or two only previous to it, and had drawn a sort of interest to the occasion; the sermon being, as I think you know, a very beautiful and profound dissertation as he delivered it; as having been re-thought and re-written since, it is much more beautiful and coherent; and along with certain notes which he has appended to it, it forms an essay upon philosophy, morals, and academical education which it will delight you to read. I have directed him to have a copy for you at the Athenæum (he is gone

---

[1] Mr Hare would have preferred *publisht;* see *preacht* on page 160.

to London), and I hope you will receive it safely and soon. It was well we had such a reserve in the way of sermon, for Peacock, who preached for us on the day, had given himself only a day or two for the writing of his discourse, and the end of it was, that it was a sort of political essay on the duty of regulating our views and feelings in accordance with the new and reformed state of things, which, though by no means devoid of cleverness and dignity, sounded rather like an article in the Morning Post than a sermon. However, we had besides this something better; for Birks, who was our declaimer, gave us a dissertation on the thesis, "that there is a moral truth which in its own way is as certain as mathematical truth," such as I really do not know any other person who could have written. The philosophy was most profound and consistent, and the views of the nature of morality of the pure and elevated kind, which I hope we shall always hear from our best men here ; and withal there was a richness and poetry of illustration, which did away with any unfitness which you might have anticipated in such a subject for such an occasion. His images often reminded me of Bacon's ;—a bright flash of ornament with a clear thread of poignant analogy sparkling through it. I think I shall ask him to print it, but at present he is busy with his preparations for being, I hope, Senior Wrangler, so I must not disturb him with such vanities. I perceive I am getting to the end of my paper, so I must tell you before it fails me, that I think it possible I may pay you a visit for a day in the course of next week. I intend to go to see Jones, who is either at Worthing or at Brighton, and as I have a sort of dim notion that all your people on the Southern border are in a connected region, I hope I shall be able to win my way to Herstmonceaux. I cannot be absent long, for I am very busy with business not pleasant in itself, inasmuch as it is book-keeping; but very pleasant in this point of view, that it is, I trust, my leave-taking of such business. I have transferred all tutorial money matters to Perry, and have grand projects of writing a philosophy, such as shall really give a right and wholesome turn to men's minds. Indeed they are more than projects; for I have written one Book out of them, and collected large materials for others. But I will not vapour about this matter any more till I see you. I hope to find that you have not discovered the necessity of being a party man when I visit you, if I am so fortunate. One obligation I have to my philosophy is, that it makes it irksome to me to think of such things. I will bring you a book of Milnes's, which, if you have not seen it, will I think, please you.

*   *   *   *

Yours ever, W. WHEWELL.

Jan. 7, 1834.

MY DEAR ROSE,

I send you this interesting document[1] with great pleasure, that I may ask for your congratulations on the occasion of its being the last of such literary essays which you will receive from me. The business of keeping other people's accounts and receiving and paying money for other people has always been a heavy and wearisome load, and I have long had a resolution formed to get rid of it before many years were gone by. I do not know that my resolution would have taken the shape it has, or have been executed at this particular time, if it had not been for some accidental circumstances; but feeling now that every stroke of my pen is filing away my chain I go on with right good will in my task. I look forwards with much pleasure to the rest of my tutorial employments, when I consider that I am to have them unembittered by that part of the business, for which I knew myself to be most unfit, and which I felt to be an abominable annoyance. I shall now probably have a little time to philosophize for myself, but whether I shall do so to any purpose, time must shew.

I am glad to hear a good account of your university[2]—for such on the whole I hold yours and Whitley's to be. I dare say we shall be able to learn some things of you. I wish most heartily among other novelties, you would some of you discover or write a system of morals which might take the place of Paley. Sedgwick tells me he has sent you his sermon; when you read it, you will

---

[1] An account of some customary payments due by Mr Rose to Trinity College.

[2] Mr Rose was at this time Professor of Divinity in the University of Durham; and Mr Whitley was Reader in Natural Philosophy and Mathematics.

see that he has declared war against both Paley and Locke. This puts them on a different footing in Cambridge from that on which they have hitherto been; for though opinions to the same effect were in very general circulation in the place, they were never, I think, clothed with any thing like an authoritative expression before. The task of writing a system of ethics is certainly not easy, for it must not only be erected on sound principles, but so framed as to bear an advantageous comparison in its logic and execution with the best of other systems, for instance, with Paley's book—which is no easy condition. I am afraid, from what your *Brit. Mag.* says of Wardlaw's *Christian Ethics,* he has not solved this problem. I have sent a parcel of books for you, addressed 250 Regent St., but having been a very short time in town I had not an opportunity of ascertaining that they had fallen into good hands and were likely to reach you. I hope Durham continues to agree with you and you to like it. Pray give my best remembrances to Mrs Rose. I hope she likes Durham and all the people there, or, if this be too large a wish, at least a great many of them; and if there happen to be any people that she likes less than others, that she contrives to extract amusement out of them : for it is very praiseworthy to pick good out of evil. We shall be right glad when you and she come into our latitudes again. Adieu, and believe me, dear Rose,

Affectionately yours, W. WHEWELL.

[To Professor Hamilton of Dublin.]

TRIN. COLL., CAMBRIDGE, *Mar.* 27, 1834.

MY DEAR SIR,

I am glad to hear you have been turning your thoughts to Mechanics, and have no doubt you will make a hole quite through them with your long analytical borer, and for aught I know bring up purer waters from greater depths than we have yet known, as they are wont to do in this country. In the mean time I, who have been long muddling at the bottom of the well, have persuaded myself that I have got the mud to subside, and have been trying to distinguish how much of the stuff comes from the clear spring of intellect, and how much is taken up from the base mud of the material world. I send you my attempt to render to the principles of intellectual necessity and empirical reality what belongs to them—to each its due. If what I have written leads you to speculate about the matter I shall be glad, and glad too to hear of your speculations from you.

I send you another little production which I think you will like, by one of our best mathematicians—indeed I think much the best of our young men, and likely to be a great analyst. The speculations appear to me very steadily conducted under all the profusion of ornament in which he indulges.

The Airys have had a series of illnesses, but are now, I hope, nearly well and likely, if one may judge from appearances, to be *more* as well as merrier. Sedgwick, you may have heard, hurt his arm very much by a fall from his horse ; he dislocated his wrist and it was not set again for 19 days. He is now recovering the use of his hand. I hope you have seen his Discourse on the Studies of the University. I believe he wanted to send it you but could not find a channel—you will wonder, I suppose, that the *Irish Channel* did not occur to him.

Believe me, yours very truly, W. WHEWELL.

Will you have the goodness to send my memoir to Lloyd ?

[The next two letters were written to Professor Faraday, who consulted Dr Whewell as to the nomenclature to be used in the exposition of his researches in Electricity : see Vol. I. page 99.]

TRINITY COLLEGE, CAMBRIDGE, *April 25*, 1834.

MY DEAR SIR,

I was glad on several accounts to receive your letter. I had the pleasure of being present at the R. S. at the reading of your paper, in which you introduced some of the terms which you mention, and I was rejoiced to hear them, for I saw, or thought I saw, that these novelties had been forced upon you by the novelty of extent and the new relations of your views. In cases where such causes operate, new terms inevitably arise, and it is very fortunate when those, upon whom the introduction of these devolves, look forwards as carefully as you do to the general bearing and future prospects of the subject; and it is an additional advantage when they humour philologists so far as to avoid gross incongruities of language. I was well satisfied with most of the terms that you mention; and shall be glad and gratified to assist in freeing them from false assumptions and implications, as well as from philological monstrosities.

I have considered the two terms you want to substitute for *eisode* and *exode*, and upon the whole I am disposed to recommend instead of them *anode* and *cathode*. These words may signify *eastern* and *western way*, just as well as the longer compounds which you mention, which derive their meaning from words implying rising and setting, notions which anode and cathode imply more simply. But I will add that, as your object appears to me to be to indicate opposition of direction without assuming any hypothesis which may hereafter turn out to be false, *up* and *down*, which must be arbitrary consequences of position on any hypothesis, seem to be free from inconvenience even in their simplest sense. I may mention too that *anodos* and *cathodos* are good genuine Greek words, and not compounds coined for the purpose. If however you are not satisfied with these, I will propose to you one or two other pairs. For instance, *dexiode* and *sceode* (*skaiode* if you prefer it) may be used to indicate *east* and *west*, agreeably to Greek notions and usages, though their original meaning would be *right* and *left*: but I should say in this case also, that right and left, as it cannot be interpreted to imply a false theory, any more than east and west, would be blameless for your object. Another pair, *orthode* and *anthode*, which mean *direct* and *opposite* way, might be employed; but I allow that in these you come nearer to an implied theory. Upon the whole I think *anode* and *cathode* much the best.

I have already said that I liked most of your new words very well, but there is one which I should be disposed to except from this praise; I mean *zetode*. My objections are these. This word being grouped with others of the same termination might be expected to indicate a modification of electro*de*, as eiso*de*, and exo*de*, or ano*de* and catho*de* do. Instead of this, it means a notion altogether heterogeneous to these, and the *ode* is here the object of a verb *zete*, contrary to the analogy of all the other words. It appears to me that, as what you mean is an element, all that you want is some word which implies an element of a composition, taking a *new* word, however, in order that it may be recollected that the decomposition of which you speak is of a peculiar kind, namely, *electrolytical* decomposition. Perhaps the Greek word *stecheon* (or *stoicheion*) would answer the purpose. It has already a place in our scientific language in the term *stœcheiometry*, and has also this analogy in its favour, that whereas your other words in *ode* mean *ways*, this word stecheon is derived from a word which signifies to *go in a row*. The elements or zetodes are two things which *go*, or *seek to go*, opposite ways. I might add that, if you want a word which has a reference to your other terms, the reference must be to the process of decomposition by which these elements are obtained. You might call your zetode, an *electrostecheon*, especially if you had occasion to distinguish these elements obtained by *electrolytical* processes from others obtained by *chemolytical* processes, that is, the common analysis effected by the play of affinities. Elements obtained in the latter way might be called *chemostecheons* in opposition to *electrosteckeon*. But I am afraid I am here venturing beyond my commission and out of my depth; and you must judge whether your stecheons or zetodes, or whatever they are to be, are likely to require the indication of such relations. If you were to take anode and cathode and adopt stecheon, I think *anastecheon* and *catastecheon* might indicate the two *stecheons*. If you stick to zetode, *anazetode* and *catazetode* would be the proper terms; but perhaps *zetanode* and *zetocathode* would be more analogous to zetode, which is a word that, as I have said, I do not much like.

My letter is become so long that I will recapitulate: *anode, cathode, zetanode, zetocathode* fulfil your requisitions; *anode, cathode, anastecheon, catastecheon* are what I prefer. With great interest in your speculations, and best wishes,

Believe me, yours very truly, W. WHEWELL.

TRINITY COLL. CAMBRIDGE, *May* 5, 1834.

MY DEAR SIR,

I quite agree with you that *stechion* or *stecheon* is an awkward word both from its length and from the letters of which it is composed, and I am very desirous that you should have a better for your purpose. I think I can suggest one, but previous to doing this, I would beg you to reconsider the suggestion of *anode* and *cathode* which I offered before. It is very obvious that these words are much simpler than those in your proof sheet, and the advantage of simplicity will be felt very strongly when the words are once firmly established, as by your paper I do not in the least degree doubt that they will be. As to the objection to *anode*, I do not think it is worth hesitating about. *Anodos* and *cathodos* do really mean in Greek *a way up* and *a way down;* and *anodos* does not mean, and cannot mean, according to the analogy of the Greek language, *no way.* It is true that the prefix *an*, put before *adjectives* beginning with a vowel, gives a negative signification, but not to substantives, except through the medium of adjectives. *Anarchos* means *without government,* and hence *anarchia, anarchy,* means *the absence of government:* but *anodos* does not and cannot mean *the absence of way.* And if it did mean this as well as a *way up,* it would not cease to mean the latter also; and when introduced in company with *cathodos,* no body who has any tinge of Greek could fail to perceive the meaning at once. The notion of *anodos* meaning *no way* could only suggest itself to persons unfamiliar with Greek, and accidentally acquainted with some English words in which the negative particle is so employed; and those persons who have taken up this notion must have overlooked the very different meaning of negatives applied to substantives and adjectives. Prepositions are so very much the simplest and most decisive way of expressing opposition, or other relations, that it would require some very strong arguments to induce one to adopt any other way of conveying such relations as you want to indicate.

If you take *anode* and *cathode*, I would propose for the two elements resulting from *electrolysis* the terms *anion* and *cation*, which are neuter participles signifying *that which goes up*, and *that which goes down;* and for the two together you might use the term *ions*, instead of *zetodes* or *stechions*. The word is not a substantive in Greek, but it may easily be so taken, and I am persuaded that the brevity and simplicity of the terms you will thus have will in a fortnight procure their universal acceptation. The *anion* is that which goes to the *unode*, the *cation* is that which goes to the *cathode*. The *th* in the latter word arises from the aspirate in *hodos* (way), and therefore is not to be introduced in cases where the second term has not an aspirate, as *ion* has not.

Your passages would then stand thus :

" We purpose calling that towards the east the *anode*[†], and that towards the west the *cathode*[‡].......... I purpose to distinguish these bodies by calling those *anions*[§] which go to the *anode* of the decomposing body, and those passing to the *cathode*, *cations*[¶]. And when I have occasion to speak of these together I shall call them *ions*.

† *ἀνά*, upwards, *ὁδός*, a way; the way which the sun rises.

‡ *κατά*, downwards, *ὁδός*, a way; the way which the sun sets.

§ *ἀνιόν*, that which goes up (neuter participle).

¶ *κατιόν*, that which goes down."

I am so fully persuaded that these terms are from their simplicity preferable to those you have printed, that I shall think it a misfortune to science if you retain the latter. If, however, you still adhere to *dexio* and *scaio*, I am puzzled to combine these with *ion* without so much coalition of vowels as will startle your readers. I put at the bottom of the page the explanation, if you should persist in this [1]. I would only beg you to recollect that even violent philological anomalies are soon got over, if they are used to express important laws, as we see

---

[1] *δεξιός*, on the right hand, and hence, the *east;* *σκαιός*, on the left hand, and hence, the west. [This note is Dr Whewell's.]

in the terms *endosmose* and *exosmose;* and therefore there is little reason for shrinking from objections founded in ignorance against words which are really agreeable to the best analogies. The existing notation of Chemistry owes its wide adoption and long duration to its simplicity.

I am afraid you will think I am fond of playing the critic if I make any further objections, otherwise I would observe on your Article 666, that if you are not sure that you will want such words as *astechion,* it is throwing away your authority to propose them. If what I have written does not answer your purpose, pray let me hear from you again, and believe me,

Yours very truly, W. WHEWELL.

P.S. If, adopting the term *ion* for *stechion,* you do want the negative *astechion,* I do not think there will be any difficulty in devising a suitable word.

UNIV. CLUB, *July* 11, 1834.

MY DEAR MRS MURCHISON,

Though I have been so long without acknowledging your kind letter, I hope it has never occurred to you that I was not both very grateful and very much delighted with it. In fact it made me moralise in a way, which I think you would allow to be very sagacious and enlightened, concerning the relative value of one's gentleman and lady friends, and the result of my reflexions was not at all to the advantage of the former ; for Sedgwick, on whom I depended for giving me currency among his Scotch friends, has not given me any credentials. In truth, however, I believe that if I were to grumble at him for this I should do him injustice, inasmuch as he has induced you to give me your advice and assistance, which I value much more than his. I should have written sooner for the purpose of profiting by your kindness, if my plans had not been too indefinite to dwell upon, but at present they are a little more clear, and I look forward with great pleasure to the prospect of executing them. I shall make my way to Glasgow early in August and shall try to find your friend Lady Mary at Bonnington.

I shall then move on to Arran and to various parts of Argyleshire by land and water, as weather and opportunity may tempt me, and so to Mull. I doubt whether I shall extend my tour to Skye, and rather think that I shall travel the whole length of the Caledonian Canal to Inverness; and from that place I shall turn my face southwards to Edinburgh, taking such of the lochs and hills and other "seeingsworthinesses" (as the Germans call them) as I can pick up on the way. You have now such an outline as I can give you of my present intention, and I shall be really much gratified by any hints of yours as to what is worth notice. I shall probably take a hammer with me, but rather as an excuse for rambling up and down the rocks than for any other purpose, for my main object is to enjoy the sight of hills and lakes and coasts and arms of the sea ; without some such feast the summer leaves me quite empty of agreeable recollections, and when I sit down in the flats of Cambridge in the autumn I have a painful feeling of inanition. So you see I am going to bolt your Scotch hills whole, without all the ceremonies of slicing and cooking which your husband and Sedgwick bestow upon them in order to make them suit their artificial and vitiated tastes. Mr Lonsdale, who has promised to convey this to you, informs me that you have joined the geologues. I hope you fell in with them in time to prevent their turning their fratricidal hammers on one another, which I feared would be the result, if they would not agree about the dovetailing of the two portions into which they have partitioned the unhappy principality.

You will hear that we are in a political ferment here. The ministry is dissolved, and they are trying to patch it up again so as to last out the Session, but at present nothing is known. The report was yesterday that Lord Althorp was to be premier, but he denied it last night to his friends. Anything which you may be kind enough to send me have the goodness to send to Friar Street, Lancaster, where I shall be in the end of this month. I should like to know when Sedgwick will be in the West of Scotland and where. Does Murchison accompany him there, and do you do so ?

Pray give my regards to your husband and to Sedgwick, if he

is at the surface of the earth and near to you. I hope his hand has learnt to rejoice in its ancient exercise. Believe me, dear Mrs Murchison,

Yours faithfully and obliged, W. WHEWELL.

[Extract of a letter written to M. Quetelet of Brussels, Aug. 4, 1834.]

You will find that the Statistical Section, which sprung up under your auspices at Cambridge, is grown into a Statistical Society in London with many of our noblemen and members of Parliament for its members. Our Committee has had several meetings, but we are still somewhat embarrassed by the extent of our subject. I have no doubt, however, that we shall do something. The transactions of the last meeting of the Association are just published ; you will probably receive them through some channel, and you will find a good deal which will interest you. Among other things you will find an account of the observations of the Aurora Borealis, made at different places and tabulated in a manner which seems likely to give a good view of the phenomenon.

I have heard two or three times from Sir John Herschel. He is delighted with the country and with his situation at the Cape of Good Hope, but does not find the climate very favourable for observing. His mirror became tarnished, and the stars are not well defined. He says that the Barometer is certainly higher at the Equator than in high latitudes.

FORT WILLIAM, N. B., *Aug* 21, 1834.

MY DEAR JONES,

I am brought to a standstill by the weather, which I am so far from wondering at, that I wonder it has not happened sooner. I have been turning to good account the sunshine we have had, meandering up and down steep glens, and sailing about grand lakes and bays in the middle of ever-varying lines of mountains, thrown one behind another by every shade of atmospheric grey and blue. I want to go to Glencoe, but the weather says "you

shan't," so I shall give you the benefit of its churlishness.   I have
been thinking several times about my philosophy; travelling often
alone one has many happy moments of such speculation.   One
conclusion which I have come to is, that a letter which I wrote to
you about this same philosophy, I think from Lancaster, would
not give you the slightest notion of the general character of the
part of my book which I wanted to explain to you; so I shall try
again.   I shall in the *First Book* give a history of some of the
principal sciences, marking the *epochs* when the great steps were
made, the *preludes* and the *sequels* of these epochs, the way in
which each was essential to the next, and so on.   This will be
varied narrative and therefore, I hope, not dull.   After this I expect to shew that in all great inductive steps the type of the
process has been the same.   And I have in the *Second Book* to
explain what parts this process consists of, what conditions it
requires, what faculties it calls into play.   I expect to shew *clearly*
that in order to arrive at knowledge or science we must have, besides impressions of sense, certain mental bonds of connexion, ideal
relations, combinatory modes of conception, sciential conditions, or
whatever else you can help me to call them: they are what I
called *Ideas* in my former letter.   Thus *space* is the ideal relation
on which the science of geometry depends; *time, cause, likeness,
substance, life,* are ideal relations on which other sciences depend.

Now when I have shewn distinctly how these ideal relations
are the conditions of physical sciences which have already made
a great acknowledged progress, I shall have to try to discover the
nature of the analogy which exists between these sciences and
our knowledge respecting morals, taste, politics, language, and
generally, all hyperphysical knowledge.   I suppose my course of
proceeding must be something of this kind.   In the physical
sciences my philosophy points out what the ideal relations are on
which the existence of the sciences depends, shews that these
relations are different from the impressions which they connect,
and that they cannot be resolved into impressions of sense, though
of course they can only exist with impressions.   The philosophy
points out also the nature of each of the inductive steps of which
the progress of *physical* science has consisted, under what con-

ditions it could be and was made, and especially how it rendered the next step possible by its influence on current ideas and on language. Now I shall want to do the same thing with regard to some of the *hyperphysical* sciences, and thus shall have to give a criticism of their past history fashioned upon a general type. For instance, in the case of language I shall have to point out that *meaning* or *significance* is the ideal relation on which all knowledge must rest; that certain axiomatic principles, flowing from our idea of meaning, govern, and must govern, all our philological speculations. I shall have to give a sketch (only a sketch) of the inductive steps hitherto made in the philosophy or history of languages, and thus to shew where we now are. I shall have to do the same thing with respect to morals; shewing that the ideal relation that there are things which we *ought* to do, is inevitably involved in all our thinking on this subject, and tracing the fundamental principles which have been assumed in the principal moral philosophies; and so of the rest. This kind of criticism ought to be interesting. I believe you will fear that I shall lose myself in so wide an undertaking. But that you may not think me absurdly presumptuous, observe *first*, that I shall not take all branches of hyperphysical speculation under my management, but only those on which I have something of my own to say, and *next*, that I shall be far more bold and dashing in my judgments in this part of my work than in the history of the physical sciences: in *that*, every word is to be founded on close examination and proved by reference; but in morals and taste, such a proceeding is neither possible nor necessary. I must take advantage of my own philosophy, which, as it points out that all knowledge comes by induction and that induction is guessing, allows us to publish guesses, acknowledgedly imperfect, as contributions to knowledge. In short, I shall write reviews of the progress of morals, taste and the rest, much as any other critic would, only keeping in view the analogy of the true type of the progress of science. Now I wonder whether I have made myself at all intelligible, and if I have, what you think of my scheme. I will add that when I have shewn, as I hope to do, that the relations of duty and of the affections are as fundamental a part of man's

thoughts as the relations of time and space, and direct his will
in the same inevitable manner, I think I shall have assigned man
a moral and social constitution on firm grounds. The first and
far the most frequent exercise, and apparently the object, of all
the ideal relations which the mind possesses is *action ; theory* is
a secondary use of the faculties. The principles which operate
invisibly in action are detached and exhibited by speculation.
The speculative moral principles, so far as they are the true ones,
are thus the destined guides of our will ; and the principles which
they imply are as substantial as the axioms of geometry. I hope
to get thus to a satisfactory point of view respecting moral truth.
The same observations apply to taste, art, language, and the
rest.———Are you not asleep ? I dare say you are ; but
some day, when you are very wakeful, pray read this and make up
your mind what you think about it, which tell me when you can.

I hope your book goes on. Or is your lecture in the way ?
If so, get that job off your hands as soon as you can, before the
occasion is forgotten. Pray write to me at Edinburgh. I do not
mean about my philosophy but about yourself. My best regards
to Mrs Jones.

Yours always W. W.

DUNKELD, Aug. 31, 1834.

MY DEAR FORBES,

I consider myself as drawing near to Edinburgh, and expect to be there in a few days ; but I have still a little *giro* to
make before my tour is completed, and shall hardly meet you
before Wednesday or Thursday. I hope that will be early enough
to do you any good I can do by being with you before the general
gathering of the philosophers. When I arrive I shall enquire for
you at the College ; so if you will leave directions for me there
I will regulate myself accordingly. I should have finished my
ramble a day or two sooner, but I was misled by the error of the
maps to expect a road where no road is to be found. I think that
this is a case to bring before the geographical section ; but we will
see about that when we meet. I wish also that your meteoro-

logical committee could have arranged it so that I should not be
wet through more than twice a day upon the average, but I am
now become accustomed to the existing state of things, which gives
a higher average, and do not much mind it. I hope to find everything promising well for the Association, though I am told that a
bank is broken in which our Edinburgh friends had deposited a
magnificent dinner for us. We shall easily get over this loss and
shall philosophize none the worse.

Believe me, Yours very truly, W. WHEWELL.

KELSO, Sep. 19, 1834.

MY DEAR AIRY,

I have not been able to make out even yet with certainty
when Arago will come to Cambridge, though he assents fully, as it
seems, to my representation that it is impossible for him to leave
England without seeing our Observatory. He has a double will,
one part being lodged in his own breast and the other in that of
Pentland who is his leader in his travels, so that it is not always
easy to tell what will be the resultant volition. The most probable
event at present appears to be, that they will arrive in Cambridge
in the course of next week. I suppose I shall be there when they
arrive, but it would be much better if you could be present.
Arago is in admirable good humour and has been a grand *cheval
de bataille* in our Association, talking and speechifying on the
slightest provocation, and always very well. Upon the whole our
meeting has gone off in very good style and with good report.
We had more than plenty to do in the sections, and no special
absurdities or quarrels among the performers. The most difficult
business was the arrangement of the evening meetings; for
in our lack of matter, or of willing and able men, we allowed
Dionysius to tyrannize a whole evening concerning Babbage's
machine, which was universally declared to be a very heavy
infliction. Moreover, the room for our evening meetings was
much less than it was at Cambridge, and the company much
greater, so that the heat was something quite awful. I wonder it did not melt all the philosophers together, and produce

a new sort of Corinthian brass. We are to meet at Dublin next
year with Hamilton for our secretary, so if we follow any rule
or order it will be very wonderful. The tryst is for the 10th
of August. We have legislated many wise and good laws to
guide our future proceedings, and are altogether a very prosperous and hopeful crowd, as I will explain to you more at
length when we meet. We had more Cambridge men than I
expected, but everybody cried shame upon you for not being there,
and lamented your absence without ceasing. The Edinburgh
people made much of us, and barring the heat and the bustle
the whole affair was very delectable.

I shall be in Cambridge on Tuesday as I expect. I hope
you have gone on climbing mountains with good legs and fine
weather. If you are at Edensor pray remember me very kindly
to all the family from one end to the other; for numerous as
they are, I would make a regular enumeration if I supposed
you capable of encouraging any of them to forget me. Sedgwick
has been as usual complaining of being ill and talking more than
any seven men who are well. My kind regards to Mrs Airy
and your sister.

Yours always W. WHEWELL.

[To Mr Rickman.]

DURHAM, Sept. 20th, 1834.

MY DEAR SIR,

In the course of the meeting at Edinburgh I was
accosted by a gentleman who told me he was your brother-in-law.
But in the bustle in which we were I lost him before I had
much conversation with him, and did not find him again. He
informed me, which I was very sorry to hear, that you are still
unwell; I am willing to believe, however, that you are improving,
and that you are well enough to like to hear of my Scotch
Architectural Observations. If you are not, you can easily lay
them aside for a better time, which I hope will soon come.
In truth, I have not given much time to architecture during the
few weeks I have spent in Scotland, but I will tell you what my

impressions have been, and should be very glad to have an opportunity of correcting them by yours. In the first place there are in Scotland clear well-formed examples of Norman, of Early English, and of Decorated, exhibiting very slight differences from the English examples of the same style. For instance, Iona, Dryburgh and Kelso are good *light* Norman; Holyrood, Iona, Glasgow, part of Elgin, Dumblane, and Dunkeld are Early English. The Chapter-House at Elgin, the whole of Melrose, parts of Iona are good Decorated. The main differences appear to be that the E. E. is seldom so free from attempts at geometrical tracery as ours—for instance, as Salisbury. You will notice that in Germany and great part of France the formation of a complete Gothic style is almost always accompanied by geometrical tracery. Also, though the E. E. bases are quite regular, the capitals are not exactly so; the division into three members being here less clear than with us, and the forms more clumsy. I think, moreover, that the E. E. toothed ornament is continued in the Dec. work. It begins in the Norman, and is sometimes used very large, but never so well executed as our best. The flowing tracery of the Dec. work is rather French than English in its character, both by the want of subordination in the tracery lines, by the forms of the openings, and by the *screwed* flow of the pattern (I think you will know what I mean). But one thing which I have observed is that vaulting was much less generally practised than with us. Glasgow, which is fully developed E. E., is not vaulted, nor Elgin, which is E. E. and Dec., nor Dumblane (E. E.) nor the decorated part of Dunkeld. I do not mention Kelso, as it is Norman, and therefore its not being vaulted is according to rule. I have seen no stone vaulted roofs over the main spaces (the aisles are often vaulted) except Holyrood, Melrose, and (doubtful, as it is a restoration), Dunkeld. From the arrangement of the clerestory windows, wall-plate, and end-windows, I should suppose there had been a waggon vault at Glasgow, Elgin, Dumblane. As to the question of Glasgow, I have not examined it historically, but this one may say on mere architectural grounds. The date of the foundation is about that of King David (A.D. 1136), to which period belong

Kelso, Dryburgh, Jedburgh, and the foundation of Melrose. The whole existing building at Melrose is of the time of Robert Bruce (1320), and Jedburgh I have been obliged to leave unseen. But the oldest part of Dryburgh is light Norman with shafts, repeated mouldings, chamfered dripstones, and the like. Kelso is of the same character, and this appears scarcely to have been touched by repairs, and may, I conceive, be taken as a type of the style of King David's time. I believe also that Jedburgh belongs to the transition from Norman to E. E. Any one who maintains that Glasgow belongs to the same period must assert that different styles have no definite relation of time to each other. For Glasgow is E. E. in all details, and, though some parts deviate from our E. E., they deviate much more from Norman. Even the crypt is clearly of E. E. character, and some of its vaulting is curiously complex. I did not see Mr McLelland (I think that is the name of your friend), for I was in Glasgow but one day, and was then tempted to run off by the prospect of a tour in a steamboat which I performed very agreeably. I then rambled about the Highlands till the meeting called me to Edinburgh. The meeting went on very prosperously and is to reassemble next year in Dublin. You will hear the details from some of your Birmingham friends.

I beg my kind regards to Mrs Rickman, and am, my dear Sir,

Yours very truly W. WHEWELL.

I have heard from E. Sharpe, who desires me to tell you so, but I cannot at present get at his letter easily. I shall be in Cambridge in two days.

TRIN. COLL. *Oct.* 6, 1834.

MY DEAR JONES,

I wonder whether there is any chance of my seeing you before my term sets in. I had some intention of going to London, or perhaps further in your direction, but now that I am here and have done with the fellowship examination, I feel little inclined to move. I can get a little of my Induction done, and am at work upon it with great activity and comfort, which is more than I can expect when the days of lectures begin. I have never been able to do much in term time, and shall perhaps be at least as little as ever my own master, for Perry is fallen ill, at least too ill for hard work, and I fear I shall have a lecture extraordinary thrown upon me. In the mean time I am going on very prosperously, and getting through a good deal of my hardest work. The history of mechanics and astronomy is so important and instructive, that I must be tolerably full. It has never been written according to any philosophical view (how should it?) and I think I can make it amusing to others —it is excessively so to me. I expect too that by means of it I shall be able to stop the mouths of all gainsayers of my philosophy. For you are to understand that I still hold my philosophy of the sciences to be one of the most valuable parts of my speculations. *D'abord,* it is essential in order to give anything like consistency to my views of the method of philosophising, or to Bacon's or any body's else. However, you shall doubt about this if you like—till the time comes when you are ashamed to have doubted. By the way, you are not to judge of my philosophy by the scraps and snatches which I send you of it: my palace is not to be appreciated by specimen-stones. I have only given you sketches of the general notions which I had most difficulty in getting into a tangible form. I dare say you will think the difficulty remains still unsurmounted. In the mean time I have found an excellent device for keeping all these cramped and puzzled chapters out of the narrative part of the book. I write at the same time two Books, one of history, and one of philosophy, and when I find myself, in the course of my historical researches, becoming metaphysical and transcendental, I open *Book two,* in which all these things fall into their places, and will in the end make the most beautiful system that can be imagined. Like D'Alembert, I am tempted to exclaim, "*Ah, que je fais de belles choses que personne ne lira pas!*"

*　　*　　*　　*

Tell Mrs Jones that I should delight of all things to hear her music box, and hope to do so before long, but that there is another lady, "philosophy, heavenly maid," who keeps such an incessant din in my ears that I cannot hear anything else, day or night, and who will hardly let me stir to the other end of my lime-tree walk without accompanying me. I hope to get rid of her some day or other, or, if not, to make her better company.

*   *   *   *

Adieu.   Yours always W. WHEWELL.

TRIN. COLL. CAMBRIDGE, *Oct.* 7, 1834.

DEAR PHILLIPS,

I send you a short statement of what I said about Herschel's theory, which is too conjectural to be of any permanent value, but may serve to shew that we undulationists do not conceive that we are inferior in many points to our adversaries, that is, if we have any, for I do not know where they are to be heard of.

There are only two points about which I have any thing to say with regard to our publication. In the first place, pray do not omit to mention what the Liverpool people, the members of the Athenæum and the Corporation, have done for Tidology; I do not see that any of the public prints have noticed it. I gave you a written memorandum about it. In the next place, either print Mr Hodgkinson's researches on collision in full, or let me know what you are going to do about them, that I may inform him. They appear to me to be well done and of value, and, as he seems to have been moved by my expressions to undertake them, I consider myself as specially bound to watch over their interests, and see justice done them somewhere.

I think I have devised an anemometer, which will answer the object I have always considered the main one, of determining the total amount of the wind in any time; but I have not yet had leisure to think about the execution of it. Arago, who was here, says that in respect of your York meteorological observations your zeal and diligence are admirable, but your apparatus indifferent. Believe me

Yours very faithfully, W. WHEWELL.

Sedgwick has not been heard of here. I have sent your letter to Dent for him. Who is Mr Badnall?

TRIN. COLL. Oct. 19, 1834.

MY DEAR HARE,

*   *   *   *

There is another point about which you enquire, your Coleridge prize. I do not think the scheme will answer as you state it. *That* subject is, I think, an unsuitable one for a prize essay, on which we have no system of opinions which is generally accepted as right, while the publication of opinions which were considered to be wrong would involve the writer in the most odious charges. A subject so vast, so important, and so unsettled as the philosophy of Christianity should not be tossed over to a few ardent and, very likely, fearless young men, to make their theories on for the sake of a prize. You and Sterling[1] perhaps think that Coleridge has drawn the outline of such a philosophy, and that it only remains to develope the parts of it on his principles. I do not know that this is so; I am sure it is not generally allowed to be so; and I hold it certain that candidates would not write in such a spirit, except it was made an express condition. So much of the scheme itself—but as to its reception here, if you think a moment, you will not dream that it will be tolerated. The very phrase, "the philosophy of Christianity," or any equivalent one, will frighten our dignitaries from their propriety. They will see in it nothing but a trumpet-call to heresy and extravagance. Where will you find judges? In short, I do not see what can be made of the plan. I should think it unadvisable in proportion as I think the subject itself important in the highest degree. At the present moment

---

[1] The scheme of founding a prize at Cambridge to commemorate Coleridge was started by John Sterling, who at this time was acting as curate to Mr Hare at Hurstmonceux.

some coherent system of views, *capable of being admitted into the English minds of the present day*, and bearing on the great problems which the phrase suggests, would be of inestimable service to men's religious and intellectual stability. You must not think me narrow-minded because I make the condition which I have underlined (though I dare say Sterling will think so) and you must allow me to say that I think Coleridge's views somewhat fail in this, and all the German philosophy that I have seen fails still more. It would be a long and dim discussion, if I were to attempt to give you a detailed *how* and *why* for this; but you will not contest with me, I think, that the next step which our public can take in abstract speculations must depend on the steps they have taken already. The meanings which words and modes of expression have acquired, the connections and generalisations which it is possible to call up in men's minds, must depend on the past progress of literature and speculation among them; and truth is not truth, if you alter the discipline which this progress exercises. Coleridge appears to me to assume, and require, for the understanding of his religious speculations, an intellectual discipline different from that which the English have hitherto had ; Schleiermacher and the best of the Germans undoubtedly do so. I conceive therefore that the truths, which may be found in the writings of these men, must be taken up in the mind of some genuine Englishman and given out in a suitable form, before they will take a national hold upon us. Do this, if you can and will, and you will be an immense benefactor to England; and when this is done, you may with more safety and profit set youths to write prize essays on your subject.

I had the pleasure of seeing Coleridge a few weeks before his death, and, what was also a great pleasure, took Worsley to him, who had never seen him before. He talked wonderfully well; among other things expressed the deepest sorrow at Thirlwall's letters. I spent a day with Wordsworth with great satisfaction; sailing on Windermere and wandering on its banks all day with him.

*   *   *   *

Yours ever. W. W.

MY DEAR HARE,

The Coleridge prize, as you present it in your last letter, is a more feasible scheme than as I had conceived it. I still think there is a good deal of difficulty about the plan, but it is worth considering. My reflexions upon it are of this colour. If you would be content with essays which, in their subjects and in the style of treating them, should not differ materially from our present race of Hulsean and Norrisian Essays, I do not think there would be any insurmountable difficulty in setting it on foot; but then it may be questioned whether the addition of another such prize would be any very valuable boon to the University, or any very lofty monument to Coleridge's memory. Still I would not at once look slightingly at such a project; for the mere name of Coleridge would, to a certain extent, tinge the speculations of the writers, and the establishment of the thing with its name would bring him more to the notice of our divines, young and old, as a religious philosopher; and certainly they could not acquire this knowledge of him without great benefit to themselves. But for this purpose I think it would be better to avoid such a phrase as "the Philosophy of Christianity," and to describe the subject in some way wider in extent and more modest in expression. I am not sure that it might not be best to follow your suggestion, and make it a prize for metaphysical dissertations on speculative subjects generally, or with only a recommendation that they should by preference be such as illustrate the history of Christianity, and the connexion of the Christian scheme with views of man's place and nature borrowed from other sources. The Philosophy of Christianity is a good phrase now, and would suit Coleridgian speculations well; but it may easily become a technical phrase, and in ten years may be fastened upon some system of extreme absurdity or error; or at best the subject would lose its interest for want of variety. If you still adhere to the essay, there is a good deal of discretion and knowledge of the land requisite in fixing the conditions of the candidates, judges, competition, &c., and undoubtedly the cast of the institution in future would probably depend much upon the subjects of the first one or two prizes. In my opinion they ought not to be too wide. A vast field encourages the vain and foolish, but puzzles or misleads the clever, and destroys the effect of the exercise as a discipline.

I have talked only to some of our younger men about this matter, as they have been the only ones who have come in my way. They appear to agree with me, but that may be from hearing my view first, so you had better come and get at their genuine opinions.

I am right glad that you are coming here. If you will let me know when you are coming, I will have rooms ready for you and all other things suitable. Thirlwall is not here, and Blakesley conjectures that he is gone to look at the effect of the storms on the cliffs of the Isle of Wight, a project which he had in contemplation. The Marchesa bids me say that she rejoices extremely in the thought of seeing you again. Adieu.

Yours always, W. WHEWELL.

TRIN. COLL. Dec. 19, 1834.

MY DEAR JONES,

I was right glad to read the way in which you talk of your speculations, except that I have some doubts how far your older work may come out sooner from your having to write a new one first. But I do not much mind this. If you will put your views together in a form, that you know to be sound and right, and publish them with a tranquil mind, confident as you well may be of their value, and leaving the world to appreciate them when it has a lucid interval, as it will, you cannot fail of being all you would desire. Townsend brought me some absurd story of your having a fancy for publishing sermons, which I spurned with a declaration that anybody could publish sermons, not much I think to the satisfaction of Rose, who was with us at the time. I want extremely to see in what state your labours are, and will come to you sometime during the vacation that you may tell me, so pray arrange that it may be so. I hardly know when I shall start, for I have not been able to do much during the term, having had, contrary to my expectation, two lectures a day, and besides a deal of tide work; but I have been my own master almost three days, and am getting on famously. I am the more unwilling to break up my camp yet, because I want to have done with astronomy, which is of course my pattern science, and must have its story told with some detail and completeness. I think I am within a week of the end; certainly within a fortnight, if I can escape Christmas junkettings. My work becomes clearer and firmer before me as I go on, and I have great confidence that you will admire even my philosophy when you see its bearings. But we can talk of this anon. I have had a letter from Herschel dated Sep. 21. He is in the most charming spirits and hopes—has a new daughter quite well and very quiet—and is doing and enjoying as much as possible. Hare has been here and is just gone back to your neighbourhood; that is, if you are at St Leonards, as I shall suppose in directing this. So has Rose, and I lament to say I think in worse health rather than better; but I may be mistaken. The Coleridgians want to establish a Coleridge Prize Essay on a metaphysical subject as a memorial of him; and we have taken for granted that you are Coleridgian and metaphysician zealous enough to join us—so you are on the list, if the Heads consent. Adieu. I must work hard tonight and tomorrow morning for the books are to be sent back to the Library. I hope Mrs Jones is well and well pleased with your manifold migrations.

Always yours, W. WHEWELL.

[To Professor Phillips.]

TRINITY COLL. CAMBRIDGE, Dec. 28, 1834.

MY DEAR SIR,

I send you a sketch and description of my anemometer, which will, I hope, make it intelligible, and enable you to set up one of the same kind, if you approve of it. I am fully persuaded that the only anemometrical indication which is of any value in meteorology is that which I have here attempted to obtain, namely, the quantity of wind transferred upon the whole in each direction. It is true that, even if we obtain this, we do so only for the lowest stratum of the atmosphere, which may be the least important; but this is all we can do at present. I should be very glad to know what you think of the contrivance and its probable utility, and whether you foresee any difficulties and defects in the machine.

I am glad to hear that Mr Hodgkinson is pursuing the subject of his paper. I am not sure whether I interpret aright a passage in your letter, where you seem to suppose that I have the paper in my possession. This is not the case; I left it in what I supposed to be the official hands for transmitting it to the printing press, when I was at Edinburgh—if not in yours, I presume in Lloyd's, the secretary of our Section.

I know nothing of Mr Badnall or of his communication, nor did I hear anything of Mr Saxton's Thermometer. Perhaps this was something which was brought forward in the mechanical sub-section. I will tell Miller what you wish me to do.

My communication on the action of the eye in vision I did not find time to make, which was very fortunate. I only volunteered it at first because they were reported to want matter in that Section, and should probably have edified them very little if I had given it. It need not be referred to again.

Prof. Sedgwick is at Norwich, and, I should suppose, has rather more leisure than usual ; so if you write to him again you may possibly get something out of him.

I am glad you are looking forward to the arrangements at Dublin. The evening meetings are undoubtedly the great difficulty, but, if people will look at them in good humour, there is nothing which may not be well managed.

The observations at Monk Wearmouth Pit are very interesting. I do not see how you are to disprove the existence of chemical action in the subterranean regions. But then, to produce such a diffusion of temperature as we find, the chemical action must be on a scale so large that its effects can hardly be distinguished in our conception from central combustion or the remains of combustion. Is this sound doctrine ?

Believe me, Yours very truly, W. WHEWELL.

[To M. Quetelet.]

TRINITY COLLEGE, CAMBRIDGE, *Feb.* 3, 1835.

MY DEAR SIR,

Allow me to press upon your government, through you, the claims of Science, to which they may at present render an essential service by a small expenditure of money and trouble. Your countrymen have the power of conferring a great benefit upon astronomy by instituting and carrying on a series of tide observations on your coasts. The laws of the tides have never yet been determined from observation, and the time seems now to be come when this strange and shameful deficiency is on the point of being remedied by the governments and mathematicians of Europe. The English observations and calculations are now in progress, which will solve the problem so far as our coasts are concerned. I have good reasons to hope that the subject will soon be prosecuted to the same extent in France, and I shall shortly cause a representation to be made to the government of Holland for the same object. The Belgian coast at Nieuwport and Ostend would supply an important addition to the knowledge of those seas which we require—the more important, inasmuch as the course of the tides in that region is very curious and complex. According to the usual statements on this subject, the tide wave is at the same moment travelling in two opposite directions on the shores of Belgium and of England: and the way in which the tides on the two coasts are related to each other is certainly such as can only be determined by good observations in both countries. Besides these local peculiarities, a continued series of good tide observations on the Belgian coast would be of great service in supplying materials for the general theory of the tides, which theory cannot be completed without observations at many places. I trust that your government will not be slow in taking its share in that patronage of science, and especially of astronomy, which is now the pride and the glory of the most civilized nations. The promotion of astronomy has long been considered a national object in every part of Europe, but our knowledge of the tides has been left in arrear of the

other branches of astronomy. The time is now come to repair
this neglect, and the institution of observations for this purpose
belongs to maritime nations in particular. We shall soon have
a chain of tide observations along the coasts of Europe, and it
would be a matter of great sorrow to the mathematicians of
Europe to find an interruption in this chain occurring at the
Belgian shores. I trust that this will not happen, but that, on the
contrary, we shall be able to combine Belgian series of tide
observations with the best series obtained in France and England;
and I am fully persuaded that the government which procures
such observations to be made will feel itself amply rewarded
by the reputation of having contributed to fill up the only
remaining blank in our astronomical knowledge, and by the consciousness of having rendered a good service to navigation as well
as to theoretical science.

Believe me, my dear Sir,

Very truly and faithfully yours, W. WHEWELL.

P.S. I send along with this copies of my printed papers on
the subject, which I shall be much obliged, if you will present to
the Minister of your Marine as a testimony of my respect.

TRIN. COLL. Feb. 13, 1835.

MY DEAR HARE,

I do not know whether your hopes were very confident
that the Heads would embrace our proposition about a Coleridge
Essay: but, if so, it would appear they were delusive. The
V. C. informs me that the answer to your application to them
is that they are not disposed favourably towards such a proposal;
and I hear both from him and from our Master that the objection
is not to any accidental and extraneous part of the scheme,
but to the name of Coleridge. With our governors, it seems,
the vagaries of his earlier times are better known than the
Christian philosophy, which he has impressed upon so many in
his riper years. I am sorry for this, for I think the Essay might,
with good management and good fortune, have been of beneficial

influence here; but I do not see what further can be done in the present state of things. If you can suggest any thing, I shall be glad to be of use. When Townsend was here he was very zealous about the erection of a monument to C. in Westminster Abbey; holding that many in after years would in some tranquil moment or other imbibe from such a memorial some sober and philosophic thought. I told him that there were in these times few people quiet and humble enough to be benefited by such suggestions; but I did not convince him, and indeed had no very serious wish to do so. By the way, I have just received from him some verses of Stewart Rose, mainly about him, and a little Essay by him on the subject of the intellectual progress of the poor—the latter very amiable; but what sort of a man does it require to take a right hold of that subject in its whole extent?

You will be glad to hear that Jones is appointed, or likely to be appointed immediately, to the Haileybury Professorship. I am sorry to say, however, that the appointment is but a precarious one; for the scheme of abolishing the College, which has always found favour with a portion of the directors, is now more loudly urged than ever, and will probably be brought before parliament when it meets.

*　　*　　*　　*

Yours very truly, W. WHEWELL.

TRINITY COLL. CAMBRIDGE, *Feb.* 14, 1835.

MY DEAR FORBES,

I am so much interested with your experiments on the polarization of heat, that I do not know if I can wait till the first of March for the sequel of them. When I talked of double refraction without polarization, I thought it possible that you had got two separate rays of heat by double refraction, and had examined sometimes the one and sometimes the other. Such double refraction without polarization would result if the vibrations which constitute heat were longitudinal instead of transverse, and this seemed a very possible theory. I have no hesitation in saying that I wish to look forward to the general theory in collecting the laws of particular phenomena, and, with regard to heat, I do not know any experiments which promise so much in this way as yours. If you have heat polarized with double refraction, you must have transverse vibrations. I see no reason why you should not, but I should like extremely to know in what these vibrations differ from those of light. There is plenty of room for such, for all the vibrations below red and above violet are not light. Pray let me know what you have proved, or even conjectured, about the length of a vibration of heat. Can you not get your rings and brushes by letting the heat fall upon some very thin film of fusible varnish covered with coloured powder, or some such matter? At any rate, when you have time, pray let me know what progress you have made.

I sent your suggestions on meteorological doctrines to Herschel, from whom I have recently heard, and who is in excellent health and spirits. He had spent one night in tiger hunting, but seemed to think it poor sport compared with the "*chasse aux étoiles doubles.*" He is president of a Literary and Philosophical Society of the Cape of Good Hope, and I have sent them a memoir on their own tides, which I hope will set them to work. It is a great shame that you have not regular tide-observations at Leith. Pray how goes on your observatory after the ambiguous sort of encouragement that the British Association gave it? By the way, what a wonderful article is that in the *Ed. Rev.* about us! My leading reflection upon it is, that we have had a very happy escape in getting out of the track into which the author (of course Sir D. B.), would have urged us. To "obtain a direct national provision for men of science" is, in other words, to get pensions from the government for our friends: a worshipful employment truly! And then that you and I should tell people where to establish fisheries and railroads! If our present employments are not more useful, they are at least what we know something about. I have some fears, however, that this combustible article may make a blaze at Dublin, but I do not think it need be very mischievous, though it seems to be intended to be so.

I hope you have had your newspaper hitherto, but I must

tell you that I was not able to get one on the terms you mentioned, and ordered the publisher to send it on his own, having
a prospect of managing better afterwards, which has not been
fulfilled. You must tell me whether I am to continue to send
you a primary *Cambridge Chronicle*, if I cannot obtain a secondary
one. I hope your library and your booksellers have by this
time got Deighton's books; he assures me the former were sent.

Airy is gone to town to give his evidence about South's
equatoreal, and to predominate over the anniversary dinner of
the Astronomical Society, of which he is to be president. We
have got our large telescope by Cauchoix, but it is not yet
adjusted. Sedgwick is just returned from his Norwich prebend
in great preservation. Adieu.

Always truly yours, W. WHEWELL.

TRIN. COLL., CAMBRIDGE, *Feb.* 20, 1835.

MY DEAR ROSE,

My architectural speculations are somewhat like the de-
cisions of the judge, who only heard one side because his head
was confused by hearing both. I look at the features of the
architecture, and pronounce very authoritatively what the date
and history of the building *ought* to be. This does very well
when I take a large and distant field like France or Germany,
for it seems to enable me to classify and speculate, and there
is a reasonable excuse for turning aside from the labour of verifying my conjectures by historical research. But it would not do
so well for a single building in England; people would certainly
expect me to shew by some external evidence, that the period
and circumstances of the work were what my doctrines make
them. If I had any historical testimony about Barfreston, I would
try what sort of story I could make out ; but without that I think
your readers (for I suppose you want me to discourse in the *Brit.
Mag.* though you do not say so) would imagine me a consummate
dogmatist, if I gave them a hypothetical narrative and claimed
their belief. I have not at this moment time, nor do I know
where to seek materials, to make out a history of *Barson.* The

only historical fact which occurs to me, is that it must have been the residence of "goodman Puff of Barson"; who, according to Mrs Quickly's notion, interfered with Sir John Falstaff's claim to be "the greatest man in this realm." If any of your antiquarian friends will look into the history of the place and make out anything about the church, I will do my best to make the architecture *agreeable* ; and, if anything comes in my way of that kind, I will see if I can dish it up in a decent and presentable shape; but without this I think you would have little reason to be thankful for my setting up for goodman Puff of Barson, and attempting to puff the church a hundred years backward or forward with my unassisted breath.

I agree with you altogether about Jones. I think he is now excellently well placed, and that all his best friends can do is to hope his position may be permanent.

Are you surprised at the election for Speaker? I suppose not, for by this time you must have learnt to be surprised at nothing. Stanley's conduct seems to me to be as great a gain for the conservative cause as the loss of the Speakership is a loss. As for a church reform, I dare say it will be a reasonably well compounded dose when one considers the doctors; and probably when the first soreness of the application is over, we shall feel rather better than worse for it. At any rate, we must hope it will be so, as our good hopes are the best inheritance we have left.

Always very truly yours, W. WHEWELL.

TRIN. COLL. CAMBRIDGE, *Mar.* 5, 1835.

MY DEAR LYELL,

Airy has shewn me a letter of yours which, as it refers to matters touching the tides, he thinks I am entitled to answer. Now if your enquiry be whether Laplace's calculations do really give us any ground to believe the depth of the sea to be any one of the depths mentioned by him, I have no hesitation in saying that I do not think that any competent judge can really find in his reasonings any sufficient ground for any opinion whatever about the amount or law of the depths of the ocean. That this is so, will, I think, be obvious to anyone who considers that Laplace's reasonings all go upon the supposition of an earth *covered* with water; on which consequently the tide-wave might sweep round and round with no interruption such as continents produce. The existence of continents, breaking the ocean into seas of limited longitude, so entirely alters the problem, that it would be absurd to talk of Laplace's solution as an approximation. There is no more reason why the depth, or the law of the depth, should approach the results of his hypothesis, than there is why we should consider Newton's solution as an approximation, which leaves the depth quite indeterminate. Indeed I myself am fully persuaded that Newton's solution is at this moment the more trustworthy, and that Laplace's arguments against it (still founded on the hypothesis of a universal ocean) are of no weight whatever.

Laplace himself allows the inconclusiveness of his own reasoning on the grounds I have stated. Thus (*Méc. Cél.*, P. 1, Liv. IV., Art. 8) he says "the observations made in our ports shew that the difference of the two tides on the same day is small, and that therefore the sea is nearly uniform in depth. But from the difference it appears that it is somewhat deeper at the poles than at the equator. But this consequence is *subordinate to the hypothesis* of a fluid regularly distributed over the surface of the spheroid, which is not the case of nature." The same may be said of all his other inferences on the depth of the sea.

With respect to the uniformity of the depth of the sea as collected from the fact just referred to, the smallness of the difference of the two tides the same day, I not only doubt the reasoning but disbelieve the fact. All the good observations of the tides made by the coastguard last June from the Scilly Isles to the Isle of Wight shew a decided difference of the morning and evening tide: the Bristol tide-gauge confirms this very clearly. The non-appearance of the difference at London is easily explained. I have no doubt of the existence of the difference, but I do not think Laplace's theory enables us to collect from it anything about the depth of the ocean.

I have never been able to see through Laplace's analysis what
the mechanical principle is, by which he makes the laws of the
tides dependent on the depth of the ocean. If it be, as Airy
thinks, that he takes the depth to be such as will suit the rate
of travelling of the tide considered as a wave, and calculated
on the ordinary hydrodynamical theory, there is not much difficulty in the subject, but the conclusion is worth nothing, on
account of the fallaciousness of the theory of the motion of waves.
According to this theory, the velocity of a wave is the velocity
of a falling body acquired down half the depth; and in my memoir
on co-tidal lines I have on this principle stated that the depth
of the German ocean at the Nore is 90 feet, on the east coast
of England 120 feet, of Scotland 360 feet, of the Atlantic west
of Ireland 2600 feet, of the Atlantic in its middle part 50000 feet
(Sect. 4 of my paper). I do not believe these results, but they
are of more value than Laplace's, because they are deduced from
the actual velocity of the tide-wave.

I have, however, a strong persuasion that the depth of the
Atlantic may be deduced from the velocity of the tide-wave along
it, when we know, better than we yet do, the laws by which the
rate of travelling of a wave depends on the depth and form of
its channel. I believe this improvement of our knowledge must
be looked for from a combination of theory and experiments;
some experiments in canals are now going on. But there is one
point which is, I think, well worth the attention of geologists. The
velocity of the tide-wave in the Atlantic certainly *depends* on
the depth, and if there be a *secular* change in that velocity, there
must be a *secular change in the depth* of the Atlantic. By ascertaining such facts about the tides, we should have an instrument
of research for the bottom of the ocean which is out of the reach
of all other methods; a sounding line which is always out, and
which is longer than ever plummet reached. Nor do I think
it is impossible to make out such secular changes. Lubbock
has, I conceive, shewn that the tide comes later to London than
it did in Flamsteed's time. If we had tide observations at the
Bermudas and Azores 200 years old, they might betray a new
Atlantis rising gradually to-day. Is not this worth thinking

about? I could give you more reasons than I have given for not trusting Laplace's guesses about the depth of the sea; but perhaps you will have enough of this. I am in despair at not being able to get tide information about the Pacific and its shores.

Yours ever truly, W. WHEWELL.

TRIN. COLL. CAMBRIDGE, *April* 12, 1835.

MY DEAR HAMILTON,

I am glad to find you are seriously pondering the prospects and conduct of the British Association, and especially the proper course of behaviour which becomes it at Dublin. My intention is certainly to attend, in which case I shall set my foot on Irish ground for the first time in my life. I have also some supplementary schemes of prefixing or postfixing a small meandering through the land to the grand explosion of spouting which is to take place in your city, according to the reviewer's apprehensions. You talk of Brewster's article as containing hard knocks, but, in truth, my feelings in reading it were mainly those of self-gratulation, that any scheme in which I was involved had escaped the character, which, according to him, it was the object of the founders of the society (that is, himself of course) to give it. I would not for any consideration on earth be a party to a plan for "inducing the government to make a direct provision for scientific men"—that is, for bullying them into giving us pensions; and I should think myself a goose, if I were to be tempted to give practical men my advice about railroads and fisheries. Whoever it was that gave the Association another turn did the state some service.

As to the future functions of the body, they are too doubtful in my mind to speculate much about. If you work the Dublin meeting well, as I have no doubt you will, I think you may fairly leave the guidance of the affair afterwards to the permanent officers (Harcourt, &c.), and to the general body. Nevertheless, if you think this base counsel, as if it advised flinching from a difficult duty, and if you feel that you have a vocation for wielding the future fates of the Association, you may be a great benefactor to it: but I do not feel that I have any views on the subject worth talking of.

It is tolerably clear from the experience of Edinburgh that there is no fear of want of employment at the sectional meetings, and that the difficulty of management applies to the evenings. The success of the public discussions at Cambridge, as appears to me, made people perhaps expect too much from such evenings. Their smaller success at Edinburgh has probably made people think too unfavourably of them. I do not know what I should recommend, but you will easily choose a course, and we will support you. I should have no fear of going on in the old way, if the President is of that opinion.

I received your paper for which I thank you much. William Wordsworth and his wife are here, and have talked of you more than once.

Remember me to Dr and Prof. Lloyd very kindly, and believe me,

Yours very truly, W. WHEWELL.

TRIN. COLL. CAMBRIDGE, *May* 2, 1835.

MY DEAR JONES,

I ought to have written to you sooner to say that I do not think it at all likely that I can make my way to Haileybury on Wednesday. Lectures and other matters have got dominion over me, so that I hardly know which way to turn myself, and I believe I must be content to be tethered at Cambridge for the ensuing month. Among other unexpected calls, I find that I cannot get on very well without lecturing in Butler's *Sermons*, which is carrying me into the whole business of moral principle with reference to Hobbes, Mackintosh, and all the squad of speculators on such matters. This will rather interfere with my prescribed course of writing about Induction, but still I hope to manage in such a way as not to lose by it. I shall expect to manage so that what I do will come into its place at some time or other; for the beauty of my Induction is, that it is like the

Devonshire man's pie, into which he puts every thing which he catches.

*   *   *   *

Yours always, W. WHEWELL.

[To Mr Rickman.]

TRIN. COLL., *May* 6, 1835.

MY DEAR SIR,

I am at last able to send you my Norman notes. I am almost ashamed of them when I see how trivial they are, and they appear the more so after the extremely acute analysis and excellent classification which we have just had to read in Willis's book. However, I suppose they may do for beginners, and have some effect in diffusing a knowledge of the subject, so I am not sorry that I have finished them. You will see that I have taken occasion to write a preface containing a sort of review of the present state of the subject.

It is intended to reappoint the Fitzwilliam Syndicate in order that they may devise the best plan for obtaining the opinion of the University respecting the building plans. I do not at all know what plan they are likely to recommend, whether a preparatory selection of a few, or that each person should vote at once for one plan, or for an order of merit of several.

I hope you continue as well as when I saw you here. Believe me,

Yours very truly, W. WHEWELL.

TRIN. COLL., *May* 9, 1835.

MY DEAR JONES,

Pray tell Mrs Jones that, though I cannot speak, she has only to look at me attentively to see that I am quite as much pleased with her presence as she can be with mine, and that it is only the reserve and gravity which Mr Eddis has thought it fitting to bestow upon me[1], which prevents my breaking out into

---

[1] This refers to a lithograph of a picture painted by Mr Eddis for Mr Worsley, at present Master of Downing College. The lithograph was by a daughter of Mr Dawson Turner.

14—2

a broad grin by way of shewing the satisfaction her complacent looks and comfortable sayings give me. Indeed, if she looks again, I am sure she will find that, however much Worsley may have tried to have me made looking profound and metaphysical, I am in fact amusing myself at the expense of those who imagine me bestowing my thoughts on any thing except the persons I am with, and hardly bridling my smiles when such visionary people come into my head.

I had seen the London Review of Sedgwick, but had looked at nothing else, till you directed me to the note about you. I am somewhat puzzled, not being able to make out whether the reviewer is a scoundrel, who, by bringing together ferocious expressions knowingly, endeavours to excite people's passions against an antagonist, or whether he is a real *bonâ fide* example of that silliness which belongs to Benthamites and the like; and which can see nothing but moral horrors in all persons of opposite opinions. However, I do not know that it is any great business of yours or mine to settle this knotty point. I do not think Sedgwick will fight. He was applied to by the book-seller who publishes his sermon (which is now a widely circulating book), but, wisely I think, laughed and refused. It would cost him some trouble to make his view of the matter water-tight— for he has talked somewhat loosely—not, however, that the reviewer is any thing very formidable. Indeed I think he does not believe what he defends. I have been reading a little about such matters of late, and though I thought it an interruption, it has answered pretty well. I have got a glimpse, which I have long been wishing and struggling for, of the inductive history of ethics. Mackintosh's *Dissertation* is my main guide, but I think I can both correct and extend his view, besides making it merely a branch of the universal tree of knowledge. I have got another scheme of writing a short Essay, on Mathematics as a part of liberal education, to appear as the preface to a new edition of one of my many mechanical books, which is wanted. The plan at least is after your heart, and we will talk about it. I hope you work, and actively, at your printed lecture book.

I perceive that you plume yourself upon some architectural

clearsightedness of yours which, it appears, you have shewn, but I cannot make out in what you suppose it to have been exerted. If it were in conjecturing that Saracen architecture had to do with Gothic, I hope you will observe that the only feature of resemblance which Willis allows is the pointed arch, which requires no peculiar eye, and that the novelty of his opinion is, not that there is a resemblance, but that Europe derived it from Asia; a point of history, not of eye. Observe too that if Rickman and Willis have done anything, it has been to make that a distinction or resemblance in enunciated principle, which was before a matter of eye or of guess (like enough, mistaken). The architecture of other eastern countries may have borrowed from Byzantium, for it exhibits domes, the only peculiar Byzantine feature. Whether domes are Byzantine depends on the way in which they are supported. I should like to see any sketches which illustrate this.

If I go to London in the next fortnight, I will try to hear one of your lectures. Adieu.

Yours always, W. WHEWELL.

TRIN. COLL. CAMBRIDGE, *May* 26, 1835.

MY DEAR JONES,

I was thinking that of us two it was time that one wrote to the other, when your letter came. I have hardly had time to con over your examination papers, but they appear to me on a first glance to be excellent. I see many reasons for thinking that your system of examinations by the lecturers themselves has several advantages over ours. It is also, I think, in some degree the Oxford system. I am still scheming and speculating for the improvement of our system here. Among other things I lately wrote to the Editors of the *Encyclopædia Britannica* offering, if they would edit Mackintosh's *History of Moral Philosophy* in a separate and convenient form for our men, to write a preface, or analysis, or some other kind of appendage, to recommend it to our students as the best text-book on the subject. They have agreed to this, and I think I can make a good job of it, and fix moral philosophy here as one of the subjects for our class men. I also want to write, as I think I told you, a pamphlet about mathematics as a part of a liberal education, which will all be founded on the principles of the true philosophy without my telling people more of them than is requisite to be told for the purpose. I want you to come and talk about this, for I dare say we shall disagree hugely. Another business I have on hand is a sermon to be preached before the Brethren of the Trinity House on Trinity Monday—an appointment of our new Chancellor[1]. If I can get hold of you in time, I must make you be my critic for that too. Besides this, I have I cannot tell you how many re-editions of books to look after, so that I am likely to lead a very literary existence for some time. By the way, you never told me whether you agreed with my architectural speculations, but that also we can settle hereafter. I am rejoiced that you are still resolute about your book that is to be. By all means and on all accounts get on with it as fast and as well as you can. As you say, this is no world to lose time in, for we may be out of heart or out of breath in a little while, but for the present I hope, as you say, that we are both going on well and doing what we were intended to do. I have contrived to work through the history of Heat, Electricity and Magnetism during this term, which are very pretty now they are done (so, you know, are all my eggs when new-laid). And now I have got to write a report for the British Association on the same subjects, which is to be added to my literary employments proper to the next month—that too will be "very pretty," as I presume. Come here as soon as you can, and let me know when that is to be. I want to go to London, and may perhaps come and hear your lecture, but I move with reluctance and not without inconvenience at present. Lodge and Sedgwick are just returned from Yorkshire, having voted on opposite sides; Sedgwick has been exhibiting as a hustings orator. One cannot but feel that the present state of the country breaks up all those circles of early friendship which at a former period one hoped would supply a good deal of the quiet pleasures of maturer age.

[1] The Marquess Camden.

However, the business is to make the best of what we have.
My kind regards to Mrs Jones.

Yours always, W. WHEWELL.

TRIN. COLL. CAMBRIDGE, *June* 2, 1835.

MY DEAR WILKINSON,

I hold myself bound to answer all reasonable questions
concerning the tides and my own lucubrations about them, and
therefore I should not have neglected to reply to yours while
they were somewhat fresher than they now are, if I had not
expected to be able to fortify my doctrines by a paper which
I intended to publish soon, but which has only just been finished,
and which I now send you. You were in some doubt, if I rightly
recollect, about the stoppage of the tide undulations at their
meeting, and your doubts were very consistent and rational;
I wish I could get the practical men, the sailors and surveyors,
to look at the matter in a way half so agreeable to mechanical
principles. I could have told you then that the two opposite
undulations would not in fact stop at their meeting, but would
go forwards, each modifying the other, but I could not have quoted
at that time the evidence, which I now have, that this is really
the state of the case on our shores. You will find the result
stated in the paper which I send you. I am sorry that it refers
to the South Coast alone, but I could get no more work done
by my clerk at the Admiralty, and so was glad to publish what
is already done. I was the more desirous of doing this, inasmuch
as we are going this month to repeat the observations of last
June from the 9th to the 28th: and the tides are to be watched
for this period, not only at the coastguard stations in the British
Isles, but also at a great number of points on the coast of Europe
and America; for the governments of Sweden, Denmark, Belgium,
Holland, France, Spain, Portugal and the United States have
been applied to, and most of them have agreed very cordially
to co-operate. You see I shall have such a register of the vagaries
of the tide-wave for a fortnight as has never before been collected,
and, I have no doubt, I shall get some curious results out of it.

Indeed, I think you will see that the inferences from my last year's observations make this quite certain. I do not know that I shall be in your part of the world this summer. The British Association meets at Dublin, Aug. 8; if you can move then you cannot do better than join us. I shall try to hang on an Irish tour as an appendage to this meeting before and after; and shall put myself in motion hence as soon as our own hubbub is over. By the way, you will perhaps come here to help us to install, in which case I shall have the pleasure of seeing you in July. I hear, not unfrequently, from Herschel, who appears to be enjoying his residence at the Cape exceedingly, and has got over some of the difficulties which he found at first in the way of observing. Jones is become a sort of distant neighbour of ours, having been appointed to the Professorship of Political Economy in the East India College near Hertford, which was vacated by the death of Malthus. Sedgwick is just recovering from a bad cold which he brought with him out of Yorkshire. I suppose he caught it by turning hustings orator in the rain and cold, which according to his own account, as well as that of others, he did while he was in Dent. I hope you were edified by his doctrines.

Give my kind regards to Mrs Wilkinsons both, and remember me to the children.

Always truly yours, W. WHEWELL.

I send you a lithographic effigy of myself, about the resemblance of which opinions are divided.

16, SUFFOLK ST., July 15, 1835.

MY DEAR MURCHISON,

You will be grieved to hear that Sedgwick is prevented joining you by a misfortune in his family. His brother, who came from the Isle of Wight to enjoy the amusements of the Installation week at Cambridge, on Thursday morning, without any warning, received the sad news that his wife had died suddenly of a premature confinement. He was overwhelmed, and Sedgwick is gone back with him home. When this is done, he returns to Cambridge and stays there till the meeting at Dublin, which he still purposes to attend.

As for myself, I cannot even yet tell you certainly whether I shall be able to join you at Milford Haven. The Admiralty are to give me a clerk to work the tides for me. I saw Lord Auckland on the subject yesterday, and want to set my man to work before I leave London. It is possible I may yet overtake you at the Haven on the 20th, but, if I do not, you will of course not think of waiting for me. Only leave a line at the Post Office, Milford Haven, to mention your line of march, that I may judge of the chance of falling in with you.

Our festivities at Cambridge went on very prosperously, the weather being fine and the arrangements all successful. Your two foreign friends, Teleki and Clary, were there and appeared to be amused and pleased, except that the Prince lost his hat in the great dinner at Trinity. I believe, however, he recovered it next day.

If I do not reach the Haven in time to go with you, I shall perhaps sail direct for Cork, which, I think, was not your plan. In that case we may possibly come together on the banks of Killarney. I shall expect to find you waking the echoes with your hammer, and Mrs M. propitiating the genius of the place for such a violation of his sanctity by putting a flattering like-ness of him in her sketch-book. With my best regards,

Yours truly, W. WHEWELL.

Lieut. Drummond is appointed Secretary for Ireland, and starts for Dublin immediately.

16, SUFFOLK ST., *July* 15, 1835.

MY DEAR FORBES,

I rejoiced much to hear from you, and hope that I am not too late to catch you at Montpelier. As you anticipated, your letter arrived just at the moment of the eruption of our installation festivities. These are now over well and prosperously, and I am on the point of starting for the south of Ireland, where

I am to make a little *giro* previously to the gathering at Dublin. I was much interested with your account of the reception of your doctrines about heat, and glad that you arrived so opportunely as to prevent their being rejected untried; for though the suffrages of people, who judge after Biot's fashion, are not worth much, they have their consequences. I am glad too the tides are a printing, but I fear they will go on slowly. Our Admiralty here have undertaken the reduction of the June observations, and I am in London now in order to set them to work. I had a few days ago a letter from Herschel, who begs me to thank you for your notes about meteorology. But he says he cannot understand why you have assumed such a formula as $A \ (cos \, . \, lat)^{\frac{1}{2}} - B$ for the diurnal variation. He has been making experiments with his actinometer, which convince him, he says, that invisible vapour stops more heat than air does.

We are going to suffer a grievous loss at Cambridge, no less than the loss of Airy, who is going to replace Pond at Greenwich. I have no doubt that he will make Greenwich much the best observatory in the world, but in the mean time what is to become of us and of the undulatory theory? I hope some arrangement may be made by which Airy may continue to lecture at Cambridge, at least for some years; but nothing can prevent his departure being a great calamity to us. He does not go immediately; but I believe the matter is settled as to essentials. You lose nothing by not having my book with you, for there is only part of Normandy in it, and nothing south of that. I have seen prints of the buildings you mention, but will look at them again. I am glad to find you have found such objects interesting to you.

Gay Lussac is in London at present, ordering steam engines to be employed in a communication between Marseilles and Constantinople. I have not seen him, and I am told that he does not intend to be at Dublin at the meeting. I am going to try to see Poisson's book about heat to-day, as I have to make a report on that subject among others.

Sedgwick was to have gone to Ireland with Murchison and me, but has had to change his plan in consequence of a very melancholy event—the sudden death of his brother's wife while her husband was among the amusements of our installation. He, poor man, was quite stunned by the blow, and Sedgwick is gone home with him. Perhaps I shall hear from you at Dublin. In the mean time a pleasant tour to you.

Yours always, W. WHEWELL.

16, SUFFOLK ST., Monday, July 20, 1835.

MY DEAR JONES,

As to the important matter of my autograph, I suppose it is happily terminated already, the universal Mr Magrath having made me write my name for the purpose the other day. I am still here, having staid to finish my preface to Mackintosh, which I have just done. I should have liked much to shew it to you, all inert critic as you are, but I do not know whether I can manage to do so, as I want to have it off my hands before I set off for Ireland, which I intend to do to-morrow or next day. Yesterday in the Athenæum I saw a German-looking man, and charged him with being Von Raumer, which he forthwith confessed to. He told me that he had sent the new edition of his little book to Hayward—for you—for me. This circuit of communication has not been completed, and I have not seen Hayward, so that I do not know whether it is broken at you or him. I wish I had seen the book, for Von Raumer tells me that he has given his judgment of Bentham in the new edition. I hardly know how you are to send it me if you have got it, as I shall probably be gone before it arrives; but I will thank you to try. Send it here to-morrow (always supposing you have it), and as I shall probably not depart till Wednesday I shall receive it; and, if not, it will stay here. There is lying here for you Cathcart's translation of Savigny's <i>History of the Roman Law</i>, Vol. 1, and a missive from the Statistical Society. The latter I have taken the liberty to open, and it turns out to be a summons to a meeting on Wednesday next, "to consider the Report to be drawn up for the information of the British Association." If I were sure you would come here on Tuesday or Wednesday,

I would put off my journey till Thursday, but with such dispositions as you talk of I imagine there is no chance of that. I am very glad to hear that you work with so much zest. I have a great belief in the merit of anything so produced; but pray, good as your project of Indian lectures is, do not let it interfere with your book. You must get that out, and then, you know, it will be working while you are abed. I am sorry to hear that you are out of order—pray get your leg into reasonable trim with as little physic as possible. I am glad Mrs Jones is coming back to take care of you. My letter will reach you just in time to meet her with kindest regards.

I have been imprisoned in Mrs Mould's black back room for the last three days writing my Preface. Dr Holland volunteered to overhaul it for me, which I was glad of, for it is well to be criticised before publication, especially in a matter which half London will fancy it knows as much of as I do, and Holland is an acute man disciplined by much intercourse with the world. So my papers are now in his hands. I am now going to look after my tides at the Admiralty, my Trinity House sermon, and my wind measurer, and when I have set these matters in order I am off for Ireland. If I do not hear from you here, write to me at Dublin. You may depend on my being there till the 17th of August. Adieu.

Yours always, W. WHEWELL.

TIPPERARY, *Aug.* 5, 1835.

MY DEAR AIRY,

I have been meandering about the wilds of Kerry and dipping myself in Bantry Bay, with so little regard to science that I even allowed my watch to stop, but now, finding myself approaching Dublin, I am collecting my thoughts and considering how I shall conduct myself like a philosopher when I arrive among the wise men. Among other profound ideas bearing upon that subject, it occurs to me that I shall be expected to know whether you have seen the comet and what is your account of the length of its tail and the weight of its head. I beg therefore that you will set me up in these points by sending me a bulletin to Dublin (care of Prof. Lloyd). I hope you have caught the vagabond and impounded him, so that any man may know where to have him. Another thing, which I should be glad to be informed, is whether you have any definite plans for magnetic observations at Greenwich, and whether you want any special instructions or suggestions from those to whom such things are a care. Sabine and Cap. J. Ross, whom I have seen at Limerick, appear to be taking a great interest in the subject. Ross, in particular, has been determining intensity, dip, &c. at Limerick to form what he calls a magnetic base for Ireland, and is full of sorrowful indignation at the lack of English observations. I think it was supposed to be settled that there were to be in future magnetic observations at Greenwich, and, if so, it will be well that the plan should be adjusted so as to suit those who are likely to use them. You see I am talking of Greenwich, as if it were certain that you are going there immediately. I sometimes persuade myself that some delay may occur, but this I fear is not likely to any material extent. I am as much discontented as ever at the thought of your leaving us, and have quite satisfied myself that all the reasons are worthless for the change, except the consideration of not working you too hard. I suppose no more is known about any Cambridge arrangements, consequent on the change, than was known when I left you.

When I had got the din of our installation out of my head, I felt discontented at having had to go away without bidding you and your visitors good bye, and without talking a little about the hubbub after it was over. I heard from Sheepshanks in London, that Mrs Airy was well again, and that you (but how many *yous* I did not make out) were gone into Suffolk. I was detained in London so long that I missed Murchison, and have been rambling about without looking beyond the surface of the ground. Killarney is pretty, but, except one or two views, seems to me to gain little by the comparison usually made between it and the English lakes. There is too much flat or nearly flat ground about it. Nor are the sea lochs of the Irish coast which I have seen at all equal to those of the west coast of Scotland, but then I have not seen the best. I am come here from Limerick in order to see the cathedral at Cashel to-morrow, and, if possible, an abbey about five miles further off. Next day up the Shannon with Sabine and Capt. Ross. Give my kind regards to Mrs Airy. I hope to find you still far from your departure when I return to Cambridge.

Always truly yours, W. WHEWELL.

MARCUR, SLIGO, Aug. 21, 1835.

MY DEAR JONES,

You have, I suppose, received a letter of mine written from Dublin in some discontent at the thought that you had procrastinated so long in giving me your opinion of my preface that I should not be able to write to the publisher from Dublin. Indeed this happened, for after the first day the bustle became so unremitting that I could attend to nothing else. I was very glad however to receive your criticism a day or two afterwards. But I am rather puzzled with what you say. There is nothing in the Dissertation or in my preface, which appears to me to be an attempt to establish a parallel between the moral sense and our physical senses. That our moral judgments belong to a peculiar *faculty* indeed, and cannot be resolved into judgments concerning mere pleasure and pain, is Mackintosh's opinion and mine, but is far from new. The arguments which I have given in favour of this opinion are put in my own way rather than his, and were intended to be the main value of the essay; for in the part about selfishness I have done little but try to arrange the arguments given by preceding writers; and the main use of that division of my subject is to shew that we must grant motives which go beyond personal gratification, and thus to prepare the way for the establishment of a sense of duty as a motive. Mackintosh's analysis of conscience I do not entirely assent to or ascribe much importance to, and I would say the same of the corollaries he deduces from his theory; but I thought that I had explained that part of the subject in such a way that I must appear only an expositor. Indeed throughout I am far from considering

Mackintosh's view as containing the whole truth; but that is to be remedied only by writing a new treatise. The great merit of the book in my eyes is that it corrects some prevalent errors; and I certainly hoped that the arguments against the utilitarian view were likely to produce some effect. However, I do not know that I can mend them; so they must go to press as they are. As to the passage about Bentham's treatment of the ancient and modern philosophers, you may leave it in its place; I shall probably strike it out when I see it, as I have outgrown my fancy for it in a great measure; but you do not give me your own judgment about it, and therefore I am no nearer a decision than I was.

Send the MS. to Adam Black, Publisher, Edinburgh. As soon as I foresee when I shall be stationary for a week, I will write and tell him to send me the proofs. Probably this will be in London or Cambridge, for I am disposed to return as soon as I can and try to do some work. Our Dublin meeting went off very well. The trial of the vitality of the thing will be the next meeting, which is to be at Bristol.

I grieve very much at what you say about the intended abolition of your College. I hoped the storm would blow over. Some daylight may still break through it. Do not let it disturb the progress of your book. My kind regards to Mrs Jones.

Adieu, yours always, W. WHEWELL.

CAPEL CURIG, *Sept.* 1, 1835.

MY DEAR JONES,

It is perfectly true that I am here, at the foot of that very Moel Shiabod on which you sprained your ankle about seven years ago and mended it by some Opera Dancer's recipe; and after climbing to the top of Snowdon by a much finer path than we took, and with a day a good deal better, though still by no means so bright and brilliant as I could have desired. I thought I could spare a day to retouch my recollections of the places we visited together. Some of them are marvellously changed. The road between Capel Curig and Llanberis, which was then hardly passable, is now a road as good as the Simplon, traversed by four coaches daily: and besides the inn where we had the bottle of port which excited your enthusiasm, there is, close to Dolbadern castle, an enormous hotel with three sides of building and three stories, an establishment of eight chambermaids, a separate stable court, handsome carriages standing and liveried coachmen lounging before it, and everything as unlike as may be to the ancient days.

To-morrow I go on towards London, and I shall be there about Friday. I have directed Black to send me the proof sheets of my preface, which I suppose you have sent him. I told you I was puzzled by your saying that Mackintosh and I between us had tried to make out a parallel between the moral sense, as it has been called, and the physical senses. I have been trying to make out any passages which you might think to have this aspect, and only two occur to me. In one passage it is said that it must be expected to be a difficult matter to make out whether our moral judgments are peculiar perceptions or are capable of being resolved into others, since even in the senses, as for instance, that of sight, the same difficulty occurs. If this be what you mean, I might perhaps remove all ambiguity by saying this.—" It is difficult to shew clearly of any judgments that they do, or do not, depend on a peculiar faculty; for instance it has been maintained that we have a peculiar faculty by which we judge of beauty, and that our perceptions of the beautiful, and its reverse, cannot be resolved into any other perceptions (this is Brown's doctrine); while on the other hand many eminent writers have attempted to make such a resolution. All these attempts are declared unsatisfactory and hopeless by the most recent writers. And even in the physical senses the same difficulty occurs. It cannot be considered as yet established to the satisfaction of readers in general whether we can or cannot *see* figure." Surely this is not attempting such a parallel as you mention, but only guarding against a hasty and confident rejection of the trouble requisite to a discussion of the question. Again, it is said somewhere in the preface that the moral sense implies a memory of our judgments and a power of anticipating what they will be in given cases, as the other senses imply such powers. This was said in consequence of some absurd quibbling of Mill's, to which I will not refer more distinctly, in which he says that the judgments of which Mackintosh speaks, being judgments of past actions or dispositions, cannot direct future actions. If this be what you mean, it will not be difficult to avoid the implication which you speak of, and which I should not have expected anyone to collect from the words. You see I want to get something definite in the way of criticism, and then I can make it useful. Write to me at the Athenæum, if you write before Saturday. I shall get to Cambridge as soon as I can, for my book haunts me, and I shall have no peace of soul till I am at work at it again. I have almost resolved not to leave Cambridge again, except so far as is necessary, till I have it in the press; for it pursues me like the image of a mistress, and everything brings it to my mind. The only difference is, that I shall never rest till I have *parted* with it.

I hope you go on writing your book. That must be the best plan on all accounts and whatever betide, besides which I cannot understand how you can reconcile yourself to the enormity of imprisoning your full grown truths any longer. They have plenty of muscle and only want feathers.

My regards to Mrs Jones.

Yours always, W. WHEWELL.

TRIN. COLL. CAMBRIDGE, Sept. 9, 1835.

MY DEAR JONES,

I am grieved to find that you are out of health and out of heart, and hardly know how to dispose of my anger towards you, for I have been irate and am still rather so. I believe the best way of getting through my ire will be to tell you of it. You must know, I think, as well as I, that such general criticisms as you have sent me can be of no use. If you had pointed out any passages to which your remarks apply, I might have profited by them, and this is precisely what

I wanted you to do. Without this, such warnings can only torment a poor author, if he pays any attention at all to them. It was for this purpose that in my letter from Capel Curig (which I suppose you got) I had referred to the only two passages I could think of as giving room for one of your objections. I begged to know whether, in what you said, you meant those passages; and of this request you take no notice. Another of your objections, which I also replied to, you repeat in your last letter in the same vague and useless way.

I am the more vexed at this, because, for my own part, I consider this mutual criticism one of the important offices of friendship between people who happen to publish. I have always given my best thoughts and care to it, not only for my more intimate friends, but even for slighter acquaintances, whenever it was asked. And the subject being one on which you have thought much, on which a critic with such a qualification is not easy to find, and on which a little good advice may save a great deal of misapprehension in general readers, I certainly thought I might depend upon you.

Every evil has some remedial circumstance; and while I find that other persons have not the same impressions with myself of this obligation, I grow more confident in my own power of judging my own productions; and as I have never cared much about immediate success as an author, I can bear to trust to myself when I find I must do so. And there I leave that matter.

*   *   *   *

Yours always, W. WHEWELL.

TRIN. COLL. CAMBRIDGE, *Oct.* 2, [1835].

DEAR PHILLIPS,

I send you a statement of the purport of what I said in the geological section, and, as it may save you trouble in copying, I send it on a separate sheet, taking for granted you will charge the Association with the double postage. You may abridge my account if you find you can do so, and if you are pressed for room; or, if you find it necessary, you have my permission to omit it altogether. I will write to Quetelet and tell him that he is to appear in our *Transactions.* Of course he will be allowed a certain number of separate copies, about which I will ask his instructions and communicate them to you.

I am glad Willis has set you a thinking and scheming for the benefit of Gothic architecture. I have by me a plan for the tabulation of diversities, which I drew up with a good deal of thought some time back, but I have never published it, and I do not know that I shall. I begin to think it is both officious and unprofitable to draw up plans and to expect other people to work upon them, and I shall hesitate about proposing any more plans of notation and nomenclature, till I have got some body of information which I can communicate by means of them. It is in that way only that systems of names and notes make themselves generally known and used. I shall be glad to see your system of geological representation, and should be very glad if you or any geologist would try to arrange a system of colours, upon such principles as I suggested, or any others which are coherent. Any one who would publish good geology in such a form would be sure to gain disciples, for people would be able to read his maps by the eye, instead of having to spell them out by the dictionary in the margin. Believe me,

Yours very truly, W. WHEWELL.

When you write again, tell me what number of separate copies are allowed to Reporters, and how soon you expect to be able to publish.

[To M. Quetelet.]

TRINITY COLLEGE, CAMBRIDGE, *Oct.* 2, 1835.

MY DEAR SIR,

I regretted very much that you were not able to join our meeting at Dublin. I did not receive your report on Belgian Science till my return from Ireland, when I found it at Cambridge. I sent it to the Secretary, and am informed by him that it will be printed in French in the volume of our *Transactions* which is now in the press. I read it with great interest, and did not think it at all too long. The views are presented in a striking and rapid manner, which keeps up the attention. When it is printed, I believe the rules of the Association allow you a certain number of separate copies of the Report (I forget how many). I shall be glad to have your directions how I shall dispose of them ; probably you will wish the greater part of them to be sent to you. You will see in the *Philosophical Magazine* the official account of a part of our *Proceedings ;* when the account is complete it will be sent to the members. The *Transactions* will appear as a separate publication. Some of our miscellaneous periodicals have also given an account of the Dublin meeting, and I send you a few numbers of the *Athenæum*, in which you will find our history. Upon the whole the meeting was a very good one, especially in our section (General Physics) ; we meet next year at Bristol, which, being the first place of meeting which is not a University, is a new experiment, but I have no doubt of its success.

I am much obliged by your exertions with respect to my Tide Observations. When I was last in London, a month ago, none of the foreign observations had been received at the Admiralty, except those from Holland and Spain, both sets apparently excellent. I shall be at the Admiralty next week, and hope to find the Belgian Observations.

I received your work " L'Homme " with great pleasure, both as a mark of your kindness, and for the interest of the work itself. Your researches are as curious and interesting as anything expressed by means of numbers can be. The Statistical Section of the Association met at Dublin, and had some important papers, but I was not able to attend their sittings. The Statistical Society in London was engaged before the meeting in drawing up a large collection of questions for circulation. They will, I have no doubt, obtain a great deal of information, but my opinion at present is that they would go on better if they had some zealous *theorists* among them. I am afraid you will think me heterodox, but I believe that without this there will be no zeal in their labour and no connexion in their results. Theories are not very dangerous, even when they are false (except when they are applied to practice), for the facts collected and expressed in the language of a bad theory may be translated into the language of a better when people get it; but unconnected facts are of comparatively small value.

TRIN. COLL. CAMBRIDGE, *Oct.* 30, 1835.

MY DEAR LUBBOCK,

I have spoken to Peacock on the subjects to which your letter refers, and both he and I think that Hamilton's Conical Refraction appears the most promising paper for the medal. He, however, speaks of giving the subject further attention, and intends to get Jerrard's papers and read them again carefully. At the meeting at Dublin a sum of money was voted to Sir William Hamilton to employ in procuring calculations to be made in order to test the application of Jerrard's method in particular cases, as the best way of ascertaining the value of it. Till the result of this examination appears, I think it would be hasty to give our decision upon it. I will look at Hamilton's memoir again, and will see Murphy as soon as I can. I shall probably be at the first meeting next month. I will come if possible. Peacock says that he knows nothing of the Duke's intentions[1].

I have been working at your results of the Liverpool tides, and have, I think, got the lunar inequalities into very pretty order. I cannot get any satisfaction about the solar inequalities, and have no confidence in any thing we have yet done about them. In order that you may see my grounds for mistrusting our previous modes of proceeding on this subject, I send you some leaves of the draft of my paper, which I will thank you to return. I am persuaded that you will have to alter your solar correction for the London tides, and should expect that your Tables would be improved by leaving it out altogether. Since we have got the correction for lunar declination and parallax

---

[1] The Duke of Sussex, President of the Royal Society.

with a considerable degree of exactness, I do not doubt that we might obtain the solar correction by correcting the tides for the moon's effect, and then tabulating the *residual* quantities with reference to the sun. I propose therefore that in the recalculation of the Liverpool tides, which Stratford is going to undertake, the solar correction be altogether omitted. The errors will then enable us to get this correction, or at least offer the best promise of it. I hope you will agree with me in this, and shall be glad to hear what you think of these views. Believe me,

Yours very truly, W. WHEWELL.

TRIN. COLL. CAMBRIDGE, *Nov.* 11, 1835.

MY DEAR JONES,

*   *   *   *

I had a note from Drinkwater[1] with a translation of a German play, which I suppose he has executed to shew that he is not overworked. I receive it with great joy as a symptom of vitality, and I dare say I shall find that he has made the Jungfrau talk in a very pertinent way when I have time to read it. At present I am somewhat overwhelmed with tutor work, and I have besides a great tide-paper on the stocks, which is to be a considerable advance on all that we have yet done. All this smothers the Induction, and I have not been able to do a stitch at it all the term. Still I am trying to struggle into daylight. I have almost resolved not to interrupt myself in any way I can help, till I am ready for the press, and then I may, I think, finish in about a year. Of course I cannot cast off tuition and tides, but I must allow myself no holidays. If I could bring my work to Brighton, I should like very well to come and work with you there, but I do not think that will be possible.

*   *   *   *

Yours always, W. WHEWELL.

[1] Mr Drinkwater afterwards took the name of Bethune. In 1848 he went to India as a member of the Supreme Council. See De Morgan's Preface to the *Treatise on Problems of Maxima and Minima*...by Ramchundra.

[To Mr Rickman.]

TRIN. COLL., *Jan.* 7th, 1836.

MY DEAR SIR,

I was glad to hear from you, as I always am, and should be happy if I could give you any useful information on the points on which you ask it. I am not sure that I shall be able to do this, for on many of the details which you mention I have not formed any opinion of my own, and do not know what the opinions of others are. And there is another difficulty, which it is best to state distinctly. Of course, if I were to tell you what I think as to the questions on which you are in doubt, my judgment could go for no more than the opinion of a private person, which may be bad or good, but which could not be a sufficient ground for your acting, and still less could be quoted, or in any way referred to, in justifying or explaining your plans. I may go further, and say that I know, from circumstances which have recently occurred in another case, that any suspicion that an individual had taken upon himself to offer suggestions to an architect, employed in preparing designs for the University, would produce great disapprobation and indignation, which might be prejudicial to both parties.

I tell you this, merely in order to shew the caution with which I am obliged to give my views, and not at all with any wish to avoid answering any enquiries, so far as they regard my own judgment of what is best. But I conceive that, this being the state of things, all additional instructions as to details, like those to which most of your enquiries refer, cannot be obtained in any useful way but by application to more official persons than myself. If the Vice-Chancellor were applied to on such points, he might get the Syndicate together and consult them on the propriety of such additional instructions. With respect to the rate of increase of the library, which is certainly an important point, I think the best way would be to write to the librarian, Lodge, who is the best authority on these heads, and would, I suppose, feel no difficulty in replying to that or any other essential questions respecting the accommodation wanted for the library.   I believe the annual rate of increase is 4000 or 5000 volumes.

As to the general bearing of your enquiries, I should be disposed to say this : If a design proposed provided for the immediate erection of considerable additions to the library, and two or three museums (the Woodwardian to be one); and if it admitted in the sequel of turning to similar account (library, museums, and lecture-rooms) the remainder of the old court of King's, it would be no objection, *in my eyes*, that it was defective in the provisions for removing the existing façade of the library, and putting a better in its stead[1].   For I consider the increase of our accommodation to be the main point, and I no longer look upon the removal of the existing front as a necessary condition, or think of it as a thing which will certainly happen.   I believe that several other persons think with me, and that the re-issuing the instructions to architects with so little alteration, and the informing the architects that the old plan was resumed, was a step adopted principally to avoid the difficulties which would have encumbered any other course of action.

But then you must consider, on the other side, that the instructions require a design for an entire new building, and that, if the design does not come to us in this form, the Syndicate, which will have to report upon it, may very possibly report that it does not fulfil the conditions ; and then it would be of no avail that it was approved of, even by a majority of the University.

I believe the main reason for fixing originally so short a time for the architects to prepare their designs was, that it was supposed they would send in their former designs without any fundamental modifications.   But this was merely a conjecture, as I conceive, and by no means a condition.   Still, I imagine that a change so great as the substitution of a Gothic for a Grecian design would be an unfavourable circumstance to begin with. I do not say it might not be got over.

I am afraid I have not helped you much by my explanation,

---

[1] For an account of all the transactions connected with the building of the University Library at Cambridge, Mr Bashforth's pamphlet should be consulted : see Vol. I. page 58.

and it is easy to see that, with the difference of opinion which will probably arise in the University as to the object and the means of accomplishing it, the problem must be a very perplexing one for the architect. I have stated my own views, which must, as I have said, be looked upon as strictly my own, and as possessing no authority whatever.

I hope your health continues good. The best wishes of the New Year to you and Mrs Rickman.

Believe me, my dear Sir,

Yours very truly, W. WHEWELL.

TRIN. COLLEGE, CAMBRIDGE, April 9, 1836.

MY DEAR HERSCHEL,

I do not know if it appears to you, as it appears to me, a longer time than usual since I wrote to you. There have been various causes of this; but having just got your letter to Dr Adamson, which I have read with great interest, I will give you one or two of the reflexions which occur to me upon it. To begin, after our ancient fashion of disputatiousness, with the points on which we differ, I am inclined to think that you dispose rather too lightly of the ancient languages, as instruments of giving a philosophical civilization to the mind; or rather perhaps I ought to say, that I should think this, if your scheme were proposed as one for European education. For I am fully persuaded that the study of the Greek and Latin authors, in their original mode of presenting themselves, can never cease to be an essential part of the liberal education of the present age, till the habits of thought and taste, of intellectual analysis and imaginative enjoyment, of the age cease to depend in an esssential degree upon that which we have, through various channels, near and remote, direct and tortuous, derived from them; and this I am quite sure will never be. To suppose that man will ever become, in these respects, independent of the history and antiquities of literature, seems to me to be equivalent to supposing that he will cease to keep his footing in the path of intellectual progression, of advancing civilization, of the mind and feelings, which began with the philosophical and poetical age of Greece. It must, I think, always be the peculiar character of a liberal and intellectual education, that it leads people to seek and find grounds and reasons for the tacit assumptions, rules, and methods, which are diffused and accepted wherever literature, in the best sense of the term, is cultivated. And this process in education nothing, I am convinced, can supply, except a study of literary antiquity.

So much for my criticism; to which, after all, you would probably not be far from assenting. There are other things in your letter where I go along with you still more decidedly. I am quite delighted to find that the ancient subject of our earlier liking, the Inductive Method of Philosophising, is still so prominent an object of your affection; although, in truth, I needed no new evidence of this. I have always doubted, however, whether Bacon's book could now do much more than inspirit its readers with an admiration of the dignity and vigour which the prosecution of this object infuses into the mind. I do not think it complete and methodical enough to teach much to students; and it is clear, I think, that it requires both to be accommodated to the present state of thought and knowledge, and to have its vast vacuities gradually supplied. You have done no little yourself towards this object; but I still think that much remains to do; some part of which, from what you have sometimes said, I still expect from you. In the mean time I am myself working in the same career; and hope, in no long time, to be able to offer you some contributions. My first must be a sort of survey of the History of Inductive (Physical) Philosophy up to the present time; to correspond, as to its place in the development of the subject, with Bacon's Advancement of Learning. This I expect to finish in a few months, and then I shall print forthwith; so that, except you sweep up your stars very speedily and come home, I shall have to send it you. I do not know how you manage to carry on so many speculations at once; but for my own part I begin to find that I have set myself a task, which is hardly consistent with my other employments here; and as it appears that the reform of our Philosophy is the work to which I have the strongest vocation, and which I cannot give up if I were to try, I have very serious thoughts of resigning my share in the active business of the College, and of giving the rest of my life to the formation and exposition of a Philosophy and Logic such as we ought to have. I shall do this with some regrets; not only from long habits of kindly and grateful association, but also because I think I perceive that any improvement in our academical studies (and of course a reform of philosophy ought to improve *them*) may be introduced with greater advantage by a person actively engaged in them than by an insulated spectator. But I have tried for several years, and I cannot combine these two employments to my own satisfaction; and I think it is more wise and right to transfer to other hands occupations, which I am conscious of being unfit for, and duties, which I discharge imperfectly, than to go on with an impossible struggle, and to endanger the attainment of a great object (you know how we imaginary Bacons talk), which I feel considerable confidence I can materially contribute to. I have not quite worked myself into such a condition, with regard to worldly affairs, as to make this a very prudential course, but at the same time I believe I shall be able to manage it without any inconvenience which may not reasonably be incurred for such an object. So in a year or two I expect to be a philosopher and nothing else.

So much for philosophy. When I talk of giving the rest of my life to it, I always reserve to myself the *tides*, as a corner of physics which I shall go on and work at, till all is done that I can do. I have made so much progress that I have great hopes I shall leave that subject in a condition very different from that in which I found it. One reason why I have not written to you sooner has been that I have been looking forwards to the time when I should work up the observations you made for me last June. I have good observations all along the coasts of Europe from the North Cape to the Straits of Gibraltar, and these are now undergoing reduction; yours among the rest. The results will be very curious. I have also *long* series of observations at London, Liverpool, and Bristol, which give very interesting laws with reference to the theory; but these I will tell you of at some other time.

In the last letter I had from you you augured that the Ministry, that now is, would be likely to continue the existence of the East India College, where Jones is so happily placed. I fear this is too good to be true. They expect to be dissolved, and at any rate the Government have refused to make his appointment permanent. In the mean time he employs himself rather with public matters (drawing Tithe Bills and the like) than with his theories, which I hold it is his duty, under any circumstances, to complete and publish. I often see him, and we never fail to talk of you. From some things which you say, I have strong hopes that your return is not far distant; the sooner the better, for, without any disparagement to the world at large, we want such people as you in England.

I hope Lady Herschel and all your young things are well. I shall be quite delighted to see what effect an African sun and the habit of standing at right angles to us (to say nothing of several years of time) have produced in their appearance and figure.

Believe me, dear Herschel,

Affectionately yours, W. WHEWELL.

TRIN. COLL., Ap. 23, 1836.

MY DEAR MURCHISON,

My friend M. Rio, a French literary man, was admitted as a foreign member of the Athenæum three years ago, when he was in London for a short time. He is desirous of obtaining the same privilege again; and, as his former friend, Sir John Malcolm, is gone, he has begged me to try to make him a new friend who will render him the same service, and I do not know any one who can do it better than you. I suppose an application to the committee will produce the effect with little trouble. Rio is worthy of the distinction, being a person of great learning, eloquence, and accomplishments. He was Professor of History in the University of Paris before the "three days"; and, refusing to take the oaths to the new government, was stripped of his appointment. He is, in fact, a high royalist and catholic. After travelling in Italy, where he studied the history of painting, on which he is publishing a very striking work, he came to England, and married a Miss Jones, of Llanarth Court near Monmouth, a catholic family, one of the oldest *Silurian specimens;* so you see he belongs to you by right.

I shall be glad if you can render him this service, as he is an excellent fellow, and a great friend of Sedgwick's and mine.

* * * *

TRINITY COLL., CAMBRIDGE, *May* 27, 1836.

DEAR HARCOURT,

There are some points which I should like to urge at the next meeting of the British Association at Bristol, and I should be glad to know your opinion about them beforehand. There is a man there, by name Bunt, who has been working very hard at the Tides with a view to constructing Tide Tables; and has become so much fascinated with the subject, that he has given more time to it than will be repaid him by any return he is likely to get. His results appear to me to be valuable as an addition to our knowledge of the subject, and I should wish to propose to purchase his calculations from him, or in some other way to make him some recompense. He is likely to go on with the subject, and to do more than he has yet done. If I were authorized to employ a certain sum upon him, it would be exactly what Lubbock has done in the case of Dessiou.

Another matter connected with the tides is the execution of our ancient scheme of fixing the relation of land and water, at least at one or two points. The observations of tides at Liverpool and Bristol have been so multiplied and so well analysed, that the position of the measuring scales, with regard to the sea level at those places, may be considered as very exactly determined. I would therefore propose that these scales should be well connected with the natural features of the country in some clear and permanent manner, so as to afford a point of comparison in future times. I think this is the only way of executing the scheme, formerly recommended, of ascertaining the permanence of the level of land and water.

Nothing has been done respecting the Antarctic magnetic expedition. At one period Capt. Beaufort, the Hydrographer, sent me a message that there was a ship which we might possibly have if we asked for it. But I found that there was no chance of this, and that, if sent at all, it would go to the north, as it has done. I found also that nobody, except Murchison, came to the meeting of the committee held on this subject, so that I thought it better to wait, than to urge the point weakly and ineffectually. Next year, with Sabine's Report in circulation, we shall have, I hope, a better chance.

I have not yet heard of the time of meeting at Bristol, though it is now late for it to remain uncertain. I hear too that Phillips is ill, which I regret very much.

There is an election to take place soon, in which the Archbishop of York is an elector (one of four), and in which we Cambridge men are somewhat concerned; I mean the Mastership of Downing College. Of the candidates, Starkie[1], though in other respects an excellent man, would, I think, be objectionable as a non-resident lawyer; Dawes has taken a living, which would also take him away. Worsley is, as appears to me, a candidate every way unexceptionable; and being a person of very conciliating character, very great accomplishments, and very desirous of making the College more effective for the purposes of good education than it has yet been, he would, I think, be a great gain to us. There are few persons of whose religious principles and right intentions I think so highly, and he is a very general favourite both for his literary and conversational endowments. You will not think me superfluous in saying so much, for, of course, it must be the wish of the electors to choose the person most likely to manage the College well; and you can communicate what I have said to the Archbishop, or not, as you think best.

I shall be in London soon, where I may perhaps meet you.

Believe me, dear Harcourt,

Yours very truly, W. WHEWELL.

---

[1] Mr Starkie, senior wrangler in 1803, was Downing Professor of Law; Mr Dawes, fourth wrangler in 1817, afterwards became Dean of Hereford.

TRIN. COLL., CAMBRIDGE, *May* 28, 1836.

MY DEAR FORBES,

I ought to have written to you long ago, but I have been more than usually busy; among other causes, from the illness of one of my assistant lecturers. I read with great interest your account of the eclipse, as did Airy, who was here at the time. He has been giving us a parting course of lectures, which, with some difficulty, we persuaded the Lords of the Admiralty to allow, and has just left us; leaving the impression of our loss in him stronger than ever. Challis, however, buckles to his work with great zeal and courage, and I have no doubt will work the Observatory well. He gives Plumian lectures for the future. I was much interested by your account of your optical experiments on the eclipse; although, in truth, Brewster's conjecture appeared to me so arbitrary that I never attached any importance to it.

I thought Chevallier's answer to your "philosopher" well enough, but rather too hot. I cannot set to work with that subject in special at present, but it will find its place in a book which I hope to write by and bye, and which I am ashamed to find growing in its scheme larger and larger. It is partly in consequence of this scheme that I cannot yet speak very positively about my visit to Scotland this year; for I want, if possible, to execute a definite portion of my task before I go a gadding, and I have not been able to write a sentence for some months on account of other employments. Among the rest, I have had a large paper on the Tides to write, the result of observations made all over Europe last June, which appears to me to be curious and important. However, I have now nearly finished this, and I think a month of leisure, if I can find it, will take me through the work I have set myself. If I manage this, I should like much to visit Scotland again in July, taking Edinburgh in my way, if you are then there, looking again at some of the noble scenery of the West, and working on to Orkney, if I have time. The exact period of the meeting at Bristol could not be settled the last time I met the managers, in consequence of some uncertainty about Assizes, &c., but it appeared to be understood that it would be late in August. I intend to be there, for I want, among other things, to get some more tide work done. I hope you will come. If I could have travelled north at present, I should have been glad to have joined you in your Highland tour, but that is out of the question.

In reference to the undulatory theory, I see that Cauchy has been professing to prove, from the theory, the laws of elliptical polarisation by reflection from metals, as established experimentally by Brewster. I am sure you cannot do any thing better for the credit of your university than give this theory prominence, for in a few years people will wonder at the slowness of the world in admitting it, as they now wonder about gravity.

My Mechanics, new edition, will be ready in a week or two. We are at present trying to introduce some mixed mathematics (Mechanics, Hydrostatics, Optics) into the general examination for degrees, as I recommended; and I think I shall have to publish a little book, containing about thirty simple propositions in Statics, for this purpose. I will send it you, when I have printed it.

I hope you go on polarizing heat. I do not know at present any subject more promising or interesting. Biot is writing a series of articles about Newton, &c. in the *Journal des Sçavans*, in the way of a review of Flamsteed's letters. He is very temperate and reasonable. I have also received from Rigaud some papers throwing some light on Newton's matters, which I hope he will publish. Brewster's inaccuracies come to view at every turn. I will write to you again when I know my own movements better. In the mean time, as always, believe me,

Yours truly, W. Whewell.

[To M. Quetelet.]

Trinity College, Cambridge, *May* 31st, 1836.

My dear Sir,

* * * *

...I have been much engaged of late in discussing the observations of the Tides made last June, and I think that I have obtained some results quite as satisfactory as I could expect. I am just going to communicate these to our Royal Society. The course of the Tides in the German Ocean in particular appears to me to be extremely curious, but at the same time I think I have established it beyond all doubt. I am glad to hear you have been discussing tide observations, but you ought to get calculators to do all the laborious work. The construction of Tide Tables is with us undergoing rapid improvement. In particular the calculators of such tables at Liverpool and Bristol are improving them by introducing the *Diurnal Inequality* by which the morning and afternoon tides differ. This inequality is of large amount both for the heights and the times. I have also with some difficulty extricated from the Liverpool Observations the Solar Inequality. It is the subject of a memoir which I hope to be able to send you very soon. You have perhaps received my Report on Heat, etc., and you may see there that I shall look with much interest to your thermometrical observations at different depths.

The Cambridge Observations for 1835 are delayed by the reductions of the Observations of the Comet, which were very numerous, and, we hope, very valuable.

You are aware that we have now lost Airy from the Observatory here, and a very grievous loss it is: but his successor, Challis, is a zealous and intelligent mathematician, so I hope we shall still be able to hold up our heads. The Duke of Northumberland's large telescope (12 inch object glass) is not yet mounted, but is making progress. When it is mounted on its Equatorial Axis it will be one of the finest instruments in the world. In the mean time Airy is putting the Observatory at Greenwich in a state of vigorous activity, of which your astronomers in other countries will, I have no doubt, soon see the results. You have probably seen an account of the recent discovery made by Forbes respecting the polarization of heat. This appears to me extremely curious, and shows us that we have not yet done with the undulatory theory. The bearing of the properties of crystals upon the laws of heat now becomes an extremely interesting and important subject for investigation, for since heat can be polarized, it must be affected in some way or other by its relation to the axis of crystalline form.

I have to thank you for your *Annuaire*, both on the part of our Philosophical Society and on my own.

Among other reforms in the Observatory at Greenwich, I believe it is intended to establish good magnetic observations, and if you have any suggestions to offer on that subject, I have no doubt that Airy will be very glad to receive them.

The British Association meets this year at Bristol, I believe about the 20th of August. Is there any hope of seeing you there? I wish there were, for it is long since we met, and I do not think I shall travel on the continent this year. With kind remembrances to Mad. Quetelet,

Believe me, my dear Sir,

Yours very sincerely and faithfully, W. WHEWELL.

UNIVERSITY CLUB, LONDON, *June* 10, 1836.

MY DEAR HERSCHEL,

I should have written you more frequent and orderly letters than I have done of late, but that I could not get into order the tide observations of last June, yours included, so as to see what they led to. It is wonderful what an immensity of reduction these things have required, for I had observations all the way from the mouth of the Mississippi to the North Cape of Norway, and I have reduced them to as much regularity as possible both by your favourite method of graphical interpolation, and by taking means, expending on them a great deal of labour both of my own and of certain calculators which the Admiralty have given me. At last I have got my results and am finishing my paper; and though I will not now trouble you with all my general inferences, there are some points which it may be worth while to mention. It has appeared by the discussion of long series of observations made at London, Liverpool, Bristol, that the actual tides exhibit the features of the equilibrium tides, both in the heights and times, with wonderful fidelity; namely, the semimenstrual inequality, the corrections for lunar declination and parallax, and a solar correction, and a diurnal inequality. Of course the epochs of these equilibrium tides are to be determined by observation, and depend upon what I called formerly the *age of the tide*. I remember that you did not quite concur in this way of considering the subject, and I am not sure that you will not turn out to be right in your doubts. At least there is a very odd difficulty which I must explain to you. The diurnal inequality of the heights is very marked both on the coasts of America and on the west coast of Europe. On laying down the heights in means it is strikingly regular and uniform, but with this difference. In America it corresponds to the *actual* declination of the moon, that is, it is greatest on the day when the moon's declination is greatest, and vanishes when she crosses the equator. But in Europe it vanishes two or three days later, and attains its maximum at about the same interval. Well—this may be accounted for by supposing our tide to have a day or two of age more than theirs. But then how old is your Cape tide, which, according to my former speculations, was one of the youngest which exists? Why, by this criterion, it is as old as that on the coasts of Spain and Portugal. For, by your observations of the heights at Table Bay, the diurnal inequality was very strongly marked on the 16th, 17th, 18th of June, and on the 2nd, 3rd, 4th of July, when the moon was nearly in the equator. This point appears to me very curious, and though I am loth to give you any trouble, I should be very glad if you could get it attended to. I do not want observations of the times, but of the *heights, both day and night,* for some time; in order to see where the moon is when the diurnal inequality is the greatest, and when it is 0. It will be very curious, if it is really true, that your tide is the equilibrium tide of two or three days previous, while the American one is that of the very day. I have not been able to use the Simon's Bay observations with reference to this point, as the night tides were not observed. I hope Mr Maclear has received a tide-machine. I pressed his application on the Admiralty, and was told an order had been given for complying with it. We are going on so prosperously with our tide-researches here, that I am quite greedy of materials which may tend to complete our knowledge. I am not, however, so unreasonable as to wish to take up your time with it; still less to be willing to have you washed off the pier at Cape Town in the service, as I am told happened on the former occasion. I have been in town for above a week, finishing the account of the observations of last June, and their results; but I am impatient to get this subject off my hands, for I want to go to Cambridge, and to spend the halcyon days of the vacation in completing my History of Inductive Science, which I hope to get into the press soon, and which I am willing to believe you will like. I fear it will be a horrid big book in spite of all I can do. Jones is so much immersed in Tithe Bills and the like that I can get no general philosophy out of him at present, but I shall try in quieter times, if such ever come. Adieu for the present. I am trying to find the German pamphlet, *John Herschel's Neueste Entdeckungen den Mond und seine Bewohner betreffend*, that I may send it you.

We have had the Airys at Cambridge for a month since they went to Greenwich, which was a great happiness. My kind regards to Lady H. I am glad to hear from Lyell that you are all well.

Yours ever, W. WHEWELL.

[To J. Phillips, Esq.]

LANCASTER, *Aug.* 10, 1836.

MY DEAR SIR,

My apparatus sent to Bristol consists, I presume, of two anemometers of the construction which I described last year. I have fixed up one at Cambridge, and intend to offer those two for the use of two persons who will engage to make careful and regular observations with them. If you can find persons in remote parts of England who are likely to do this work well, I shall be much obliged. I had no intention of lecturing about these to the evening audience, nor would the subject, I think, bear it. I have also some notices about the progress of my tide researches in general, and those respecting Bristol in particular, to give; but these also will not, I think, even including the local interest, be entertaining to the public at large, and I intended them for Section I. It has always been difficult to fix upon lecturers for the evening. If Faraday comes to us, as I suppose he will, it would be desirable, I think, to get a lecture out of him, so good a lecturer as he is. His last lecture at the Institution, on the Nature of the Ultimate Analysis of Bodies (or some subject like that), might be well suited to the occasion.

I have written to Mr Greenough proposing that, instead of a single notch, we shall make a level line of notches all the way from Bristol to the Land's End, or at least for some considerable distance, which will settle more than one interesting question. If you find any engineers or trigonometrical surveyors, ask them with what accuracy two places forty miles distant can have their comparative level determined, using as many intermediate stations as may be convenient. If this can be done within a foot, (as I suppose it can), it will be an object worth spending the Association's money upon. I also want to apply for some money to spend upon the Bristol tides. There is a most zealous and intelligent calculator, called Bunt, who lives in Smallstreet Court, Bristol, who has been working at them for some time, and whom I wish to keep at work.

I should rather like to bring forward in the geological section my plan for colouring geological maps, but I fear I shall not be able to work it out, or to exemplify it, so as to recommend it properly.

I shall try to visit Bristol for a day, the 15th or 16th, but, if I do not, I shall be much obliged if you will secure me tolerably good lodgings not inconveniently situated.

I am glad to find you are already on the ground. Believe me,

Yours very truly, W. WHEWELL.

16 SUFFOLK ST. *Oct.* 6, 1836.

MY DEAR SEDGWICK,

I have just seen Murchison who has given me such notice of your whereabouts as he thinks will enable me to hit you at a long shot. He tells me that you go on setting the strata of

Devonshire to rights, which is good; but he tells me that you
have had bad health, which I grieve much to hear. Pray take
care of yourself and avoid hard work, bad weather, and Dr
Sedgwick's physic. Consider that the two former cannot be
expected to agree with a rosy prebendary as they did with a
lean geologist, and that the doctor in question knows too much
of other things to make his prescriptions good for any thing.
I want very much to have your advice and opinion on a point on
which it will have great weight with every body, and with me
will be decisive. Your leading geologists think they see strong
reasons for not taking Buckland for their president next year,
and are at a loss whom to pitch upon. Now it appears to me
that if in this embarrassment they are tempted to turn their eyes
to a person who is not a geologist, and who cannot be supposed,
by any one acquainted with the subject, to have any detailed
geological knowledge, they would lose that eminence and respect
which has hitherto belonged to the office in the geological world
of Europe. And if they were likely to do this, I think that no
private regards would prevent your giving your opinion against
such a step. The case in short is, that Murchison and Lyell have
proposed the office to me; and the extreme pleasure, which
I cannot but feel at finding it possible that such a proposition
should have entered any body's head, is dashed by my own consciousness of my want of the qualifications which I think are
requisite. I am not going to trouble you with disqualifying
speeches, nor need I tell you, I think, what I think of the
Society and the Office. I hold it to be a dignity much superior
to that of President of the Royal Society. And I want to fortify
myself by your authority, that it demands a person more professedly and notoriously versed in geology than I am. I should
be very much obliged to you to write me about this as soon as
you can: and if you think, as it appears to me that you must
think, that I am not the proper person for your generalissimo,
you will of course make no scruple of saying so, and I will then
put an end to all further thoughts of this scheme. You need not
fear hurting my vanity by telling me that I ought not to have
the presidentship; for I am quite sufficiently gratified by the

opinion which the proposal implies in such men as Murchison and Lyell to need no humouring in the mode of telling me what I know to be true. So there is my case for you, and now let me have your advice.

There is another matter more connected with the geology of the field, on which I have been requested to ask for your opinion, and you may send it at the same time. A gentleman in North Wales has found lead and copper ores in his estate. The place is Pembeddo, a mountain between the village of Penmachno and Lynn Conwy, south of the road between Cernioga and Bettws y Coed. He wants to know if you can put him in the way of finding any one who can examine the place and its produce, and give an opinion of the probability of making anything of it. He wants to know also whether you think A. Wright of Keswick, who has offered his services, and boasts of your good opinion, a likely person to execute such a survey to a good purpose.

I am going to Cambridge in a day or two, so write to me there as soon as you have leisure. And take good care of yourself, and come back to us as soon and as well as you can.

Always yours truly, W. WHEWELL.

TRIN. COLL., CAMBRIDGE, *Dec.* 4, 1836.

MY DEAR HERSCHEL,

It appears to me a long time since I wrote to you, and I think it is still longer since I received a letter from you. During the summer I was running about from one part of England to another, and since I returned into College, and took to my usual employments, I have hardly had time to breathe, having more than my usual quantity of work on my hands, as I will tell you by and bye. But first of other matters. All your registers of the tides came to hand; both your letters and the original records. I have made some use of them, and could have made more, if I had had intermediate points on the East and West shores of the Atlantic, to connect your tides with ours; for the longer I attend to the subject, the more cautious I become in generalising. I dare not, at present, identify the tide which takes place on the coast of N. America with that which takes place on the coasts of Spain and Portugal at the same instant. For I find that their features are different. The *diurnal inequality* of the tides happens two or three days later on the coast of Spain than on that of America. This is briefly shewn in my memoir, which is to be sent to you along with this. This same diurnal inequality is a most odd business; and I should be very glad if you could get me observations made relating to it either in Table Bay or Simon's Bay. It will not be very difficult; for I do not want the time, but only the height, of every high water (and low water, if convenient). The point to be determined is this—How many tides after the moon's being in the equator does the diurnal inequality vanish? The differences in different places in this respect are curious and perplexing, but the dependence of the general course of the inequality on the moon's declination is clear in all. In some parts of the Indian Seas (at Singapore), the diurnal inequality of the heights is very much greater than the difference of spring and neap tides ; and at some places one tide vanishes altogether (King George's Sound, as well as the old story of Batshan in Tonquin). I shall be able soon to write a curious memoir on this subject. But at present I am, as I said, very busy with another matter. I am printing a book on which I have long been employed, and which I hope you will like. It is a history of all the physical sciences from the earliest to the most recent times, from Tubal Cain to Faraday ; and as it does not deal in generalities merely, but is thickly set with quotations from the original authors of discoveries, you will easily imagine that it must have cost some labour. It is, however, only the beginning of labours ; for it is, and professes to be, only a survey of the present state of knowledge, in order to learn from it the best method of philosophizing, and the right view of philosophy. As you have formed and executed a plan of the same kind yourself, I hope you will not be horrified at the rashness of my plan. I think I cannot be mistaken in believing that in most parts of it I have got my subject into its true shape, although new. The project is one which has been assuming a more and more definite form, and growing closer to my heart, ever since our undergraduate days; and few things have encouraged and animated me more in the execution of it than your book on the *Study of Natural Philosophy*. We shall not, however, have much ground in common. My scheme is so wide that all the life that is allotted me will be little enough for carrying it through. Still I feel confident that I have a large portion of the subject under my hand. The book I am now printing—the History— I shall, I hope, be able to send you in February, in three goodly octavos. They tell me you talk of returning, which I rejoice to hear, but I suppose you will hardly make your move before next summer.

I have a good account to give you of Jones, if you have not heard it already. After doing great service to the ministers and the bishops by drawing a bill for the Commutation of Tithe in England and helping to carry it through, he was appointed one of three Commissioners whom it created with a salary of £1500 a year. As he retains his Haileybury Professorship (at least for the present), which is £500 more, he is well off:—is just settling himself in a house in town, and is in great force. I am going to stay with him in the Christmas vacation. The only misfortune is, that he is less and less likely to write the books he owes the world. He professes that he shall still do much in that way, but I confess I doubt it: and I doubt with grief, for in certain branches of Political Economy I am persuaded he is a long way a-head of anybody else, and might give the subject a grand shove onwards.

I have been constructing an anemometer to measure the *total quantity* of wind which blows in any time, and its direction. My persuasion is, that this mode of measurement gives me what is the most important function of the wind in meteorology. The machine works very well. Green, the balloonist, has offered his balloon to the R. S. for experiments. I hope we shall use it on a grand scale. Perhaps we may come your way. If you keep a sharp look out, you may see us "hang in the clouds", like Milton's fleet. You will have seen that Green went from London to Coblenz easily. I have no room left to tell you of the British Association at Bristol. A letter of yours, written for Dublin and too late in arriving, was read by Hamilton with great effect, and delighted everybody.

I hope Lady H. and the children are all well, as they have been reported in the most recent accounts of which I have any knowledge. Darwin, who was with Capt. Fitzroy, and who visited you, is come home. He has made great natural history collections, and is become an extreme Lyellist in geology. Adieu for the present. Do not let Lady H. and the children forget us, and bring them home as soon as you can.

Yours ever, W. WHEWELL.

5, HYDE PARK STREET, LONDON, Dec. 26, 1836.

MY DEAR HARE,

I think you will not dislike my giving you an opportunity of making a well-bestowed new-year's present. Perry, like many other persons, has seen with great pain and grief the destitute state of Barnwell as to religious instruction; but with more courage and heart than others, he has set about trying to remedy it. He has bought the advowson, giving a large sum for that which is worth nothing; * * * and now he wants to build a church which may be in some measure nearly adequate to the necessities of the place. You will see by the list that he has not spared himself, and I think the spirit in which he has set about his task, as well as the object itself, will induce you to do for him what you can.

I often wish for you—on other grounds—but especially as a philologer. I am in the midst of my second volume in printing, and have many orthographical and grammatical perplexities about which I should like much to consult you. Shall I write show or shew; criticise or criticize; neoplatonic or neoplatonician; Alexandrine or Alexandrian? But I am especially puzzled about my title. "A History of Inductive Sciences"—that is too indefinite. "A History of the Inductive Sciences"—that implies "all the" and is too presumptuous. "A History of the Principal Inductive Sciences"—that is too narrow, for though I do not wish to say so, I have taken all which are, properly speaking, at present Inductive Sciences. I am mightily embarrassed with this dilemma.

I am staying here with Jones, who tells me his Tithe Commission is working well. Worsley is in town waiting for the appeal[1], which is not yet lodged, but is to come soon. The delay of the opposite party, as well as their other proceedings, appear to me inexcusable.

Have you heard that our Greek Professorship is to be vacated by the resignation of Scholefield[2], who has got another living, and that we expect Thirlwall to be a candidate for it? Except indeed, *you* will come and live with us half the year teaching Platonism, and leaving Herstmonceux to other hands for that time. I shall be here for about a week longer. Then at Norwich with Sedgwick, and at other places, but always attainable through Trinity. Adieu—

Yours always, W. WHEWELL.

Can you tell me whether *business* is Teutonic (byseg, A.S.) or Romance (besogne, bisogna).

16, SUFFOLK ST. *Feb.* 20, 1837.

MY DEAR FORBES,

I do not know whether you are discontented at the long interval since my writing to ·you—whether you are or not, I am so myself: but I must throw the blame on that which of late has been the cause of a derangement of all my habits—the writing and publishing of my book. I was very much obliged to you for your remarks on the subject of the sheet which I sent you; and, if I ever reach the felicity of a second edition, I will try if I can express more precisely the facts of the case, without giving more room to the subject, which I can hardly afford. But alas! what book on science in three thick volumes ever reached a second edition? While I was writing my book, I thought it could not fail to be very entertaining; but now that the glow of composition is over, I begin to perceive that

---

[1] See Cooper's *Annals of Cambridge*, iv. 612.

[2] This resignation did not take place.

it is too crabbed for the general reader. However, you shall have the means of judging of it, if you choose, in no long time, for I expect to be out in less than a month.

I hope you find my anemometer work well. The workmanship is certainly very much better than that of Newman was, but I am sorry that this advantage influences the price so much. I have put up one at Cambridge Observatory, and it appears to work well. I shall, as soon as I have time, write a paper describing the construction of the instrument, and comparing the results in various places; so that, if the one which you have is in action, I shall be glad to have a copy of the register to compare with the other places.

I hope, when you go abroad, you will take Cambridge in your way, that you may tell me more about your plans and intentions. I shall be very glad to hear how your thermomultipliers tell their story about the subterranean temperature. Challis has just been reading us at Cambridge a memoir on the temperature of the upper regions of the atmosphere, but I cannot yet give you any exact account of his results. Mr Monck Mason, too, has offered to the R. S. to navigate Mr Green's balloon into those parts, in order that we may see what the case really is. I hope we shall have a scientific expedition to go 20,000 feet high; for, as you know well, corresponding observations of the barometer, thermometer, and hygrometer, carried up to that height, would be quite invaluable.

You would be much grieved to hear of Dr Turner's death, as everybody that knew him must be. Besides our private sorrow, he is an immense loss to science; I do not know how he is to be replaced. He was buried yesterday, and a sort of funeral oration was made by one of his brother professors over his grave—a new thing in England.

The Geological Society have made me their President for the next two years, which I fear you will think no great argument of their wisdom. I suppose your expedition is to be a geological one, *inter alia;* so take care and bear due allegiance to me as your rightful head, and let me know how you go on in your researches.

I hope you escaped this influenza. I had it, and for a fortnight lost my voice, and got a severe cough—a bad exchange. Remember me kindly to your sisters, and believe me, dear Forbes,

Yours always, W. WHEWELL.

TRIN. COLL., CAMBRIDGE, *Feb.* 26, 1837.

DEAR LORD NORTHAMPTON,

My ballad[1] will be happy to appear in the good company of which you think it worthy, notwithstanding the change you mention. I can hardly answer very exactly your enquiry about its foundation; I wrote it some years ago, nearly as it stands now, and am not certain whether the Vienna cross has any definite story belonging to it. I think it only amounts to a tradition of a maiden who had lost her lover and used to spin there. I am very much pleased that you find nothing to object to, till you come to the last line but one. I am rather disposed to stick to my "point." I think this usage of the word is common in English Poetry. Thomson occurs to me :

> How many stand
> Around the death-bed of their dearest friends
> And *point* the parting anguish !

You will perhaps say that Thomson is a Scot, and not great authority for English : but what say you to Milton ?

> Mount Hermon, yonder sea, each place behold
> In prospect as I *point* them.

This seems to me justification enough for a ballad, where laxity of construction is tolerated. Except, indeed, you are afraid the reader will understand the word in the mason's sense, and suppose that the Viennese ladies employed themselves with mortar and trowel in beautifying the structure ; which is not a recognized way of showing sympathy and remembrance.

As to the " beloved," I believe it had better be marked as you

---

[1] The ballad is called *The Spinning Maiden's Cross;* see Vol. I. page 168. The line to which the letter relates is

"And point the fane beneath the wall."

It was suggested that the sense required *point to* which did not suit the verse.

propose, "belovèd." I do not like "belov'd," because "beloved" in prose is often spoken as a disyllable without any apostrophe being used. And though I think that people who read verse ought to have ear enough to keep it verse, when the author has made it so, without the assistance of the eye, it is very unwise to reject any easy means of avoiding mistakes.

I am very willing to look forwards to the pleasure of visiting you at Castle Ashby, but I hardly know when I can promise myself the gratification. I am busy, at present, with a history of science from the earliest to the present times, which has given me a great deal of trouble, or at least of work, and is likely to do so still. Believe me always,

Yours very truly, W. WHEWELL.

TRIN. COLL., CAMBRIDGE, *May* 13, 1837.

MY DEAR JONES,

I am right glad to hear you have defeated your internal foes, and are so stout about beating your external adversaries when the fight comes. I do not doubt you, and the less because I know that you will wait to let them assume the offensive, and then shew how much you have been doing in a proper, quiet, and judicial spirit.

I should certainly like well to see a fair review of my book in the Edinburgh Review, if it were only for the sake of the ancient times, when it used to contain the judgments of Playfair, Mackintosh, and such people. And though I have little respect for its philosophy or philosophers at present, I know no other periodical in which sound philosophy would appear with a better grace. But I do not want to disturb myself with any anxieties about the immediate reception and circulation of my book. I have, I think, pretty successfully kept myself in a state of fit tranquillity with regard to the immediate impression it may produce, and am quite willing to let it wait its time, if the worshipful the Reviewers are not yet ready to take their share in its operation. I hear nothing but what convinces me it will make its way among those who shape the opinions of future generations. For instance, Clark, whom I consider as one of the first living authorities in his own department, speaks with enthusiasm of the physiology—perhaps the most hazardous and difficult of my tasks. If you can, without putting yourself in any inconvenient position, write a review of me in the Edinburgh, it will be a great pleasure to me to see the impression it has made upon you, and to owe an increased currency to your pen; but I shall be quite content, or at least I shall try to be so, if they take their own way of dealing with me. As for the British Critic, I think I am sure of fairness from them ; and, that being so, I should have a curiosity to see how the matter strikes an average independent judge. But I have got other work in hand. I am writing a little book on University Education (entirely on the educational question, and not with any reference to gentlemen commissioners like yourself), which I want, if possible, to get out in a few weeks. I think it is much more likely to have a popular currency than either of the other books; and of course it is founded, actually and professedly, upon the History, as all good books in future should be. I shall be in town on Wednesday. Dine with us at the Geol. Soc. if you can. I believe I shall stay some time, as lectures &c. are drawing to a close. I forgot to say about Lockhart, that, if he asks you to write a review, I should much like to see you do it; and I dare say you would get through it in time in spite of old habits and new employments. But I have seen too much (though in fact not much) of him, and of his caution and shifting, to wish you to make any advances to him. A man has just sent me a pamphlet about Inductive Logic—meagre enough—but it shews that people's heads are seething.

Adieu. Yours always, W. Whewell.

Athenæum, Tuesday, June 6, [1837].

My dear Murchison,

I believe you must call your beast *Tretaspis* ($\tau\rho\eta\tau\dot{\eta}$ $\dot{a}\sigma\pi\dot{\iota}s$, a perforated shield). This word is as convenient as the one I told you, and is better Greek.

Yours very truly, W. Whewell.

MY DEAR MURCHISON,

I should still be disposed to keep *aspis* for the termination, when the shield of your trilobs is to be described, and to call your new genus *acidaspis: ἀκίς* is a sharp point. If you want to change *tretaspis* to something which contains a reference to the *rim* of the shield, you may call it *tretantyx, a perforated rim,* (*ἄντυξ*). And in that case you might call the other *acidonotus, a sharp point on the back,* or *rhamphonotus, a beak on the back;* but I think *tretaspis* and *acidaspis* are best. Yours to the end of the dictionary and beyond.

W. WHEWELL.

CHARTRES, *July* 12, 1837.

MY DEAR JONES,

How does your intention hold of snuffing the air of the continent in the beginning of August? I am on the point of leaving Paris, and will regulate my course in any way that will promise to bring us together. I shall go to the Rhine in the first place, picking up a few churches on my road, and shall then look for my letters at Bonn, till I hear something of your plans. I would have staid at Paris a week or a fortnight longer, if I had had you with me to enjoy the wonders of the fine arts at the Rocher de Cancale and Tivoli, which still preserve their attractions; but eating and drinking are social pleasures, and, for want of fit assistants in the task, I have made few attempts to ascertain whether the ingenuity of the Parisians has advanced or receded during the fifteen years that have elapsed since our initiation into its miracles. I fear that they are no longer the great nation in these respects that they formerly were; but perhaps this is only a melancholy fancy, generated by the less favourable circumstances under which I have contemplated them. In other respects too, I do not find that they have made any decided advance. I was curious to know what had been the result of Cousin's attempt to introduce among them a sort of Scoto-German philosophy.

The results of his efforts appear to be a complete failure; for he has had no successors in his own peculiar path. The German element of his metaphysics seems to have flown away like the smoke of a tobacco-pipe, and the only residue I find is a little Scotch metaphysics among those who profess to be metaphysicians, who are, however, very few and little attended to. I expect that you will hold this up to me as a warning, to shew the folly of engaging in a similar undertaking in our own country, and there is no denying that it offers a lesson which it would be very rash to disregard. But in truth it only shews, what I always supposed to be the case, that it is hardly possible to introduce foreign metaphysics in the lump, and that we must read German writers for some other purpose than that of substituting German metaphysics for English. I shall try to profit by this piece of wisdom when the time comes.

What will you do ? Come to Aix-la-Chapelle and Spa, and we shall be sure to find something to amuse us : and besides, if you are disposed for a longer flight, we are then within reach of the Rhine, Holland, and Belgium, of which you may take your choice. I shall be glad to see any of them with you, so tell me when you will meet me at the *Hôtel des Etrangers, Aix-la-Chapelle.* Lodge talked of joining me also in the beginning of August, and I am now going to write to him to propose a congress at Aix-la-Chapelle. His time, like yours, (if he come at all, which I conceive to be far from certain) is the beginning of August, or a little later. I do not know whether he is at Cambridge or at his living in Lincolnshire. I suppose at the latter, but I shall write to him at Magdalene with a direction to forward the letter.

I hope you keep your inner man in better order than you did when I was with you. I am quite persuaded that you would be much the better, stomach and head, for a few draughts of German wine and Spa water ; and for forgetting during a fortnight that there are such people as country parsons and farmers fighting about tithes. You shall not even, except you wish it, be put down in the books at the inns as *Herr Professor Zehnte-Commissionär Jones.*

How is Mrs Jones ? If she comes with you she must appear

as the *Frau Professorinn Zehnte-Commissionärinn*, which is a well-sounding title. Remember me kindly to her—and by all means jump across the channel as soon as you can. Write to *Poste Restante, Bonn*, as soon as you can. Good bye for the present.

Yours always, W. WHEWELL.

TRINITY COLLEGE, CAMBRIDGE, *Sept.* 6, 1837.

MY DEAR JONES,

I found no letter here on the subject of Haileybury, and indeed, notwithstanding all you said, I should have been much surprised if I had. Our interview yesterday was so short that I had not time to make all my enquiries. Do you still persuade yourself that you are writing a review of my History? If so, in what condition is it, and what are its prospects? I see by the last number of the Quarterly that Lockhart and Murray appear to have none but shallow articles, and moreover that they are at present very cross and very coarse, so I apprehend your chance there is small. Have you seen Macaulay's article on Bacon in the Edinburgh? I rejoice to see how little people yet see the philosophy of Induction, for Macaulay is no bad example of "the general *thinker*"; and yet how scanty and superficial are his views—happily expressed and well illustrated of course.

I found here a letter which, though I do not assent to it, somewhat disturbs me. It is from Sir Charles Bell, complaining that I have treated him with slight and injustice in page 425 of my third volume. I can easily reply that I never professed to arrange physiologists in order of merit, or to give a full account of their labours; but I doubt whether I can so well defend myself for saying "Sir Charles Bell *and* Mr Mayo." I had some misgivings before, for though Mayo may deserve to be mentioned here, I believe he cannot be considered as having gone halves in the discovery, and at any rate I had not examined the subject sufficiently to entitle me to give such a sentence. I have written to Bell defending myself, but, if I come to a second edition, I must consider well the propriety of modifying. I must however take into the account that it will be unwise to modify lightly, for I am not going to run after the rainbow like Buckland, by trying to keep up with the progress of things. " What is writ is writ."

I left a piece of Spa work for Mrs Jones at your office yesterday, which I beg you to give her as a memorial of the place where you and I did *not* meet.

Do not forget, at some moment of your philosophical moods, to do what I mentioned yesterday. Put down on paper, as clearly and strongly as you can, the reasons which you can find for the opinion you held a little while ago ; namely, that the simplest mechanical truths depend upon experience in a manner in which the simplest geometrical truths do not : that the axioms of geometry may be self-evident, and known *à priori ;* but that there are not axioms of mechanics so known and so evident. I am very desirous of getting this opinion in its best and most definite shape, because the negation of it is a very leading point of my philosophy. This tenet separates me from the German schools as well as from the Scotch metaphysicians, and is the basis of a long series of results both speculative and practical. The whole *art* of induction depends upon it.

Adieu. I am off for Liverpool to-morrow for a week of bustle and speechifying, and then I come home to philosophise.

Yours always, W. WHEWELL.

[*October,* 1837.]

MY DEAR JONES,

You will-probably have seen the Ed. Rev. of my History by the time this reaches you ; but, if not, get it and read it, that you may tell me whether there is anything which I need to answer. I hope not ; for I do not see what I can say without being ill-natured to Brewster, who has made the article for the most part an angry remonstrance in favour of his own rights unjustly withheld. Thus you find that I have offended him by not quoting his Life of Newton—by not quoting his Edinburgh Journal of Science —by not giving more credit to his arrangement of crystals—by

17—2

doubting his refutation of Newton's prismatic analysis—and by not adopting his own peculiar monomania about the rewards which should be given by the public to men of science, as well as by referring to his controversies with the French discoverers. I think this personal feeling in the review must be so obvious to those who know anything of science, that it must disarm it of much of its force; and even those who know nothing must laugh, I think, at the absurd fancy of my having a spite against Scotchmen, and passing over them and their exploits out of pure malice. All this is foolish enough, but it will not be easy to talk away the impression in those who have no more sense than to admit it. For what am I to say? I have not mentioned Mrs Somerville. There was no pretext for mentioning her in a history of original discovery; and I am not to compliment away the character of impartiality which alone can give value to a history. But it is harsh and pompous to say this, and I had rather leave people to think it. In short, I am disposed to stand upon my character and hold my tongue, till I can write my Philosophy, and then I can set all to rights that is really wrong. I hope to set to work as soon as my sermons are preached, that is, in December; for by that time my other book will be nearly printed.

The injustice (for I fear it is so) of that one word *and*, which I have put between Bell and Mayo, is the only thing which really disturbs me. If I could find any mode or channel of modifying this, I would do it.

In other respects the review will be useful to me. It is whimsical enough to see Brewster taking between his teeth the hardest part, almost the only metaphysical part, of the book—the causes of the failure of the Greek philosophy. I have said plainly enough that the account there is incomplete, and he might have waited on that account; but that would have made little difference in his understanding it. He is resolved that there *shall* be no inductive philosophy; and the observations he makes on that subject, coming from him, who is really a genuine discoverer and a man of genius, are instructive, though full of error. As to the example which he attacks, the round shadow of a square hole, it would be a good one for shewing that he is wrong; but then all such discussion would be a brick out of my philosophical house; it would stop the progress of the building, and give a very incomplete notion of the edifice.

I cannot help thinking of the folly of his *Scotch grumble*, as my publisher, Parker, calls the key note of the article. By the way, I was somewhat afraid that my Reviewers might complain, as Drinkwater did, that I had been hard upon Galileo; but Brewster gives him over to the bigots far more decidedly than I have done. As to the Scotch grumble, Brewster has taken special care to overlook all that I have said of his rival Forbes's discoveries on the polarization of heat. And his uproar about steam boats and gas and railways shews that he has not at all comprehended the nature of the book.

There is hardly anything that I should have to correct in the History of Optics, except, perhaps, the question between Ptolemy the Greek and Alhazen the Arabian, and I shall be glad if I find the Arabian discovered nothing, for I have a strong *European nationality*.

Adieu, and write to me, though I do live behind iron bars[1] which the light of knowledge has not been able to penetrate.

Yours always, W. WHEWELL.

I will bet you a penny that the cry of *injustice to Scotland* will prevent Lockhart taking your Review if nothing else does—and if you write it—of which I have my own doubts.

I send "the Tribute" with my love to Mrs Jones.

TRIN. COLL., CAMBRIDGE, *October* 14, 1837.

MY DEAR AIRY,

I have so little chance of seeing you at Greenwich for some time, that I must put on paper what I have to say, all little though it be. For ever since I took to going to London every fortnight, as President of the Geological Society, I have lost all power of locomotion in the neighbourhood of that awkward town, since I cannot spare any holidays in addition to these dutiful

[1] See Vol. I. page 116.

visits.   One of my pieces of wisdom that I wish to impart to you
is a strong recommendation to you touching your magnetic observations, namely, that you will adopt some contrivance by which
you can have *self-registering* observations of the daily changes of
diurnal variation.   With a great poker of a needle such as you
have got, which is strong enough to do any work, and will
otherwise be idle nine-tenths of its time, this cannot be
difficult.   A little clock-work would be well bestowed upon
keeping it out of such an absurd condition of uselessness.   You
may depend upon it, the bar will grow lazy and conceited if it has
nothing to do, except when you are looking at it.   I am sure that
with regard to all these half-formed sciences, such as terrestrial
magnetism is, we shall never come to any knowledge worth
having, except we have the whole course of the facts.   Self-
registering machines are the only inventions worth having for
those changes, of which the laws are as yet all in a mass of confusion.   And all that I have seen of magnetic observations leads
me to believe that *with* a self-registering apparatus, you may obtain very important results.   You have seen probably what I saw
a little while ago at Quetelet's, the statement of the contemporaneous changes of the variations observed at different places.   They
are utterly astonishing, inasmuch as, being altogether different on
different days, they correspond exactly over the whole of Europe,
and are so strictly contemporaneous, that people begin to talk of
obtaining the longitude by means of them ; so pray do not let
your magnetical gear be dispensed from the task of recording all
their strange vagaries, that we may some day or other know something of their whereabouts.

*   *   *   *

Ever very truly yours, W. WHEWELL.

TRIN. COLL., Oct. 28, 1837.

MY DEAR AIRY,

    I have got the Edinburgh Review upon me again about
my History of the Inductive Sciences.   The reviewer is Brewster,
and he writes very *cantankerously*, but especially about the Optics.

I was for the first day after reading it very peaceably disposed, and resolved to stand upon my character and hold my tongue, and perhaps this may be best after all. But I find myself waxing wrathy, like Major Downing, and do not know if I shall be able to avoid griping after my hickory, and I want you to drive a nail or two into the cudgel, for I have found already with these reviewers, that there is no use in not hitting hard if you smite at all. I would not give you any unnecessary trouble, but I want you to tell me a thing or two which you can without trouble, because I myself am a busy man at this present, inasmuch as I am re-editing one book, writing and printing another, and constructing four sermons to edify the University in the month of November, besides all common, and more than common, operations in the way of tutorizing, and the like. I will put down what I want to know from you.

First, of Brewster's asserted refutation of Newton's prismatic analysis of the spectrum. I have not yet read his paper (Edinb. Trans. XII. p. 125). Am I safe in asserting that Newton's analysis is not overturned by any analysis effected by means of absorbing media, and that these are the only source of Brewster's objections?

I think Arago distinctly gives the first place in the discovery of the undulatory theory to Young, not to Fresnel. Do you recollect if this is so?

Do you know who discovered the irrationality of the spectrum, who applied it to the improvement of the telescope, who examined its phenomena in many solid and fluid media? Brewster says I do not mention the man, because he was a Scotchman. He says that I am bent on doing injustice to Scotchmen; that the only successful generalization in my book is the neglect of the claims of Scotchmen; that I even hate Scotchwomen. I think this last charge is hard; but what can I say?

I have not thought much more about your magnetic apparatus, but it seems to me that some method of marking, which requires no force that you would miss, must be feasible. What do you think of this? Let the needle carry a bristle or hair at its end, pointing downwards; and let this bristle just brush a sheet of paper covered with fine dust. This would make a good mark, and would be quite permanent enough to be copied. Of course you might have a bristle at each end for greater accuracy.

Of Brewster's new fringes he speaks in his review. Of course they are yet *too* new to form a final opinion about; but he upbraids me also with knowing nothing of the *transverse fringes of grooved surfaces* and of the *phenomena of crystallized surfaces.* The last accusation is true, so far as I recollect; the former, I suppose, refers to those papers of his which appeared a couple of years ago, which I certainly never well comprehended. Do they go into the theory cleverly?

I have only yet found one word of three letters in my history, which I am seriously sorry for. I have talked of Sir C. Bell *and* Mr Mayo in reference to a certain discovery, which I believe is not a just account of the matter.

I hope Mrs Airy is with you by this time. I am right glad to hear you have thoughts of coming here. Pray let me know in time, if you conveniently can, because, otherwise, it is certain that you will be conveyed away in a hundred directions all the while you are supposed to be here, and I shall see nothing of you. We are going to lay the foundation stone of the Fitzwilliam Museum this week with great array of scarlet robes and dinners and so forth. My kindest remembrances to Mrs Airy and your sister. Also to the small bodies which revolve round them.

Yours ever, W. WHEWELL.

N.B. As you are as likely as myself to be busy, the first of my queries above is the most essential.

[To M. Quetelet.]

UNIVERSITY CLUB, LONDON, Nov. 30, 1837.

*   *   *   *

You will be glad to hear that the Royal Society of London have awarded me their Royal medal for my researches on the tides. I received it to day, and as you perhaps know, it is a very pretty plaything, even for its own sake, being a piece of gold worth 50 guineas, of very beautiful workmanship. This event gives me an additional admonition to send you the sketch which I promised you of the results of my Tide Researches. I will do this soon. I have in fact already begun to write it. I received a letter from Herschel a short time ago. He is winding up his labours at the Cape, and has already sent home the reduced observations of more than 600 nebulæ and 400 double stars. One of the circumstances of recent occurrence, which has given him most pleasure, is his having been able to observe the eight Satellites of Saturn during several revolutions. Nobody except his father has ever seen this object, at least so I believe. We are at this present time erecting the great telescope at Cambridge, which was given us by the Duke of Northumberland. It is the same size as South's (an object glass of Cauchoix of 13 inches), but is differently mounted, according to an invention of Airy's. I have no doubt this telescope will soon be in active operation, for our present astronomer, Challis, is very zealous. He has published his last year's observations, which you have probably received. We think it a great proof of his skill and rapidity that he has published the Cambridge Observations of 1836 before Airy has published those of Greenwich.

* * * *

CASTLE ASHBY, NORTHAMPTON, *Dec.* 26, 1837.

MY DEAR WILKINSON,

I am sorry, on looking up your letter to answer it, to find that I have procrastinated till no answer will serve your purpose. The fact is that I have been so busy with cares of authorship and others, that I have deferred everything which appeared to admit of delay, and I had a notion, an erroneous one as I now find, that your question was of that kind. I had left Cambridge before your letter arrived there, and have not been there since; and, as you very probably are there at this moment, you will not benefit much by this reply, which I intend to send to Sedbergh. But I shall do so, that I may, as far as is yet possible, quiet my conscience, which has been very reproachful upon the discovery, that I have neglected your letter till it is too late. I do not doubt that conscience is right in thus taking up your cause, but at the same time I think it very hard she should bully *me* in this way, considering that I have just been preaching four sermons in her honour, in which I have said more fine things in her praise than people of late have been in the habit of saying. But that is one of her odd ways. She never thinks of setting off good actions against bad ones. I have just finished my sermons—indeed I think it probable that they will be published at Cambridge this week; so that you may chance to see the course which my wisdom has taken if, as I have already conjectured, you are at present revelling your twelve days at St John's.

This writing and printing of my sermons has been one of my recent literary labours; another has been, or rather is at present, writing and printing a mathematical book, which I intend should be my last in that department, a treatise on the Doctrine of Limits and its Applications[1], in which term, however, I include all our second year subjects except Mechanics, about which I have already teazed the world sufficiently. I would not write this book, if I did not think I could treat the subjects which it contains in a more logical and philosophical connexion than has yet been done; moreover, I shall write a preface shewing that I have done so: and then I shall depose my mathematical boxing-gloves, and go on with arms of wider range as well as I may. I grieve much to hear that Mrs Wilkinson is unwell, but I trust it is only some passing illness, such as you heads of families must often have to disquiet you by way of balance to the blessings that you enjoy. I shall possibly see you at Cambridge before you read this, which makes considerable confusion in my notions of tense. For according to epistolary chronology no communication can intervene between writing and receiving a letter. So I send you my best wishes, to you and yours, hoping that they are already superfluous by repetition, and that you have answered them before you have read them.

With my kind regards to the ladies and the girls,

I am always truly yours, W. WHEWELL.

[1] See Vol. I. page 120.

MY DEAR HARE,

I have said very truly in my dedication[1] to you that it was a great pleasure to me to put in that place a memorial of our friendship. Literary pursuits and solitary speculations, if they do not unite us with our friends, sever us from them; and I hardly know how any literary or philosophical success could console me under the thought of having lost my place in the remembrance and affection of the friends of the earlier part of my life. And besides your place among these, many of the others whom I think of with most pleasure are twined in various ways among the recollections of our intercourse. I sometimes think she is rather a hard task-mistress, this Reforming Philosophy of mine; for she carries me into regions where I can hardly expect any contemporary friends to follow me—at least for a long time; and if a younger generation learn to sympathize with me, (of which I hope very boldly), they cannot quite replace my *æquales.* However this may be, my allotted task appears to me to be clear, and the sermons are a small part of it, though it must, I fear, be years before I can bring the connection into view; and before that time I hope I shall see you more than once. I am sorry that at present this cannot happen. I am so engaged that I cannot come south of London this vacation.

*　　　*　　　*　　　*

As to the name of your dwelling-place, I did not intend any irony, but I could not bring myself to your spelling, since it appeared to me to be at variance with your own principles, and all other good principles, of orthography. Herstmonceux cannot come in any way but by a blind submission to former usage (probably accidental), to the neglect of all analogy and etymology. I really had put into print your way, and then could not bear the look of it. I should like to discuss the other matters on which you remark, if I had room. I do not think, taking all St Paul's argument together, one can doubt

[1] See Vol. I. page 245.

that he held the Gentiles to have a knowledge of God's righteousness, and that such a belief was present to his mind in the passage to which you refer. About conscience, observe (1) That I have not asserted anything about its origin, but only about the independence and certainty of its conclusions. (2) That I do not adduce Scripture as proof of this, but only as in harmony with it. (3) I think that the Scripture is good, as proof *against* those who construct a system of morals in which the reference to conscience must be considered as an evasion, which is the case with the utilitarian systems.

I am going to see Sedgwick at Norwich in a week. You ought to come to Cambridge to see the College before all your friends become Prebends, Bishops, Rectors, and the like. We have a good living vacant, which ought to take away some of us (Dickleburgh, in Norfolk). Adieu. With all good wishes,

Yours most truly, W. WHEWELL.

TRIN. COLL., CAMBRIDGE, *Ap.* 2, 1838.

MY DEAR FORBES,

I was very glad to hear from you, for I believe it is, as you say, some time since we have exchanged letters. This, however, I fear is rather my fault than yours, for I am often very remiss in my correspondence; and, this being so, I feel really obliged to you, who, with a conscience generally free from such sins, are ready to overlook them in other people. I regretted much to learn from several quarters that you have been ill, and had hardly quite recovered at the date of my last authentic information. I hope this refers to a state of things now so entirely past, that you look upon it as quite obsolete. I am glad you intend moving this way early in summer. You will find me here till the second or third week in June, whether longer I can hardly tell, for I have not yet made up any plans for the summer which I consider as at all fixed. My schemes, so far as I have formed them hitherto, have only been those *instantaneous hypotheses* which metaphysical writers tell us are so useful in discovery. I am scribbling away at a new book, which is to be the sequel to my History. For, however odd your neighbour the *Edinburgh Review* may think it, I do not consider my History as annihilated by their critique, or less fit to get a scientific moral out of than I from the first expected it to be. But this process must be a work of some time. In the meanwhile I am threshing away at the tides, and expect still to get good stuff out of them. I have a paper now on the stocks, in which I expect to shew how each year's observations may at once be made to add something to the accuracy of the existing tide tables of the place where they are made. This being shewn, it appears to me that it is the business of all civilized governments to maintain tide observatories and tide calculators, just as much as observatories and calculators whose work appears in any other page of the *Nautical Almanac*. When I have written and published this, I shall be desirous of wrapping up my tide papers for a while; for there is really no end of the work to which they lead; any more than if any one were to undertake to reduce all the astronomical observations that exist in the world. As for the hydrodynamical theory of the subject, we are nearly ripe for that, but I confess that now, long unused to the heavier weapons of analysis, I do not dare to attack so formidable a problem. I shall leave that to bolder and stronger mathematicians. I am glad to hear you have been going on so well with refraction and polarization of heat. I have no doubt that you will end by putting that subject in its proper and final shape. I have not yet seen Melloni's paper of which you speak. What do your Edinburgh folks intend to do with respect to the vacant professorship? You have two good candidates in our two men, Gregory and Kelland, but Gregory does not appear to think it certain that the choice will fall upon either.

* * * *

Airy has just been reading us a curious and very laborious paper upon the theory of the rainbow treated undulation-wise; the trouble of the integrations is portentous, but some odd results come out, especially as to the supernumerary rainbows.

* * * *

Yours always, W. WHEWELL.

MY DEAR HARE,

I was very glad again to see your handwriting, and doubly glad to receive from you such an account of your pupil and mine. I hope, and from what you say do not doubt, that you will receive a good account of him when you come here. It is a great satisfaction to think that you are going to shew yourself among us again, and still more that you are going to preach to us; for I know well that you will set some right thoughts and good feelings a circulating among us. I shall be your disciple, not your teacher, in February, for I shall not give any lectures till the Easter term. The subject is large and serious, and I must wade gravely into it, not plunge headlong; and this, not only because I am growing older and more leisurely in my movements than I used to be, but because, to do any permanent good, I must produce a consistent system, or at least the elements of a system. Even if this be not the best way, it must be mine, for all that I have done, and am doing, has no value, if it will not hang together. If it will, I think I have laid my foundations pretty wide. But you shall have the means of judging if you like, for I shall publish my Philosophy (of Science) before I make morals my main object; this I shall do in the course of next year, if possible. It is certainly very encouraging to see on all sides strong tendencies to a reform of the prevalent systems of morals. The article in the London Review is an indication of this, and appears to me to be in many important points right, and at any rate right in the vigorous rejection of Bentham's doctrines and keen criticism of his character. But I confess I do not look with much respect upon a body of writers, who, after habitually showering the most bitter abuse on those who oppose Bentham's principles, come round to the side of their opponents, without a single word of apology, and with an air of imperturbable complacency, as if they had been right both before and after the change. Nor do I see any security, in their present creed, against a change of equal magnitude hereafter. I cannot understand how Sterling, who, I suppose, is a conservative, can write in a journal of professed political objects, these being, to destroy the church and democratize the nation— at least I have seen no retraction of these purposes. You will think I am ill natured; I suppose I am; for certainly, when I see a knot of young men attempting to impose violent and destructive opinions upon their readers as the maxims of a new and wise philosophy, and then a few months afterwards telling us that these extravagant doctrines were some of the circum- stances of their own self-formation, I think them little more excusable than the Marquis of Waterford breaking windows by way of sowing his wild oats, and vastly more mischievous. I had read Maurice's[1] 'philosophy' (without knowing it was his) with some interest, but thought it, as you say, fanciful; you know that is what we system-makers are apt to think of each other. But there is plenty for all of us to do, if we can in any way call up the elements of good from any quarter.

*   *   *   *

I suppose that, if Bunsen is in England, you see him; pray send him down to us here, for we want to make acquaintance with him.—Live recollective of us, and believe me,

Always yours, W. WHEWELL.

[Mr Hare in a letter dated October 8th, 1838, had drawn attention to an article on Bentham by J. S. Mill as "most masterly". On October 25 Mr Hare returned to the subject, and, in reply to Dr Whewell, defended J. S. Mill with great earnestness. I extract two sentences from his letter. "It must be six or seven years since Wordsworth told me how much he had been struck by him, and that he was one of the most remarkable young men of the age. Of his writings I have not read much; but all that I have has been most masterly in execu- tion, shewing a logical power and a metaphysical subtlety very rare in England, united with a high tone of moral feeling."]

---

[1] This refers to the article on Moral and Metaphysical Philosophy in the *Encyclopædia Metropolitana*, which Mr Hare commended as "brimful of genius and subtle thought", though fanciful.

TRIN. COLL., *Oct.* 30, 1838.

MY DEAR HARE,

I find Sandys's translation of the Psalms in Trinity Library, and herewith it comes to you. I shall be glad to see your collection of better versions than the ordinary ones; only take care that the same thing does not happen to you in this instance, which has come to pass once or twice before, that, beginning by being the editor, you have ended by being the author of two-thirds of the book. Or rather, look upon such a result as possible without caring to prevent it: for versifying psalms is a very good and wholesome employment, and such a collection as you meditate forming, whether borrowed or home-made, is likely, I should think, so far as a mere academical man can judge of your country parish wants and feelings, to be very useful.

I am glad that John Mill is such a man as you describe, and such that you think it right to defend him, and I have no disinclination to believe his good qualities. My quarrel is with the Westminster Review, which, after being destructive and outrageous on one side of the question, becomes vehement on the other side, remaining destructive, and using language which implies a claim of consistency and identity. If it were not that you *are* so resolute a vindicator, I think you would see how necessary it is in such a case *not* to single out John Mill, as John Mill the individual of the present moment, disconnected from all other Mills and all former times. He does not write as John Mill. He does not give his name. He does write as a London and Westminster Reviewer. That Review has been, and, as we country folks imagine, is still supported by one or two of the most thorough-going theoretical and practical destructives and utilitarians, avowedly for the purpose of promulgating their doctrines and carrying them into effect. It still claims to be Benthamite. If John Mill be a Coleridgian or a Conservative in any intelligible sense, he must be the most infatuated of men to imagine that he can do any good by promulgating his opinions under such auspices. If he think his views of morals consistent with the objects of the review, I should like to see the connexion made out with any reasonable logic; for certainly up to the present time the objects and schemes of the destructives have been supported upon views, the very opposite of those which you ascribe to him. So you see I am where I was. I may be ill-natured, but I cannot make out that I am unjust.

By the way, talking of young men, I suppose you have seen Milnes's poems. It appears to me that there is great beauty of thought, and very considerable skill in the execution, in a great number of them; though I cannot quite assent to Landor's extravagancies of expression on the subject. I shall be in town next week; I am afraid there is no chance of seeing you there then.

Believe me yours always, W. WHEWELL.

TRINITY COLLEGE, CAMBRIDGE, *Dec.* 3, 1838.

MY DEAR HERSCHEL,

I hope the time is come now when I can induce you to look with a favourable eye upon a proposal which I spoke of to you some time ago, and which I did not wish to obtain your final answer upon at that time. Now that you are secure from being plagued about the R. S. and the British Association, I hope that our good-humoured, quiet, easily-satisfied Geological Society may put in her petition for a little of your time. It will not be much that she will want; not more upon the whole than the British Association would have required, seeing that you must come to London often, whereas to Birmingham you must have gone expressly. Nor is it necessary that the president of the Geol. Soc. should attend every meeting. You would be considered as doing all that could be expected of a non-resident if you attend a plurality of times. I know that it would be a matter of great delight to the Fellows of the Geological Society to have you at their head; and most of all to those whose opinion you would most value. It would, I am sure, be very pleasant to you to see those persons frequently, and to witness and take a part in their debates. Indeed, with your interest in the subject, I hardly see how you can help being present at their meetings almost as frequently as the president need be. You would find it a good thing to become acquainted and familiar with the views which are now current among them; and they would be extremely gratified to hear your judgment when you formed one; and above all to have a view of the present state of the science, as it strikes you, in your annual address. I need not tell you what agreeable people to associate with the geologists are; but, what is more to the purpose, there are absolutely no quarrels or heartburnings or internal difficulties in the Society; at least none that I ever became aware of. Our only difficulty at present is to find a President for next February; and from that I trust you will deliver us. It is the earnest wish of every member whom I have consulted (and I have consulted most of the best judgments) that you will do so; and, if you do not, I declare I am at my wits' end to know what we shall do. The older members very much desire not to have the cycle of old presidents repeated; but indeed I do them great injustice if I represent their wish to induce you to take the office as depending on any consideration of negative advantages, for they wish it too much positively to look at the alternative.

I hope you will not think the sacrifice too great, when you consider the merits of the society to which it is to be made, and of the men who desire it. To shew you how perfectly reasonable I think it, I must tell you that I was on the point of writing to Lady Herschel, to beg her to be our advocate with you; which certainly I should not have thought of doing, if I had not believed that the petition was both a very fitting one, and one that would tend to your gratification. I forbore making this application to her, but pray tell her that I trust she will be our friend. I say nothing about the honour of this office (although I very sincerely think it the most honourable scientific office in London) because your friends never think of such things in regard to you.

I shall be here for the next fortnight, so pray write to me here, and pray do not reject the proposal. If you do not, I shall be considered as having done the Society my greatest service by my last act, and I am sure you will never regret it. Believe me, my dear Herschel, yours always most truly,   W. WHEWELL.

MY DEAR M. QUETELET,

*　　*　　*　　*

It gives me much pleasure to see M. Mailly's curves of the semimenstrual inequality of your Belgian Tides, and his remarks upon them. I hope you and he will find the subject so interesting that you will pursue it. It is plain from the curves that some of your observations are not very good, although I do not doubt that the result which M. Mailly has obtained is nearly correct; it might probably be improved by introducing the corrections for Lunar Parallax and Declination. You may have seen that a M. Chazallon has published Tide Tables for several French ports, under the title of *Annuaire des Marées des Côtes de France.* He asserts that the formulæ of the *Mécanique Céleste* give, for Brest, a better coincidence with observation than we obtain by our tables in London. I am glad to hear this, but I should like to see the calculations given more in detail. However this may be, you have, I think, sufficient evidence, even in your own ports, that the same formula will not suit different places.

I shall have another memoir or two to send you soon on the effect of Solar Parallax and Declination, and on the mean height of the Sea. We are endeavouring to induce our Government to send out an expedition towards the South Pole to determine the present position of the Southern Magnetic Pole, and also to establish several permanent magnetic observatories in order to observe the simultaneous changes discovered by Gauss and his friends. I suppose you have attended to this subject already. Airy has just put up his Gaussian apparatus in the Observatory at Greenwich, including a *Bifilar,* and will begin his observations soon. He has adopted the principle of the Collimator in order to observe the variation. The other stations at which we wish to have Magnetic Observatories are the Cape of Good Hope, Van Dieman's Land, Ceylon, and some places in Canada. Sabine will go to Canada, and I hope Capt. Ross will conduct the Antarctic Expedition. This expedition was recommended by the British Association at its Newcastle meeting, which I was sorry you could

18—2

not attend. We meet next year at Birmingham, when I hope
we shall see you. We have now a railroad from London to
Birmingham, so that the journey is as easy as it is from Brussels
to Antwerp.

*   *   *   *

TRIN. COLL., CAMBRIDGE, *Feb.* 9, 1839.

MY DEAR LUBBOCK,

Wishing to get Hopkins's paper off my hands, I wrote my
report upon it and left the papers and report at the R. S. The
purport of my report was, that I did not recommend the *first*
paper to be printed, except the author thought it more necessary
to his general investigation than I conceived it to be; but I recommended the *second* to be printed without hesitation. My
reasons were, that I thought it very desirable to work out the
precession for the extreme hypothesis of a rigid crust for a fluid
interior, which, even if physically very improbable, is still a sort
of limiting case in geological speculation. I dare say that the
investigation respecting precession contains a portion which is
familiar to those who have studied the best investigations on that
subject, but I was not sorry to see the result, in a case so much
out of the common range of physical astronomy, worked out
from tolerably elementary principles. I did not think that the
mathematical working could have been much abbreviated with
advantage, as the kind of motion which he considers (a fluid
revolving about a curvilinear axis) certainly does not fall under
very simple calculations without some management. If the crust
be rigid and filled with fluid there will be mutual pressure, and
the problem then appears to me far from easy, and I still think
that it is desirable to have the results of this simple hypothesis
investigated. He, or any one else, may take the problem with
other circumstances of expansibility, &c., but I should think it
best to solve it in its simplest form first.

Under these views I think I cannot retract my recommendation. As a student of geology, I was really curious to know the
result, and I think geologists in general will be so; which cannot

be known to them except the paper be printed; or at least which cannot be examined by others till then.

I was very glad to receive your Treatise on the Tides. I have no immediate intention of writing any general view; for, even if I had time, (a very baseless hypothesis) I do not see my way well enough. There are several points which I want to decide more satisfactorily than I have yet done. As to the epoch, I am now disposed to believe that the time half way from transit A to B (the transit at a meridian 90° from Greenwich) will be the best; but I shall be able to decide this, I hope, before long.

I am afraid I shall not be in London at the next meeting of the R. S. Council. I shall be very glad if we can come to an agreement in our recommendation respecting Hopkins's paper.

Believe me, yours very truly, W. WHEWELL.

TRIN. COLL., CAMBRIDGE, *Mar.* 1, 1839.

MY DEAR FORBES,

It was a great pleasure to me to receive a letter from you, for I had been grieving that our correspondence had been so long interrupted, and had a persuasion that the fault was mine; and in short I was meditating a letter to you when I received yours. Your thermotical results are very curious and interesting, and certainly prove, as you say, what I had always believed, that the material of the earth's crust is a primary element in the formulæ for the passage of heat into the interior. In Hopkins's researches there is not much of a thermotical bearing. His first paper consisted of some speculations (somewhat vague) about the solidification of a fluid mass from the surface and from the centre, with no application of the theory of heat. The second of his memoirs was an investigation, as appears to me, very interesting, of the difference of precession and nutation in a solid earth, and in a rigid crust with a fluid interior. He finds the difference of the amount of precession and nutation in the two cases to be of imperceptible magnitude on any probable hypothesis. His assumed conditions, a *rigid* crust *full* of fluid, are hardly consistent with thermotical laws; but I think he did well to take the extreme case first. I am somewhat disappointed that the resulting difference is so small; it would have been so pretty to extract the internal fluidity from astronomical observations. I am glad you are going on with your researches on heat. It was a great pleasure to me to be on the Council of the R. S. which adjudged you the Rumford Medal. I hope you received it with a proper statement of our satisfaction in recognising your discoveries. Why did you not come to Newcastle, when we met there on purpose to tempt your northern luminaries? I hope, as you say, we shall have Harcourt amongst us again, for he can do more than any man to keep us right. He appeared some time ago at a Council of the British Association at which I was present in London. He appeared very unwell, and the last account that I heard of his daughter was that she was past hope. As for me, I am working hard (among other matters) at my Philosophy, which I want to get off my hands before I turn my main attention to other matters. You will observe that I am Professor of Moral Philosophy rather than of Casuistry. The foundation of my professorship includes both titles, and I prefer the former. Why should not we here have such a professorship, as well as you at Edinburgh? I am glad to hear that Kelland and his place suit each other so well. I hope he will do good service to mathematics, prepared as he is, and with his attention fixed upon the most promising parts of mathematical physics. Gregory is here working at various matters, I have no doubt to good purpose. I hope soon to be able to send you my final speech from the Geological President's chair. I have there given a brief abstract of the researches of Hopkins about the interior fluidity. Buckland is my successor.

Your packet reached me only yesterday, though your letter is dated Dec. 21. I have disseminated your letters. Have you any intention of travelling southwards this spring or summer? If you come before July, by all means come by Cambridge, where I shall be, or at least till the early part of June.

We hope to get a magnetic expedition sent to the neighbourhood of the South Pole, and magnetic observatories established

in various regions to observe the irregular simultaneous variations, if they really are everywhere simultaneous. Capt. Ross
to command. I trust we have, after some struggle, engaged the
Admiralty to take it up.

I every year think of visiting Scotland again, but somehow
the summer always finds its own employment. I hope it will
not ever be so.

Yours always, most truly W. WHEWELL.

TRINITY COLLEGE, CAMBRIDGE, *May* 26, 1839.

MY DEAR M. QUETELET,

*   *   *   *

I do not think I have any scientific news for you. Since
I ceased to be President of the Geological Society, I have been
very little in London, and have not heard that anything remarkable is going on. The most remarkable discussion in Geology
since my reign is one concerning the geological position of
Devonshire and Cornwall, which Sedgwick and Murchison have
entirely revolutionized.

Up to the end of my Presidentship you will find a view of
our proceedings in my speech which I send you. Sheepshanks
is here, busily employed in making out the Longitude of Bagdad
from the measurements of the Euphrates expedition; he bids me
say that he intends to write to you soon.

Peacock has received a very valuable appointment in the
Church, being made Dean of Ely. This is bestowed upon him
by the Ministry; it will not prevent his continuing to be, as
he is at present, Professor of Mathematics and Astronomy.

I send along with this a small packet which may serve to
recall me to the remembrance of Madame Quetelet, and, I hope,
all the more agreeably from being associated with the name
of Göthe, whom she and you knew and loved so much. Last
summer at vacant hours I translated *Herman and Dorothea* into
English, retaining the verse and the phraseology as faithfully
as possible, and I have printed a few copies of which I beg
Madame Quetelet to accept one. I am afraid she never did

us the honour to learn our language, but probably she knows the original, and you may testify to her that the translation is very faithful.

With my kind regards, and in the hope of seeing you,

Believe me, very truly yours, W. WHEWELL.

LONDON, *June* 19, 1839.

MY DEAR M. QUETELET,

*   *   *   *

We are just going to send out the magnetical expedition, of which I think I wrote to you. Herschel especially has given a great deal of time and attention to the work of drawing up the instructions, and the observers are now, or soon will be, with Professor Lloyd at Dublin, learning the use of their instruments under his instructions. The East India Company also assists us by establishing magnetical observatories at three or four stations within their territory. Along with magnetical observations meteorological are to be combined, and I have no doubt the expedition will be the cause of great service to Science.

*   *   *   *

NUREMBERG, *July* 14, 1839.

MY DEAR JONES,

Something was said about my visiting you again at Haileybury in July; but, when you look at the date of this, you will see that such a plan is no longer very likely. I had some intention of taking a flight to Munich when I saw you, and one or two letters, which I was supplied with about that time, determined me to do so without delay. I believe it is well that I did not consult you on the subject, or your strong conviction of the grievous harm which inevitably arises from meddling with German metaphysicians might have shaken my purpose. I believe I go among them pretty well secured by a previous resolution not to adopt any of their fancies, and

I shall see what light their speculations will throw upon mine, which, to tell you the truth, I do not expect will be much, for I separate more and more from them as I go along.

I have been employing myself this summer, since it was summer, as I did last year; that is, travelling about from place to place, and when I found convenient quarters, stopping for a day or two to work at my book. The part that I am now revising alarms me sorely by its abstract and almost mathematical aspect; and, what is worse, I can neither do without this part nor make it more popular. It is about the metaphysical principles of mechanics and astronomy, and a number of the more weighty questions turn upon it. All I can do is to put it in such a shape that idle readers may pass over it as quickly as they like. But in truth it puts me out of heart about the reception of the book; but that may be as it may. The book is written, and the world must make of it what it can; and they shall, please God, have it next spring.

How do you and your books go on? But in truth I suppose they do not go on: for you seem to have renounced, at least for the present, the idea of answering Blake; and your velleity, as the French say, with regard to something on the education question appeared to be hardly strong enough to take the shape of action. I wish very much you would do something in the latter subject; for the public opinion is very strong, while their case, as it is usually stated, is not; and if you can put this in a very good and striking form for them, which I have no doubt you can if you set about it, you must needs earn a good share of gratitude from good churchmen and good conservatives.

I wish you could be here with me to see such an odd old town of the middle ages as this is. The king of Bavaria, who is a poet in his way, has written a poetical address to Nuremberg, in which he calls it the Pompeii of the middle ages; and it is something of the kind. The population is protestant, but the churches have preserved more catholic splendour than in almost any catholic city.

I shall be very glad if you can find a few minutes to write to me and tell me of anything that is going on in the world while

I am absent. If I can do anything for you, or get you any information at Munich, where I shall be for two or three weeks, pray let me know.

Give my love to Mrs Jones, and to the Herschels when you see them. I expect to be back in the beginning of August, —at least, in time for the Birmingham meeting.

Always truly yours, W. WHEWELL.

TRIN. COLL. Oct. 11, 1839.

MY DEAR AIRY,

*   *   *   *

With regard to the other question of your Tracts, I can give you a very decided opinion, so far as my own views are concerned. I have always thought a rapid change of the books read among us to be a serious evil. The best part of the effect of such books is the students' getting to see what are the difficulties in them which require labour and thought, and then labouring and thinking till they get over the difficulties. This cannot be done in a few years. Your Tracts are hard enough to occupy several generations of tutors and pupils before they are thoroughly mastered and made familiar. They are now well established among us, and, if they continue to be so for ten or twenty years longer, they will do us far more good than any other books can do. If something else were to be put in the place of them, even by you, giving your attention and time to the subject, it would not be so useful to us as what we now have. I am therefore clearly for reprinting them with as little alteration as your conscience will allow you to introduce. There is matter enough for the best mathematicians we are likely to have for the next dozen years; and, if there rise one or two geniuses who do not find them hard enough, or sufficiently accommodated to the last results of the best mathematical inquiries, such anomalous bodies will, by the very anomaly of their nature, find nutriment and exercise for themselves.

*   *   *   *

Yours very truly, W. WHEWELL.

7, SUFFOLK STREET, *June* 12, 1840.

MY DEAR SEDGWICK,

I do not know where you are, but I suppose that if I write to Cambridge my letter will reach you in no very long time. There is a circumstance of which you ought to be apprized, and which I had rather you should learn from me than from public report. You will be alarmed at such a preface, but the matter is not very serious. I have gratified myself by dedicating my book to you, and I suppose it will make its way to your door in a few days at the utmost, as I have already got a copy. Receive it graciously, and read it indulgently when you have time.

I have a letter from Murchison written in high spirits from Berlin. I go to Rotterdam to-morrow, and have only just time to get through indispensable work.

Yours always, W. WHEWELL.

DUSSELDORF, *June* 28*th*, 1840.

MY DEAR M. QUETELET,

*　　*　　*　　*

I left a copy of my *Philosophy* directed for you when I quitted England, and I hope it has reached you by this time. When you have time to read it, you will find that, when I spoke of your finding it too idealistic, this was not because I supposed you inclined to materialism. I do not know how the case may be with you, but most of my own countrymen have been of late in the habit of understanding the maxim, that "all our knowledge is derived from experience," in such a sense as to overlook the importance of ideas, and it is one main object of my book to correct this error. I should be glad to receive from you any remarks which may occur to you in reading it, and, if there are any of your friends who take an interest in the subject sufficient to write a critique upon it, I should much like to see the article. I should be perfectly satisfied even if the critique were very adverse to my doctrines, for I am very desirous of seeing what answer can be made to my arguments.

*　　*　　*　　*

15, SUFFOLK STREET, *July* 7, 1840.

MY DEAR LUBBOCK,

I am not much surprised that you do not like my *Philosophy* at first. I have some hope that, if you read it, it may in some measure improve upon you; the second volume is I think, less at variance with the metaphysical opinions commonly current, which are probably yours. I should have been very glad if any opportunity had occurred of doing justice to your labours in physical astronomy, but I think you will in reading my book see that they did not come in my way. My object in the work does not at all include the history of Science ; and the *Philosophy* of Science, in *my* interpretation of the phrase, is the discussion of the *Fundamental Ideas* which it involves. All my references to facts in the history of science are given as examples only of the processes which they include, and it was necessary for me to take the most plain and familiar examples which I could find; the more so, as I was writing not for scientific so much as for general, or at least metaphysical readers. Speaking of Pontécoulant, have you seen the severe and contemptuous criticism of him which Arago has published ?

I do not quite see the purport of your remark about the formula for the curve of rise and fall of the tide. In the expression $A \cos (nt + \alpha) + B \cos (nt + \beta)$, do you mean the two terms to refer to the moon and sun, or to the semidiurnal and diurnal tide ? In the former case, m and n are very nearly equal, in the latter, it is impossible to obtain any general formulæ, because the epochs differ at different places. No *general* formula can be given except the one I have given. It must be modified at particular places, as I think I have said; certainly intended to say.   Always yours, very truly, W. WHEWELL.

TRIN. COLL., *Sept.* 22. [1840]

MY DEAR HARE,

Your letter reacht me a day or two since, but I have been so constantly in motion that I was not able to reply to it sooner. I am now here with the prospect of staying, with slight intervals of absence, for some time; and I shall be quite glad to receive your designs and to give you the best judgment and best advice that I can about them. I must however warn you that you must not trust too much to my knowledge of the subject. I have lately had occasion to discover—having had to give advice about some small architectural reforms—how very imperfect my skill is, when applied in detail; as, indeed, a reasonable practical man like you will naturally suppose that it must be. But if you will let me see your designs, and especially if you will tell me what are the doubts which you wish to have resolved, I can, perhaps, either from what I know myself or from what I can learn from others, help you in some degree; and it will be a great pleasure to me to try to do something in so good a cause. If you set about the task in the spirit which you express, you cannot fail to prevent mischief and to improve the character of what you transmit to posterity.

I am glad to hear of the progress of Herstmonceux church, and of your good hope of what is to be done. The wooden tracery, authorized as you describe it to be, cannot fail to be a great advantage to the edifice. I shall like much to see your church when it has undergone its reformation.

I grieved at the death of your late bishop[1], for I knew him to be a most amiable and worthy man; but mainly because I knew how much you would grieve. I somewhat rejoice at the selection of his successor, for it involved the rejection of a certain savage exclusive party spirit, which is quite inconsistent with any good government of the Church and State, and which is vehemently urged by the underlings of the governing party. If I am not mistaken, you already know enough of him to like him.

*   *   *   *

You ask whether I am idle now that I have got rid of my Philosophy. Not at all. This is not a world for people to be idle in, who see, or fancy they see, their task. I have got various plans to execute, and some which I am setting about without delay. In some of my schemes I shall come to matters where I have much

[1] Dr Otter; he was succeeded by Dr Shuttleworth.

more confidence in your judgment than in my own; and then
I shall apply to you without reserve, for you are a man full of
business, and therefore, as our dear and admirable Sir John
[Malcolm] used to say, sure to have leisure for your friends.

*   *   *   *

Yours always, W. WHEWELL.

I am here for Fellowship examinations. We have *eight*
vacancies, which is rather too much; but we want hard men to
work. I hope we shall find them.

COLLEGE, GLASGOW, Sep. 18, 1840.

MY DEAR MURCHISON,

The suggestion which you mentioned to me to-day was so
utterly different from any thing which had ever entered my mind,
that I could not at the moment put my thoughts in order to reply
to it with any statement of grounds, and could only express my
extreme repugnance to the proposal, my conviction that it would
not contribute to the prosperity of the Association, and my general
persuasion of its not being advisable. I must try, before I leave
Glasgow, to make you perceive that these impressions are very
strong in my mind, and I hope that they are well-founded.

My only pretensions to such a position are what I may have
done as a cultivator of science, and my constant attendance upon
the business of the Association. With regard to the former point,
I venture to say that I can be an impartial and exact judge in my
own case, and I know perfectly well that there is nothing of such
a stamp, in what I have attempted, as entitles me to be considered
an eminent man of science. In the study of the tides, which is
my only pretension, I have voluntarily given up all the profounder
parts of the subject, and confined myself to collecting laws of
phenomena in such a manner as it could be done with little of
my own labour. My *History* and *Philosophy* of Science are disqualifications, not qualifications, for my being put at the head
of the scientific world; for I cannot expect, I know it is impossible, that men of science should assent to my views *at present:*
and those who have laboured hard in special fields will naturally

feel indignant at having a person put at their head, recommended only by what they think vague and false general views. I believe this would do much to disgust and repel men of science from the meeting.

You spoke of my being a Lancashire man, as a recommendation to my being President, if the meeting were held in that county. It will not so operate. I have lived little in the county, have no connections in it except my own family, and few acquaintances. It could only produce failure and ridicule to have me put in a place which should be occupied by some person of great local position, influence and popularity—and that in a county so populous, rich, and scientific. It would be considered as evidence that you could not find any person coming nearer to the usual conditions, and likely to give the business its usual attractions.

You spoke of my being able to preside over, and, if necessary, to control a large body. Nobody can do that if he be not persuaded that he has the stronger part of the body with him. I should not have that persuasion, for on many points—almost all questionable points—I should be on the losing side, (for instance the encroachments of the Statistical Section—the scenes of display, &c.) I should be more likely to irritate than to calm the assemblage, when such points came into controversy.

My own repugnance would not decide me, if I thought I had a duty to the Association, but the repugnance is very strong. I have in some measure wound up my account with physical science, and turned my thoughts into another field, in which I may, perhaps, do something, but in which, at any rate, I shall probably employ myself seriously for the rest of my life. I have been on this very account gradually unwinding myself from the engagements of the material sciences, and cannot think, without terror and extreme annoyance, of being again plunged in the entanglements I have left. It is not that I do not still love science; but in it I have done all I can; I must go now to " fresh fields and pastures new."

But I dare say I am taking more pains than is necessary to convince you that I should never do for your President. I am *not* a man of science, *not* a man of business, *not* a man of popular manners, *not* a man of weight in Lancashire. What other disqualifications can you require?

You will not suspect me of undervaluing your good opinion, and that of the other friends who have overlooked all these disqualifications. I am glad I have friends who can overlook them, for they are obvious enough.

Believe me, dear Murchison, yours most truly, W. WHEWELL.

TRIN. COLL., CAMBRIDGE, *Sep.* 25, 1840.

MY DEAR MURCHISON,

I certainly hoped that my letter to you would have put an end to the project of making me President of the British Association for the ensuing year; and if I had dreamt that it would prove insufficient for that purpose, I should have taken other steps which might have been more effectual. My repugnance to the office, and my conviction of my not being the proper person for it, continue unabated. It is, however, impossible for me not to be gratified, and in some degree encouraged, by the general choice falling upon me so decidedly as you say; and having laboured without shrinking for the Association up to the present time, I am unwilling to take any step which may place the managing body in serious difficulty. I will add that, from the opinion I have of the Cornwallians and Devonians, I would more willingly take a part in next year's meeting, held in that region, than in any other.

But this resolution to meet at Plymouth is so far different from what was contemplated when I left Glasgow, that I am ignorant of the circumstances which bear upon it, and which must in a great degree determine the prosperity of the meeting. Before, therefore, any thing is considered as fixed, so far as I am concerned, you must allow me to make a few inquiries.

Was there an express invitation from Plymouth? Can they receive our Sections conveniently? Can they find a place for our general meeting without taking the Theatre? (I think a theatre much to be avoided.)

Have the Vice-Presidents, Lord Eliot, Sir C. Lemon, and Sir T. Acland been consulted, and do they *cordially* assent to their appointment? If they do, I have no doubt the meeting will be very satisfactory; but I have as little, that either Sir C. or Sir T. ought to be President.

The Secretaries for Plymouth are good men. I take for granted that they have been consulted and are willing to work. I presume also that there is no danger of any change in our permanent officers—General Secretaries, Treasurer, and Assistant Secretary.

There is another matter which I think ought not to be passed over. It was impossible to listen to the proceedings of the Statistical Section on Friday without perceiving that they involved exactly what it was most necessary and most desired to exclude from our proceedings. Is there any objection to the President declaring in his place, in the most emphatic manner, that the mode in which this Section has been conducted is inconsistent with the objects and character of the Association?

There are other inquiries, which I should like to make, as to what has been the general character and course of the Glasgow meeting; but as these are not essential to the present purpose, I shall leave them for another time.

My engagements here do not allow me to be absent between Sept. 17 and Oct. 2, which must be recollected in fixing the time of the meeting if I am an officer.

As I have said, I must beg of you to consider nothing as concluded on my part till I have information on these points. Believe me, dear M.,

Truly yours, W. WHEWELL.

Pray take charge of a book which is left for me at Killarmont. I will tell the butler there to convey it to you.

TRIN. COLL., *Sept.* 30, 1840.

MY DEAR FORBES,

You expressed an interest about Gregory's election, so I am glad to be able to give you the first intelligence of it.

I am just returned from the deliberative meeting of the Seniors, and the election will take place to-morrow morning at nine. Along with Gregory we elect seven others—Heath (a brother of our friend), Hodgson, Frere, Edleston, Mathison, Eddis, and Ellis the Senior Wrangler.

I suppose by this time you are returned to Edinburgh, so I shall send this there. I was much obliged by your letter from Glasgow. My election as President was contrary both to my expectation and my wishes; but this I say to you only: and, the thing being done, I shall make the best of it as well as I know how. I was glad to find that a good number of stars, who did not shew at first, made their appearance as the time went on. I should have liked very much to hear Airy's exposition of his theory of Brewster's "new property". Was B. satisfied?

*   *   *   *

I find myself called upon for a new edition of my *Mechanics*. If you have any improvements to propose, I shall be much obliged to you to mention them. My general notion is to try to fit the book more for practical men, and to make it correspond with a work which Willis is very soon to publish on *Kinematics*, or *Pure Machinery*, as I have called the subject. By the way, looking into my book of Philosophy, I see reason to believe that I have again done you injustice. If it comes in your way, and you find it is so, you must forgive me till I have the opportunity of repairing the wrong. Believe me,

Yours very truly, W. WHEWELL.

TRIN. COLL. CAMBRIDGE, Oct. 2, 1840.

MY DEAR MURCHISON,

It is quite clear that you, according to your irresistible way of settling such matters, have made it fit and proper that I should be the President of the British Association next year, and therefore I have nothing to do but to think how I can best fulfil the requirements of the office. I do not at all assent to your general doctrine, on which you have founded my appointment, of an alternation of men of science and men of rank; for

I think that the most proper president in every case is a man of local consequence: and the precedents of Dublin, Oxford and Cambridge are against your alternation and not for it. Still less can I imagine that, if you were to have a supposed man of science, I was to be the man. But let all that pass. You shall find me as conformable as your Russian Devonians, and shall go on triumphantly from Nijni Novgorod to the Land's End, so far as any exertions of mine can contribute to your triumph. As I told you before, I have great confidence in the Western men, and I certainly think that Plymouth is by far the most proper place of our meeting for the ensuing year. How far the men of science will shew their good will by gathering round our standard there, it is of course for them to decide. We officers can only give them the opportunity of doing so if they choose. It will not be difficult to make it appear to any body that requires to be instructed on the subject, that we have now, more than we ever had before, solid and important work to do, which cannot be done by any body else; and I have no doubt we can find the means of doing it. As to the Statistical Section scruple, I cannot get over the utter incongruity of its proceedings with all our professions and all our principles. Who would venture to propose (I put it to Chalmers, and he allowed the proposal to be intolerable) an ambulatory body, composed partly of men of reputation, and partly of a miscellaneous crowd, to go round year by year, from town to town, and at each place to discuss the most inflammatory and agitating questions of the day? Yet this is exactly what we have been doing for several years. I must say plainly, that rather than be concerned in such a wild and dangerous absurdity, in defiance of solemn professions to the contrary, I would utterly renounce the Association with all its advantages. You have made me your President, with no good will of mine; in everything else I will be instructed by you, and labour, as well as I know how, for the advantage of the Association, in any way in which I can aid it: but I will make no agreement with you that I will not denounce, in the most public and emphatic manner, this gross violation of our fundamental constitution. If we offend people by recurring to our professed principles, I cannot help it. If our Association does not suit them, when conducted on its only rational grounds, let them make one of their own.

I do not wish you to imagine that this is my sole, or even main, reason for wishing not to be the president; on the contrary, a thorough sense of my unfitness, such as I tried to convey to you, remains quite undiminished. But, if I am to be president, I must at least guard myself from being misunderstood on this point.

I do not know whether you will think I write in ill humour about this matter; but, if it be so, be assured it is nothing which will prevent my cooperating with you in the most cordial manner, and trying to imitate the spirit and vigour which you have shewn in all your proceedings. If this long piping time of peace has prevented your being a great general, as under other circumstances you would have been, it has not prevented your making a Russian Campaign more successful far than Napoleon's, and then returning, faster than he did, to consolidate the state at home—I trust with far happier fortune. So I agree to serve under or over you, as you please, and am, dear Mr Secretary,

Your dutiful President elect,

W. WHEWELL.

TRINITY COLL., CAMBRIDGE, Oct. 5, 1840.

DEAR LORD NORTHAMPTON,

I am very much obliged by your letter, and your view of the position of the British Association. I am quite of your opinion, that the success of the meeting at Glasgow was such as to promise a highly satisfactory sitting at Plymouth; although I was obliged to leave you before the business had arrived at those occasions which mainly called up the interest and the confidence of those who attended. I certainly could have wished that some other person than myself had been placed at the head of the Association for the ensuing year; and this, not at all from any doubts of the progress of the institution, but from my knowledge of my own true position. I cannot conceal from myself that I have no just claim, and cannot be considered by judicious persons as having a just claim, to be placed upon a level with the persons who, as men of eminent science, or as patrons of science, have hitherto filled the president's chair. Nobody knows better than I do, how little I have done in science; and my future exertions will probably be mainly directed to do some little in sciences which do not come within the scope of the Association. Of my *History of Science* the principal notice taken by men of science has been of a hostile kind; and I do not think that any practical cultivators of special sciences will feel any deference for a person who has presumed to speculate about them all. But though I would very gladly have avoided this position, I shall not fail to give my best attention and exertion to its duties, being once placed in it: and I am very far from insensible to the pleasure of being proposed by you—a person whom I so cordially regard, and whom all the members of the Association have so much reason to look to with gratitude. The expression of your good opinion, and that of other persons whom I esteem, have made my appointment a source of satisfaction to me, as well as of dissatisfaction.

I have no fear that, with your assistance and that of the other persons who are associated with me in the direction of the Association, we shall be able to conduct it through the ensuing year in such a way as to make it answer its purposes as well as it has ever done. There is, however, one point, on which I am far from satisfied, in which I think something should be done, and about which I cannot consult any one so properly as yourself, who have repeatedly presided over the meetings under difficult circumstances, always obtaining my admiration, and I believe that of every body else, by your temper, fairness, and judgment. It is becoming more evident every meeting that the concourse which the Association produces is used by various persons for purposes which it cannot recognise, and which I doubt whether we ought knowingly and with foresight to further. I have therefore some misgivings when I see a meeting upon the stormy question, which now agitates the church of Scotland to its very foundation, combined so closely with our assembly. About this, I say, I have doubts. But I have *no doubts* respecting the propriety of having another great question of social economy and legislation, which produces almost as great a storm as the other, agitated within one of our Sections. I cannot doubt that this is an utter violation of all the principles on which we set out, and of all the professions we have made about our objects and maxims. Of course I refer to the question of the Scotch and English poor law. If such discussions be allowed, there is nothing in legislation or politics which can be consistently excluded. Dr Chalmers made an attempt to justify or mask this impropriety by saying that it was an example of the value of *numbers*. By the same rule we might have a discussion of schedule A of the Reform Bill, or of the probable majority of ministers on any party question. The absurdity of such a plea is, I think, undeniable, and the inconsistency of such discussions with our fundamental constitution. And this is not a question of form merely. For what kind of institution do we become, if we allow ourselves to be made an ambulatory meeting for agitating in assemblies, when both *eminent* and *notorious* men (Dr Chalmers and Robert Owen) address a miscellaneous crowd on the sorest and angriest subjects which occur among the topics of the day? If we cannot get rid of this character, most assuredly I shall be disposed to make my connection with the Association as brief as I can do, without shewing myself indifferent to the good opinion of friends like yourself, who are goodnatured enough to think that I can be of service to the genuine interests of the body.

TRIN. COLL., CAMBRIDGE, *Oct. 25*, 1840.

MY DEAR FORBES,

I am only just returned to college out of Suffolk, where your last note was forwarded to me to my great content. I do not think it likely I shall visit London within the next ten days, so I have little chance of seeing you there. Can you not run down to Cambridge for one day, or two, or three? I should like to talk to you about my plan of entirely modifying the new edition of my *Mechanics* so as to make it a book for civil engineers. It is the only possible course for the present time, on grounds of the highest principle, as well as convenience. Also, as you may well suppose, I shall be glad to confer with you about necessary and contingent truths, for I am quite delighted that you take so well to the investigation of the grounds of a difference between the two kinds. I shall probably find some opportunity before long of doing what you suggest.

Have your views of teaching civil engineering, or your hopes of not needing to think of such things, undergone any change? I should like to hear that any thing reasonable had been done on that subject.

Yours always, W. WHEWELL.

TRIN. COLL., CAMBRIDGE, *March* 12, 1841.

MY DEAR FORBES,

How are you all this while that I have heard nothing of you? I was a good way towards the border in January, but not near enough to hear any thing which was passing on your side of it, except the noise of your church controversy, which makes itself heard. I went into Yorkshire to look at a College living[1] which I had some intention of taking; but on examination I thought it too laborious. It would have interfered with my purpose of employing myself about the business of my professorship, which you will see by the accompanying pamphlet I have been in some measure pursuing, and wished to pursue farther.

I hope you got the new edition of my *Mechanics* which I directed to be sent you. You see how it has been sweated down into a little book by the process to which it has been subjected. I am printing my Engineering, and Willis is printing a book about Mechanism, so we shall between us do something towards giving a form to that science. How goes on your meteorology, optical and other? I hope to be able to prove

---

[1] The living of Masham, then vacant by the promotion of Mr Waddington to the Deanery of Durham.

that the tide is much affected by the barometer of the *day before*. This will much improve our predictions.

Do you come south soon? Let us see you when you do.

Yours always, W. WHEWELL.

T. C. *Mar*. 16 [1841].

MY DEAR MRS MURCHISON,

I am very glad that I am able to do what you wish in the most important point. I had begged Professor Smyth to give me a few copies of his "Sheridan",[1] that I might be able to give it to friends who had taste enough to wish to have it; and I shall delight him much by telling him that you are one of the number. I inclose a copy along with this.

I lament very much that I cannot do *myself* the favour of profiting by your other proposal. I am engaged here the whole of next week, and must resign, much against my will, the pleasant party which you offer me. It would give me all the more pleasure to meet the Archbishop and Miss G. Harcourt, after being, as I very unwillingly was, as near them as York without seeing them, when I went down in January last to look at my Yorkshire living, that was to have been. But I could not make out to my satisfaction that it would do for me or 1 for it; so I wait another chance.

I should like, too, very much to hear your husband finish the history of his last Russian campaign before he makes another visit to Archangel. But this is not to be, so far as I can see. I believe I must remain till Easter in the region of clunch and gault, rejoicing only that there are persons who write charming Quarterly Reviews of Silurian regions far away, and give us picturesque descriptions of cities, fairs, embassies, droshkies, postboys and peasants, that we shall never see.

Believe me always, yours most truly, W. WHEWELL.

[1] Mr Smyth, Professor of History, had privately printed a memoir of Sheridan, and distributed copies of it. In a letter to Dr Whewell Professor Smyth remarked with respect to Sheridan, "it is pleasing to me to observe how strong is the hold which this extraordinary man still has on the minds of those who can at all remember what he once was."

UNIVERSITY CLUB, SUFFOLK ST., *April* 1st, 1841.

MY DEAR QUETELET,

*   *   *   *

Your system of simultaneous observations on the annual phenomena in all parts of your country must give interesting results, and will probably lead people to do that systematically and generally, which they have so long been doing in a partial and imperfect manner. But you must necessarily have a great deal of trouble in arranging and comparing the results of your enquiries. I have no doubt that, if you examine your Belgic tides, among other things you will find that they have a very marked dependence upon the barometer. I find that at Bristol the water (at high water) rises one foot for a fall of one inch in the barometer, and the same thing takes place at other ports. At first I thought that the elevation of the water corresponded to the fall of the barometer some hours previous, but I find that, so far as I can determine, the effect upon the tide corresponds most nearly with the contemporaneous barometer.

We have very excellent accounts from Van Diemen's Land. Ross has established his magnetic Observatory there, and is now sailing towards the South Pole. I am glad that your work, *Sur l'Homme*, is translated, and is going to be published here with notes. I shall look to your notes with great interest.

With my kindest regards to Mad. Quetelet,

Believe me, most truly yours, W. WHEWELL.

TRIN. COLL. CAMBRIDGE, *April* 22, 1841.

MY DEAR HERSCHEL,

I am afraid your ground for calling me goodnatured is very insufficient, for I do not know that you ever asked me to do any thing for you before: and I have hardly known enough of children to know whether I like them or not. But certainly, so long as I like any body's children, it is probable that I shall like yours; and I shall be very glad that the strange and unsatisfactory condition in which I stand to them, of not having

a god-child among them, should be removed. So pray consider
me as the sponsor of your little girl when you present her for
christening, and suppose my best wishes for her temporal and
eternal welfare to be with her, if I am not there myself.

But in truth I do not know why I should not run down
to you for the day, and witness the ceremony myself. I must
leave Cambridge and go to London about that time, within a
day or two sooner or later; and I do not see that it will make
any great difference in time, if I perform one oscillation to
Hawkhurst and back before I come to rest in London. May
the first will, I think, suit me as well as any other day. So
except I hear of some change in your plans before that time,
I shall come to you then. Shall I be in time for the ceremony
coming to you by the coach on Saturday? I hope so; for
I cannot leave Cambridge sooner than sometime on Friday.
I am glad to hear of the excellent people that are to be my
sisters on the occasion.

To know that you are reading with care my Philosophy,
makes me feel the strongest interest to see what your reflexions
will be. Whatever they are, I shall be sure to learn much from
them, (I believe you will not suspect me of saying such things
for the phrase's sake), and I shall really be relieved by seeing
some other view than my own presented with a calm and earnest
love of truth. My own doctrines quite haunted me till I had
got them put down upon paper, and then fairly shot out into
the world through the press; and even now I feel a sort of
satisfaction at having done, once for all, a great task which had
been to me inevitable. I had no rest till I had thought it all
out, and am right glad to have it off my hands. I shall be
curious beyond measure to see the other side from you. My
argument is all in a single sentence. You *must* adopt such a
view of the nature of scientific truth as makes universal and
necessary propositions possible; for it appears that there are
such, not only in arithmetic and geometry, but in mechanics,
physics and other things. I know no solution of this difficulty
except by assuming *à priori* grounds; but I am most willing
to look at any proposed solution. But we can talk of such

matters hereafter, if it be necessary. I am now busy about other things, systems of moral philosophy and the like, which nevertheless are well worth attending to, as I think you are as ready to allow as I can be to assert.

You must thank Lady Herschel for recollecting my representation of my anomalous position with regard to your babes; you must tell me whether my coming to you on Saturday, May 1, will fall in with your notions and hers of the significance of the occasion, and you must believe me always most truly yours (including hers), W. WHEWELL.

T. C. *June* 26, [1841].

MY DEAR HERSCHEL,

It has added to every joy of my life to know that I might reckon on your sympathy, and most of all to find yours and Lady Herschel's so completely with me in this. I can answer for Cordelia's readiness to love and admire both of you, as she well may do; and I hope the time will soon come when she can shew her feelings towards you by her personal intercourse.

Your review makes me too proud, as I have told Lady Herschel; but also it makes me feel as if I had been too contentious, when you differ from me so gently. I really believe that I did write a great deal of it in a spirit of needless pugnacity. I knew that my opinions were opposed to those generally current, and was prompted by that recollection to sharpen my doctrines and my arguments as much as possible. If I reach another edition I will be more calm and moderate ; and especially all that I have said about the Edinburgh Reviewer, if he holds his tongue till then, shall be erased.

As to your notion that the irresistible impulse to generalise will account for universal truths without having recourse to *ideas*, I shall weigh it well before I pronounce upon it; but at first sight I do not see any reason to despair of our coming very near each other. I think it very likely that a great number of my most important doctrines may be expressed in language consistent with your view: but, as I have said, this is a matter requiring steady thought, and at present I have no head for the "bitter metaphysics", which you very naturally express your own weariness of. So we will sleep on our arms, and think of other things at present. I hold myself in the mean time the happiest of "the Reviewed".

Always most truly yours, W. WHEWELL.

UNIV. CLUB, *July* 25, 1841.

MY DEAR HERSCHEL,

I am here on my way to Plymouth, and want to have as good a story to tell there as I can contrive. Among other matters, I shall be very glad if we can devise any good schemes for employing the money of the British Association upon worthy and promising objects; for it ought to be spent and not saved, and with good management we may get money's worth out of it. Meteorology appears to me one of the subjects in which we may labour most fitly, since the labour of discussing such observations is so great, and the progress hitherto made so small. Can you suggest any tasks of this kind which might be put in the hands of commissioners with funds for calculators? I found Birt at the R. S. yesterday, who told me he was working for you, and had sent you some curves, the results of his labours. Do you want him to go on with his work, and do you want any money for that purpose? In short, can you suggest any thing which we British Associationers can do, that is likely to be useful? I do not like our meeting without such results.

I am just returned from Ulleswater, where I have been staying at the house which you and I visited some two years ago. The lake is as beautiful as ever, and the place in my eyes the happiest in the world. I cannot easily tell you how delighted I am at the thought of Lady H. and Cordelia being friends, which they appear to have made some progress in already.

I have not yet had time to think over your Review as it deserves. About space and time I do not despair of our coming to a compromise. Your notion of cause is to me new and striking, and I will not say that I may not hereafter adopt it wholly or partly. Altogether I am enchanted with the article. Always, my dear H., most truly yours, W. WHEWELL.

CONISTON, *Oct.* 16, 1841.

MY DEAR PEACOCK,

Your letter came upon me like a thunderclap altogether unexpected. The active and generous friendship which you shew in it is beyond measure grateful to me, and I shall consider the added conviction of such a friendship on your part as a great gain, whatever the result may be. In the mean time, I shall do as you recommend, and come to town immediately with my wife. We shall probably be at Lord Monteagle's, 37, Lower Brook Street. I have written to Sir Robert Peel and to Mr Goulburn, and shall write other letters as fast as I can; but I hope to be in London to-morrow evening, which will much expedite matters. Once more, my dear Peacock, believe me that I am much touched and delighted with your cordial friendship, shewn in the warmth with which you have adopted my interests, and am more than ever,

Affectionately yours, W. WHEWELL.

16, SUFFOLK STREET, *Oct.* 19*th*, 1841.

MY DEAR PEACOCK,

I wrote you a very hurried note the other day, but perhaps did not tell you that I was about to take your advice and come to town. I did so, and found a letter from Sir Robert Peel offering me the Mastership. I accepted this yesterday, have seen Sir Robert to-day, and am now going back to Coniston by this night's mail.

There has been much in the whole of this business to gratify me :—most of all things, the warmth of good opinion, and the zeal of exertion shewn by my friends :—and of all these most of all, my dear Peacock, without any exception, your own friendly warmth and zeal. Your goodwill was so cordially expressed and involved so much of generous feeling that it made me very happy, and is still one of the pleasantest parts of the good fortune which has fallen to my lot.

I go into the North this evening, as I have told you. I have left Mrs Whewell there, and shall bring her to London about the end of the month. She and I are much obliged by your congratulations on that score, and I am always, my dear Peacock,

Affectionately yours, W. Whewell.

Coniston, Ambleside, Oct. 24, 1841.

My dear Herschel,

No words of friendship are ever more grateful to me than those which come from you, on any occasion, great or small. The events of the last fortnight of my life have been great indeed, for they determine the whole course of my future. I had had quite enough of loneliness, and look for no regrets in losing that ; but I had hardly satiated myself with speculation, and do not find myself drawn back into the bustle of active life without some misgivings. But the task appears to lie upon me so as not to be avoided, and there is much good to be done ; so I take it hopefully, and the more hopefully, if people like you hope good of me. My kindest regards to Lady Herschel. I hope it will not be long before she and Mrs Whewell know more of each other. Trinity Lodge is spacious, whenever you are disposed to reknit your Cambridge recollections, which you and Lady Herschel have not done since you were at the Antipodes. I hope all your bright constellation of little stars in the school-room are shining with their usual healthy light. Cordelia is not at my elbow, but, if she were, would send her kindest regards to you and Lady Herschel.

Always truly yours, W. Whewell.

Trinity Lodge, Nov. 19, 1841.

My dear Mrs Murchison,

I confess I almost despair of your friend's friend, who makes a new theory of the Universe, and myself coming to any common understanding. She says in the course of her speculations, that she wants an answer not in mathematical language, but in plain English; and as I have accustomed myself to think that on such subjects mathematical language is the very plainest English which can be used to those that can understand it, and that no others are likely to make theories to any purpose, I do not know how I am to find arguments which appear to me of any weight, and which are likely to convince the fair theorist. But in truth I believe I had better confess at once that I put in my book all the best thinking at which I could arrive, on the subject of gravitation, and its being a part of a wider law; and having done this, I have nothing to add on the subject of any special theory, which is not included in what I have said there.

I was very sorry I did not see you the other day when I was in Belgrave Square; but my visit was very brief, and was occupied with the "Russian Campaign" of your husband, as we had agreed that it was to be. In a short time I expect to be in London again, and then I shall try to see you, I hope with better fortune.

I took possession of my "Lodge" on Tuesday last:—a very august mansion, when compared with the single room in which you found me when you visited Cambridge, and worthy of being named even by the side of your palaces in Belgrave Square. I shall be very glad when you come to take a leisurely survey of the interior of this mansion for a few days; and so will Mrs Whewell. She begs me to present to you her kind regards, and you know that I am

Always most truly yours, W. WHEWELL.

TRINITY LODGE, Jan. 3, 1842.

MY DEAR HERSCHEL,

Health and happiness to you and yours with the new year, and with all other new years to the end of the century. Your letter and its good wishes ought sooner to have been answered, but such oughts go for 0, even more than usual, with a man who has just got into a great house and a great office and is trying to puff himself out so as to fill both the one and the other. You

will perhaps have heard that I have just had Jones here to help
me to execute one part of this design, and an admirable help he is
for such a case in every way.   And now that we are by ourselves
again, I wish for nothing so much as that some more of my old
friends would come here, and confirm the comfort of my present
condition by connecting it with earlier times.   This is the first
house which I ever possessed, and is likely, I think, to be the last,
for I do not know anything which has any chance of moving me
hence.   Or rather this is the first house which I have inhabited,
not which I have possessed ; for I possess, besides this, the house
"behind the Colleges," which you know as "Farish's House," and
which I was to have occupied before I foresaw my destiny here.
When I looked to that as my prospect, I was in no small dismay
at the impossibility of finding room for my books or place for my
writing table, and feared that I should have had to give up altogether the practice of writing.   I now sit in Bentley's chair and
listen to the ticking of Newton's clock, and write with as much
contentment as either of them can have had.   I do hope that
Lady Herschel and you will come and see us before it is very
long, but I do not say much about such a plan now, because
I suppose such movements are only likely to take place in
summer.

I am glad to hear you get your Cape stars into order.   I am
trying to do as much with the virtues and vices, but I find them
somewhat refractory—the one quite as bad as the other in that
respect.   Also I cannot get out of my head a metaphysical conun-
drum which you have put into it.   In your review of my Philosophy you prove very clearly that, in mechanical causation cause
and effect are contemporaneous.   Good :—uniform motion measures time, and change of such motion takes place instantly when
force operates.   But how are we to extend this to other causation
besides mechanical ?   A drug produces disease, not instantly,
but after a time.   What is here the direct effect which is instantaneous ?   Motive produces human action, not instantly, but after
a time.   What effect does it produce instantly, and what is the
movement which measures time when no cause of change operates ?   I do not know whether I make you see my puzzle.   I try

to solve it, and with that view I am writing a paper for our Philo-
sophical Society, but I am afraid it will not turn out worth much.
However, if I can succeed in puzzling other persons as much as
I have puzzled myself, it will be something.

*   *   *   *

TRINITY LODGE, *March* 21, 1842.

MY DEAR QUETELET,

I hope by this time you have received the copies of the
Athenæum which you wished for.   It was very wrong in me to
forget to send you them; but, in truth, events succeeded each other
with great rapidity in my history after I parted with you.   I was
at that time looking to the future with some feelings of uneasiness, for I found that moving out of college into a small house
was likely to be far more disagreeable at first than I had anticipated.   It appeared to me that I should have no room for my
books, and no place for study ; and I had been so long married to
Trinity College that it was painful to part from *her*, even for a
wife of a more affectionate character.   But all my alarms were
soon dissipated ; that is, soon after I had taken the great step of
marrying.   Within a day or two of the time of my marriage, my
predecessor quite unexpectedly resigned, and in a few days more I
was Master of Trinity.   You will easily conceive that it was the
situation of all others in the world which I most desired, and even
now I can hardly believe at all times that it is true.   I wish you
would come here and add your testimony to the reality of the
change, and bring Mad. Quetelet with you.   I have now a house
of my own and a good one, and can receive you in a manner
which may, I hope, occupy you agreeably for a short time, and
I shall think it a privilege for my wife to make the acquaintance
of yours.   You speak of our coming to the Continent.   It is not
impossible that we may do so, but I fear we can hardly hope for
such a pleasure this year.   My new position and my new relatives
together occasion a great number of new claims upon my time.

*   *   *   *

W. WHEWELL.

[The next three letters relate to the Tides; they will be found readily intelligible by a student of the subject, though they lose much interest by appearing without the replies of the Astronomer Royal. These replies are preserved among Dr Whewell's papers, and copies of them exist among the collection of documents at Greenwich. For some remarks on the main topic of the letters see Vol. I. page 76.]

TRINITY LODGE, *Jan.* 18, 1843.

MY DEAR AIRY,

I have contrived to scramble through your Treatise on Tides and Waves, though in a very incomplete manner, and must give you some account of my notions about your mode of treating these subjects. The part about waves appears to me very admirable. Although the solutions include some conjectural steps, they are so plausible, and bear comparison with facts so well, that I hold them of great value. I shall be glad when I have time to study that part of your labours over carefully. With regard to Laplace's theory, I confess that I am not yet converted to a persuasion of its value, though I am very glad to have it made so much more intelligible, as you have made it. It still seems to me that it gives us no light which the equilibrium theory had not given us before; and it is most curiously infelicitous in all its distinctive results when compared with observations. I do not include among these the importance which is assigned to the rate of motion of the luminaries in R.A., and to the difference of increasing and decreasing declination, in the effect upon the tides; for these results are obvious according to any way of applying the equilibrium theory which I have seen attempted. I had bestowed a great deal of labour, in conjunction with Mr Bunt, on the different effects of increasing and decreasing declinations, and I thought I had said something about them in some of my papers; but I suppose I am mistaken, and I never obtained any results which satisfied me. Nearly the same is the case with the rate of motion, but I gave up battling with these inequalities at last, seeing how much labour it would take to make the predictions materially better than they are.

From what I have said you will see that I do not agree with such expressions as you have used about the equilibrium theory. I will even go so far as to say that I think such expressions are unsuited to a scientific treatise. It will really make ignorant mathematicians of the present day conceited to find a person like you speaking of the best theory that Bernoulli, Clairaut and Maclaurin could devise, as "miserable" and "contemptible." Nor is this censure just any more than it is tolerant. To hold the equilibrium theory to its merely mathematical consequences is quite unfair, and, what is more, it is a gross misrepresentation of the way in which it has been used. It is quite unfair; for Laplace's theory, and your own theory of waves even, require some general conjectural reasoning to bridge over the gap between the mathematical hypothesis and the case of nature. The equilibrium theory does no more. It assumes that a fluid will always tend to the condition of equilibrium, though the circumstances of the case prevent its ever reaching that condition; a very just and reasonable assumption. And as it is in my way, I will go on to say that I have already tried to teach that that, which you say is strange, is not strange at all; namely, that a theory which is not at all true should be very useful. This is so far from being strange, that it is the universal course of scientific history.

You will perhaps be amused with my defence of the character of the equilibrium theory; but it has long been my habitual employment to do what I can to prevent people from despising what is really valuable in the history of the past; and your fierce disdain of this most useful step in Tidology is a matter which I cannot pass over.

I am not at all dissatisfied with your mode of dealing with what I have done. You cannot think more slightly than I do of the conjectures which I threw out from time to time with a view of reconciling facts and formulæ. Indeed it appears to me that in several places you attach much more importance to them than I do myself. I think, however, that if I were a young labourer in the field, and of a character to be discouraged, I might have been discouraged from going on by your assuming the manner, not only of a teacher, but of a severe teacher. In some cases, too, I think your severity is misplaced ; as in (564)[1], where you say that my determination of the relation between the diurnal tide and the corresponding position of the moon is worth little. It is at least worth something ; for it predicts the fact in the most consummate manner. I wish other theories did as much.

I have no objection to offer to your view of the Tides in the German Ocean; at present I have not time to attend to the subject : but I wish you had tried to draw the cotidal lines between Norfolk and Holland as they result from your view.

I have had some notion of writing either to Arago, or to the French Admiralty, or to our own, to request them to have tide observations made at the Marquesas, now that the French have established themselves there ; but I do not know whether I can undertake the Pacific any further. It would require a good deal of correspondence, besides the reduction of the observations.

To turn to other matters. I had a letter from young Acland, who has been with the Engineers on Blackdown, working with your Zenith Sector, and is quite charmed with the working of the instrument. He speaks with enthusiasm of one "famously long night of it." He has been calculating and printing Tide Tables for Budehaven.

And now good-bye, for I must set to work to make my paper of questions for the Smith's prizemen, and kind regards to Mrs Airy.

Yours very truly, W. WHEWELL.

I send you an extract[2] from the newspapers, which seems to excite great alarm. It has been sent to me from several quarters.

TRIN. LODGE, Feb. 22, 1843.

DEAR AIRY,

I see that you stand to your guns very stoutly, as your ancient wont was, but you have not made me think that a theory;

---

[1] The Astronomer Royal admits that his Article 564 was open to objection. He meant to express his belief that "for the reasons there assigned the determination was extremely uncertain."

[2] It is not known with certainty to what this refers, but probably to an expected high tide.

which has done so much, and been put forward by such men, as the equilibrium theory of the tides, can properly, in a grave scientific treatise on the same subject, be called "contemptible" and "despicable." Nor have you persuaded me of any great superiority on the part of Laplace's theory. I do not think that it is a step *in the right direction*. It would be a step in the right direction to determine the oscillations of a sea bounded to the east and west, however arbitrary were the conditions by which the solution was attained. But it is no step in the right direction to suppose the earth covered with fluid, or to suppose a complete equatoreal ocean. What we want is to see how the forces affect the oscillations, without suffering the protuberance to run round and round the globe,—the fault alike of the Laplacian and equilibrian theory,—and that this makes Laplace's theory inapplicable is plain, because the theory obliterates a feature of the first order, the diurnal inequality. And when you vilify the equilibrium theory as having led to success on false principles, you forget one plain matter, that we have always applied it, all of us who have applied it, *in conjunction with the laws of waves,* so far as we knew those laws. And now that we know them better, thanks to your investigations, I am fully persuaded that we shall make out a better account, and a *truer*, by combining the equilibrium theory with your properties of waves, than by following Laplace's. If people are so perverse and stupid that they will attend to nothing but algebraical formulæ, I do not think it is worth while to abuse such men as Clairaut, D. Bernoulli and Maclaurin, in order to avoid misleading *them*. You may teach them to despise their betters easily; but I am afraid your care in turning them from equilibrium to motion will do them little good; for they will continue to be as unphilosophical in their mode of conceiving Laplace, as in their mode of conceiving Clairaut.

And now I think I have said all that I want to say. I am glad you have taken to this subject, and hope much good from your having it on your mind. So I send you some tide observations of Capt. Bayfield which are curious examples of that curious thing, the diurnal tide. Perhaps they may fall in with your speculations. Do you not think we ought, some of us, to put the French upon making tide observations at the Marquesas? Such observations, continued, would tell us something about the tides of the Pacific, the great blank of Tidology.

I was much interested with the account of the demolition[1]. Sedgwick gave us a good description of it, as you may suppose. I wish, when you write again, you would tell me what is the rule for the quantity of powder with which Cubitt worked; for the adjustment of the force to the work appears to have been consummate.

*　　*　　*　　*

Always truly yours, W. WHEWELL.

TRIN. LODGE, *March* 2, 1843.

DEAR AIRY,

I think I can find a few minutes this morning before Syndicates and audits begin, to reply to some parts of your last letter; and I will not omit to do so, for I should be sorry to have you think I underrate what you have done in your *Treatise on Tides*. I think that you have made a most important step in the theory by introducing friction; and that, assuming the effect of friction to be equivalent to a retroposition of the moon's place, you have done more than any one had done to bring about an agreement of theory and observation. But, if I understand rightly, the other step which you mention (I presume your Art. 552) is in a great measure included in this, though, as you say, a very striking result.

As to the combination of the equilibrium theory with the laws of waves, I conceived it thus, and I do not think you will quite repudiate it. I found from observation what you have explained by theory, that the *laws* of the tides (not the *magnitudes*), both as to time and height, were those of the equilibrium

---

[1] This refers to the blowing up of a cliff at Dover under the direction of Mr Cubitt, for the purpose of constructing a railway. The rule was: the number of ounces of gunpowder equalled the cube of the number of feet in the depth of the mine normal to the cliff face.

theory, with a retroposition of the moon. The propagation of the tide from place to place by free waves gave a further retroposition. In every case I subtracted as much of the retroposition as I could from my knowledge of wave laws, and supposed the remainder to be the imperfection of the theory, as it turns out to be.

In my admiration of Laplace's theory I can hardly yet go further than to look with satisfaction upon the differential equations, and hope they may come to some good, as in your hands they have done. I cannot take his solution as expressing the unlimited canal. A complete equatoreal canal is *not* an *unlimited* canal; it is a *reentering* canal; no more unlimited, than a circular canal on any other part of the surface of the sphere. And if the whole surface be covered, still it is a reentering canal.

I ought to have added to my former paragraphs, in order to explain my seeming to disparage your remark about the moon's velocity in right ascension (which was only *seeming*), that when I had arrived at the notion of the tides being regulated by a moon retropositioned by an unknown and perhaps variable quantity, it was plain that her velocity in R. A. would enter into the laws, and that it would make a difference whether the declination was increasing or diminishing.

As to the *magnitude* of the tide, I never from the first referred to the equilibrium theory as accounting for that, except that I supposed each cycle of elevations would have some definite relation to the corresponding cycle of the equilibrium theory. Even in calculating the proportion of the solar and lunar tides from the phenomenon, I never put forward the results as of any weight, because they must be rendered insecure, or rather, futile, by the unknown amount of original retroposition. And even this original retroposition I gave only as a way of expressing laws of facts.

Since the laws of tides and height are the same as those of the equilibrium theory, and since this is the only result of Laplace's theory, and is obtained by him through a combination of insecure general reasoning with calculation, I still see no superiority in his theory, except in his having formed differential equations which we cannot integrate.

And so much for Tides.

I send you a report of a Syndicate, by which you will see that we are trying to give fixity to our examination in mathematics. The plan is mainly mine, and one which I have for some time been trying to engineer.

I shall be glad to have Capt. Bayfield's letter back again.

Always truly yours, W. WHEWELL.

T. L. *Feb.* 14, 1843.

MY DEAR JONES,

I dare say you will get some good suggestions from Willis if you put your case before him; but in the mean time I will tell you what occurs to me. In every machine, or rather in every mechanical work, you have three things to consider— moving *power*, trains of *mechanism* connecting the power with the work, *work* done. Willis begins his book with the example of a knife-grinder—moving power = muscular power of knife-grinder; mechanism = grinder's wheel; work = knife-grinding. It is plain that, *quoad* knife-grinding, the power is much increased by the mechanism, for the man would get on very ill without it. Now your question seems to be how much the value of the work done exceeds the value of the moving power. I reply, it exceeds it by the value of the mechanism, at least; otherwise people would not acquire the mechanism. The value of all the knife-grinding done in Britain must exceed the mere wages of the knife-grinders by at least the value of all the grinders' wheels. Now this appears to me to be general—the value of the work done must be the value of the moving power *plus* the value of the machinery. In the case you mention, a cotton mill, the value of its produce must equal the value of the moving power (coals and engine) *plus* the value of the mechanism (wheels, spindles, &c.). And the productive powers of nations must be compared by comparing the moving forces *plus* the whole quantity of working mechanism; for the latter would not exist if it did not add its own value to that of the moving force.

You may get some good lessons from Willis if you wish to pursue this subject into detail. The account of the knife-grinder's wheel as a type of all mechanism whatever in the first paragraphs of his book is excellent. He says truly, that there is always a *train* of mechanism between the moving force and the work done; and his book is the analysis and classification of such trains. In one of my last books—*The Mechanics of Engineering* — is a good chapter on the Measure of Moving Power, borrowed from the French Engineers and improved. Willis and I agreed to call the Measure of Moving Power "Labouring Force"—an expression which has considerable advantages. The French term is "*Travail.*"

You will see that we have petitioned for your Welsh Bishopricks. Will Peel yield? or will he defy Celtic wrath?

Our regards to Mrs Jones.

Ever yours, W. W.

T. L. *Ap.* 7, 1843.

MY DEAR JONES,

I suppose you got the letter I wrote yesterday, which contained an answer to one part of yours; but I may as well say over again that we shall be most glad to see you and Rutherford. We shall be at Audley End from Monday to Wednesday in Easter week, but at home the rest of the time. I shall be glad to hear what you say of John Mill's book when you have read it. I believe the part you mention is the best, for it is that which has long employed his thoughts; when he comes to Induction, he appears to me to write like a man whose knowledge is new (indeed he confesses that he had much of it from Herschel and me),—and not very well appropriated. For instance, a great number of his examples of scientific investigation are taken from Liebig's researches on physiological chemistry—*just published.* The most profound and sagacious physiologists and chemists cannot yet tell which of these will stand as real discoveries; still less can they put these new views into the true relation to the old without long thought and study. How then should Mill do it, to whom the whole subject is new? And then he takes the instance, which Herschel has made people wild about, of Wells's book on Dew. Tell Herschel he has something to answer for, in persuading people that they could so completely understand the process of discovery from a single example. Wells's place in the history of science is misunderstood, and the amount of his discoveries miscalculated, when he is detached from the general train of researches on heat and moisture. But, to speak of the part of Mill's book of which you speak, I agree with you that the Logic is fairly logical; also, that it is deadly dull. The Whateleian logicians are to me far more oppressive than the Aristotelians; far more immersed in verbal trifling and useless subtilty. When Mill comes to the question of the ground of mathematical demonstration, he certainly makes a stout fight; and his arguments are valuable as serving to shew how the true doctrine may best be put so as to be understood and accepted. * * * He is quite subjugated by one whom I think a very bad philosopher, Comte, of whom he constantly talks with a veneration which I could easily shew you is a most gross idolatry. I had written an article for my philosophy about Comte, but suppressed it, wishing to avoid unnecessary controversy. Of course, I shall not notice Mill's book till I come to a new edition, and probably then no otherwise than by altering, as I have said, the mode of presenting my doctrines. By the way, new editions of my History and Philosophy are marching upon me quite as fast as I wish. The History is very nearly sold off (1500), but the publisher recommends me to let it stay out of print a year or two, which suits well with my other plans. As soon as I am discharged of the V.C.ship I feel a great desire to rush upon Morals again. I think that next year I can both give a course of lectures and publish a system. At present nothing but business can go on.

I hope your Kentish move will be of use to Mrs Jones. Let me know when you come here.

Always yours, W. W.

TR. LODGE, *Ap.* 8, 1843.

MY DEAR HERSCHEL,

I send along with this a letter for Jones, who writes me word he is going to be your guest. Pray send it to him in London, if his visit to you fails. I suppose you have by this time lost sight of your comet. Do you get any new notions out of this fellow ? Is he self-luminous, or only a very diffuse planet? We did not see him well here ; and, for my part, not at all.

Jones will tell you of a new book by young Mill about the philosophy of science, suggested in a great degree by your book on the same subject and by mine. There is in new books of this kind a satisfaction in which both you and I may have a share. I 'mean that notions and expressions, which were new and strange when we began to write, are now familiarly referred to as part of the uncontested truth of the matter. Mill agrees with you more than with me in the parts where we differ, but he does not appear to me an ally to set much store by; for, though acute and able, he is ignorant of science and still entangled in the prejudices of a bad school.

I am grieved to hear from Jones that your household have been molested by colds. We hear such complaints on all sides of us, but we ourselves have escaped pretty well. The worst complaint I have is the Vice-chancellorship, and that is not a very bad disease, except for the old reason why Love is so—

"Love is like a dizziness
That won't let a poor man go about his business"

—at least not about any other business ; so I am longing for the end of it, that I may once more philosophize and moralize. Pray do not let Lady Herschel and the girls and boys forget me ; and my wife is equally unwilling to be forgotten by you.

Always truly yours, W. WHEWELL.

TRIN. LODGE, *June* 6, 1843.

My dear Forbes,

I was glad to see your handwriting again; for, though the Vice-chancellor's office added to all other employments has made me very slack in writing to my friends, it has by no means taken away my desire of hearing from them. I was glad, too, to have the news you give me about the form in which the results of your travels are to appear. I think you have a good prospect of being successful. I mean, so far as the form may affect your success. Most readers like personal narrative, even when the narrator has little to tell, which is not your case, including in the events of your travels the steps by which you were led to your views. I do not quite know how far it enters into your plan to tell this; though I saw a few sheets in Heath's hands, as he probably told you.

I believe I must agree with you that Hopkins's experiments do not prove much more than what was asserted before—that ice in contact with a conductive body is capable of yielding to minute moving forces. But though this always seemed to me very evident, it had, I think, been denied by some of your fighting glaciators; and the proof by experiments is a satisfaction always, and usually a means of obtaining measures and constants. I am so much a lover of experiments, though on such subjects experiments on a small scale may be very fallacious, that I have several times thought of the possibility of *illustrating*, at least, your views by experiment; I mean that which appears to me a very curious and striking principle—that a semi-fluid mass, creeping down an anfractuous valley, will have its fissures, the state of its surface, and, as I presume, the surfaces of equal pressure, in given positions with regard to the sides of the valley, while the particles advance. From your saying of your having produced your structural shells in viscid plaster, I think it not unlikely that you have done what I have spoken of, and a great deal more; but if you could exhibit the fact in a model of an Alpine valley, it would be very persuasive.

*   *   *   *

Most truly yours, W. Whewell.

TRINITY COLLEGE, CAMBRIDGE, *Sept.* 17, 1843.

MY DEAR QUETELET,

This note will, I hope, be delivered to you by my brother-in-law, Arthur Marshall. I know that you will be glad to increase the number of our common friends, and you will besides be interested in the object of his visit, since it tends to make our two countries more useful to each other in the way of manufactures. I do not doubt that you and Madame Quetelet will have the pleasure of seeing something of your friend Prince Albert during the visit which he is now making to your city. Pray give my kind regards to her, and tell her that I wish her that and much other gratification, and believe me always,

Yours most truly, W. WHEWELL.

TRINITY LODGE, *Sept.* 30, 1843.

MY DEAR FORBES,

I shall write you only a short epistle, for I have great doubts whether I have any chance of meeting you at Geneva, which is the only probable address of yours that I can hear of. I want you to consider whether you cannot take Cambridge in your way northwards, and let Mrs Whewell and me have the pleasure of becoming acquainted with Mrs Forbes, and shew her Cambridge, which your wife ought to see. We shall be here, and able to receive you at any time, so far as I at present know. I am reading your book, and am quite charmed with the life and movement which it possesses, as well as by the manner in which your narrative is all made to converge to your theory. Still I have one or two puzzles which, I have no doubt, a little talk with you would help me out of.

You will find us at last clear of workmen, and shall be able to pass from your bedroom to breakfast without traversing a moraine or having canvas flap over your head. You will excuse my being brief, if the communication is not reduced to zero by not reaching you and so being no communication at all—and believe me

Yours most truly, W. WHEWELL.

T. L., *Oct.* 6, 1843.

MY DEAR JONES,

I am grieved to hear that you are so worried and so out of order; and that we are not to see you and Mrs Jones. If you can run over here for two or three days at any time, pray do. I hope it will do you good, and it certainly will be good for me. I have been working steadily on with the historical part of my Morality, and have, I think, finished the requisite reading. It has led me to go over Hallam's books much more carefully than I had done, and I am glad to find that they have risen much in my estimation on closer inspection; for I like the man. They are, especially the English Constitution, a series of condensed, able, lawyer-like, or rather judge-like, discussions of all the principal constitutional questions which our history offers. He is an old whig, and at times rather a stern one; but then he is an old whig, not a new whig; and I see scarcely any points on which we, the constitutional conservatives, are called upon to differ with him. What a difference between his judicial gravity and fairness, and the clever sophistical advocacy of Macaulay! But then again what an advantage Macaulay has in his bright transparent style over Hallam's ill-wrought parenthetical sentences often amounting to riddles! I do not know how far I may be able to systematize the principles that enter into such discussions, but that is my next task; and, at any rate, taking the matter in this order I shall not fall into vague generalities or unforeseen dangerous conclusions. I appear to myself to be following exactly the course I did in the philosophy of the sciences; taking first historically what has been done and said on each subject, and then extracting out of that the philosophy which was in the minds of those who contributed anything to what was done.

I shall be very glad to see your new edition of Rent, and am happy to hear that you have reached that stage of your progress. I send you an examination paper, in which I have inserted, as in examining for Fellowships I usually do, a question on the subject. This time I got two or three answers from men who appeared to

have read your book.   I look forwards with great satisfaction to
the appearance of your other baby.   When shall we hear its first
squall ?

*   *   *   *

I agree with you that the love of natural scenery lasts undiminished better than most other pleasures; but I am not yet
weary of poetry, only rather fastidious, and that I suppose is what
you are.   I have, however, of late found a new way of enjoying
poetry by reading it to my wife every evening, which at any rate
prevents its putting me to sleep.   Kind regards to Mrs Jones.

Always yours, W. WHEWELL.

TRIN. LODGE, *Nov.* 23, 1843.

MY DEAR HARE,

I apply to you to know if you can give me any account of
your neighbour Bunsen.   I wrote to him above a week ago asking
him and Madame Bunsen to come here any time from the 2nd to
the 11th.   I thought he might like to fulfil his long talkt of
design of seeing Cambridge in term time.   I have got no answer,
and want to arrange my engagements, so I shall be obliged to you
if you can tell me where another letter will find him.

I have been reading Sterling's Strafford, as I suppose you
have.   There is a great deal of force and skill, and considerable
dramatic power—I mean power of making living persons.   What
I most miss in it is an English tone.   The philosophy, religion,
and polity, are not at all those of the time; nor those of English
statesmen and lawyers at any time.   He has omitted, too, many
of the most animated turns in Strafford's accusation and trial.
Perhaps this was done to avoid servilely copying history; but
Shakespeare would have been content to be servile.

The Worsleys are coming to dine with us alone to-day.
I wish you could come and make a fifth in these parties.   Whenever you can come, you will be most welcome.   I do not know
that we go away at all this winter.

Always truly yours, W. WHEWELL.

You were quite right in supposing that it was the structure, not the style, of your sermon, which reminded me of Schleiermacher.

MY DEAR HARE,

I do not know whether you will think me too persevering, but I really am desirous of discovering why I cannot get an answer from Bunsen. After I received your last, I wrote to him again (directing Carlton Terrace) repeating my invitation, and begging for a reply: but as yet I have received not a word from him. As I may at some future time wish to write to him again, and to obtain an answer, I should be glad if you could put me in a probable way of doing what appears to be so difficult.

I have been looking at some of the accounts of Strafford and his trial, and am rather scandalized at the violent injustice which Sterling has done him. The broad and offensive pantheism, which he puts in his mouth in the concluding speech, is utterly at variance with history and character; and so, according to all that I have read, is his connection with Lady Carlisle; indeed the discrepance between these inventions and Strafford's farewell letters to his son and his wife appears to me to be shocking.

      *   *   *   *

Always truly yours, W. WHEWELL.

TRINITY LODGE, *Feb.* 22nd, 1844.

MY DEAR DEAN, [Dr Peacock]

I am sorry that we should not agree about the desirableness of inviting the British Association here next year. I retain the opinion which I expressed at Manchester, that it is not desirable that the Association should go on repeating its sittings in its former haunts. I think such a course would make it an intolerable burthen to the places included in the cycle, and to those of their inhabitants who took a share in its proceedings. And in this way it would lose its most valuable office, which is to stir up the scientific zeal of the places which it visits. It would do immense good to such places as Durham (a University Town like ourselves), Leeds, Hull, Lincoln, Norwich, Portsmouth, Exeter, Gloucester, Shrewsbury; but none to us. If it cannot find new places to go to, a large share of its usefulness is at an end. As for the main object of raising funds, I do not think it reasonable that the University should be asked to lend itself to such a purpose. These are my objections as a friend of the Association, and they imply what my objections are upon the part of the University. I think it would be an intolerable burthen, not only in a pecuniary view, but in its interference with the time, quiet, and comfort of the members of the University. Many persons must take part in it who exceedingly dislike such tasks, or else it must be an entire failure. It must at any rate be a failure, compared with the last Cambridge meeting; for the novelty and enthusiasm are gone, and the burthen would be calmly calculated. I do not think that the University, or at least the Colleges, could be or would be protected against the burthensomeness of the occasion even in money concerns. I do not see how the visit to Cambridge can make so much difference in the probable existence of the Association. If it can find a place of meeting for 1846, why not for 1845?

In speaking of the burthensomeness of such an occurrence to the University and the Colleges, I have reason to believe that I express the feeling of members, both of our own and other Colleges, whose opinions would carry great weight. I have not obtained them in such a way as to implicate the Association in the mortification of a rejected application which, I think, ought to be avoided. I shall be glad if you will, so far as you can conveniently, and so far as the discussion in the Council makes it pertinent, let my views be known to the Members of the Council. Believe me, dear Dean,

Yours most truly, W. Whewell.

[The British Association met at Cambridge in 1845 under the presidency of Sir John Herschel: see Vol. I. p. 154.]

MY DEAR HARE,

*   *   *   *

I think you would laugh, though you would approve, if you were here to witness what we are going to set about to-morrow—a set of singing lessons under the direction of Mr Hullah —*we* being about forty Masters of Arts, who meet at my Lodge. Many of them want to learn, that they may be able to direct the teaching of singing in their parishes, which they have, or hope to have.

I find that Taylor's *Synonyms* is going to be reprinted. * * * He did something for the knowledge of German, in England, and might have done something for literature, if he had not been all expended on periodical review writing.

Have you copies of the translations of Göthe's Hexameter Epistles which you inserted in the *Athenæum* many years ago? I should like much to see them again. I ought to have them, but I have often sought them in vain. Adieu.

Always truly yours, W. WHEWELL.

TRIN. LODGE, *March* 7, 1844.

MY DEAR HARE,

I return you your hexameters and pentameters which I am glad to have seen again. It appears to me that all we, who are set upon writing hexameters by the Germans, make the same mistake, of constructing lines with such endings as this of yours,

Turn impatiently over the leaves and then scratching his *own pen.*

You may say that *own pen* is a very good spondee. True: too good a spondee for English verse. The English verse will run very well with dactyls and trochees, but abhors spondees. I am sure that, if we hexametrists had avoided these dimonosyllabic endings, we should have been considered as English versifiers, like our neighbours, and should not have been hooted at as perverse and irrational innovators. Still I believe the chance of making hexameters in England liked by the general body of readers, to the same extent as they are in Germany, depends very much upon a poem of considerable bulk, attractive in its subject and treatment, being written in them. As Southey says, if you could march a body of 10000 or 20000 hexameters into the country, you might succeed, but your detachments of a few at a time are cut off as fast as they appear. His "Vision" was, however, a most unhappy attempt to verify his project: besides which, he allows himself professedly licenses, which entirely destroy the character of the verse; as a short syllable at the beginning of a line. Herschel has written (and printed for his friends) a very excellent translation (with one or two mistakes, according to me) of Schiller's "Walk." Hawtrey has also translated some of Göthe's Hexameters and Pentameters. Yours are almost free from the harshness which I have noticed in the Hexameter Epistles.

Will you tell me, when you have time, what form the memorial of Dr Arnold has taken? Is it an Historical Scholarship at Oxford? I seem to recollect to have heard that it was so. Some friends of mine wish to know.

I wonder that the Etymological Society have not yet included Bunsen among their numbers. I thought he had been one of the original members. I have never taken any part in their proceedings since the day we were there together. But I have no doubt they would jump at the notion of having Bunsen one of them. They do not appear to me as yet—judging from their Proceedings, which they print and send me—to have made any great way in the philosophy of languages, or in the knowledge of what has been done.

*　　*　　*　　*

We Hullahize, as you euphoniously term it, very prosperously: 40 Masters of Arts, 40 Undergraduates, 20 University ladies, 200 Town gentlemen and ladies, 200 of the people, five strong classes.

Always truly yours, W. WHEWELL.

MY DEAR FORBES,

I intended to have written to you before I received your Roman letter, but I did not meet with any of your friends who could tell me exactly where you were. And, though I have no doubt that you are too systematic a traveller to leave your track unknown to the post office, you know perhaps how much it dulls the pleasure of writing to have your correspondent presented to your imagination as vaguely diffused over all Italy. I rejoice very much to hear that your Italian sojourn has been of so much service to your health. And I may say the same of Mrs Forbes, though I grieve to hear she had so much occasion for restorative influences. As you will suppose, I am too old a traveller to be surprised at any modifications your Italian enthusiasm may have undergone since your earlier visit. The absence of surprise and novelty necessarily do much to make late impressions of objects duller than the first — *the proofs before the letters.* But besides this, there is, I think, another thing which very much diminishes the interest of sights in after life—at least for active people. They become so much occupied with the interest of actions, external or mental, that they care comparatively little for spectacles which they cannot work into their plans and speculations. I have not yet seen your last letter on glaciers, but I am amused with your finding that a stream of lava and of ice are very like each other. I still look forwards to having a discussion with you about your ribboned structure theory. Perhaps I may put my difficulty briefly thus. The minute fissures in the ice stream must be the solutions of continuity necessitated by the different velocity of the different parts. But these must be in the general direction of cylindroidal surfaces having the axis of the stream for their axis, because the central and upper parts move quickest. Therefore the ribboned structure ought to have this form, and not a frontal dip. I think your answer to this objection will probably bring us to a better understanding on the subject. There can be no doubt that you will have all our geologists on your side as soon as they have time to get over their preconceptions. Sedgwick has been reading you lately with great admiration. We have at present a struggle going on here whether the British Association shall come here in 1845. I am against it. To have that hubbub returning upon us in a cycle every few years would be intolerable. I do not know how your Edinburgh people would like it. Oxford has refused to have any more to do with the Association. Indeed I have no doubt that the majority of the Senate here are against the plan. But Sedgwick and Peacock and a few young men are for it, and perhaps those who oppose may not like to vote against them.

* * * *

Always truly yours, W. WHEWELL.

* * * *

TRIN. LODGE, *Mar.* 12, 1845.

MY DEAR HERSCHEL,

I think I must make an effort to rescue my little book[1] from your contempt. I see that you look upon it as a kind of *Elegant Extracts* of my Philosophy, published in compliance with the prevalent love of little books that pretend to contain the essence of large ones, and daintily dressed for dainty people. This is not exactly the meaning of it. I do not know if you have seen or heard (from your letter I should think not) of a book called *Vestiges of Creation.* It is anonymous, has been circulated with great zeal and mystery, and is much read and talked of. It is considered to have a materialist tendency, and so, to be mischievous. I have been much urged to answer it, which I have refused to do, except by extracting passages from my previous books; and indeed, in the passages which I have extracted, all the arguments of the *Vestiges* are discussed and answered. So you see it is not a wanton selection of elegant extracts, but a compulsory selection of theological extracts, and by no means shows what I like best, or think most essential, in the other books, or is likely to be so taken by those who read it. As for the

---

[1] *Indications of the Creator :* see Vol. I. page 155.

binding, you are not to suppose that I treat all the world to
such elegancies. They are only for people like Lady Herschel,
for whom nothing is too good. You will soon have evidence,
quite enough I am afraid, that I am not dwindled down to a
writer of little books. I shall hope to publish, in two or three
months, a book as big as any I have yet published. It is a system
of morals.

A word about glaciers. I do not quite assent to your *crush
room* theory. The fragments are not permanent ones in the
glaciers. Old cracks are soldered; new ones made; this perpetually and throughout. To make your *crush room* resemble
this, you must suppose that gentlemen are constantly splitting
in two, one half sticking to one lady, and the other to another,
and so on. Now *I* should call this a plastic mass, and so would
you, if the gentlemen were tiny elves. Add to this the center
stream of these elves flowing out into the Haymarket faster
than the sides, and let them bear traces (in their breasts of
course) of the various rendings they have suffered, and you have
a Forbesian glacier with its structural bands.

*   *   *   *

Always truly yours, W. WHEWELL.

TRIN. LODGE, *July* 18, 1845.

MY DEAR JONES,

I am glad to hear news of your book and its progress
towards a printed existence. We intend to leave this place on
Monday, and shall be at 41, Upper Grosvenor Street for a week or
a fortnight, so you must send me your sheets there. I shall
look at them with great interest, and will tell you any thoughts
which occur to me about your way of meeting your subject:
but my interest in your sheets does not depend so much upon
any hope I have of detecting any errors for you, as from my
desire to see your views in their systematic shape. I have no
doubt of their making their way to general acceptance; if not
in this generation, at least in the next. I believe that the pro-
mulgators of long pondered truths ought to be prepared to wait

a while for the gratitude of the world, for they cannot mix themselves with popular and periodical literature, or with London coteries, in such a way as to find a set of ready made admirers when they publish. But this is not to be regretted, for truths of any broad philosophical kind do not admit of transmission through admirers so made. I have been much amused, in this point of view, with the success of the *Vestiges of Creation.* No really philosophical book could have had such success : and the very unphilosophical character of the thing made it excessively hard for a philosophical man to answer it, and still more to get a hearing if he did. How do you like Sedgwick's Article in the Edinburgh ? To me the material appears excellent, but the workmanship bad, and I doubt if it will do its work. We hardly know where we shall be for the next two months. I am bent upon seeing some of the Cathedrals, Canterbury first, and Winchester last in the beginning of September, when the Antiquarians meet there. Adieu.

Always yours, W. WHEWELL.

HASTINGS, *Aug.* 19, 1845.

MY DEAR JONES,

I wish I had told you to write to me, for I want much to know how you are, as well as what you think of my notes on your Lectures. I intended to say so, and to tell you that letters addressed to Trinity Lodge will always be forwarded to me. We are got so far in your Sussex, having spent ten days at Walmer, where the Worsleys are. This place is, to-day, almost too marine, for the sea is quite noisy close to our windows, and the crowds of people hurrying past between us and the shore are as unquiet and as noisy as the sea. But this is on account of some regatta, and is not, I suppose, the usual habit of the place.

Both the *Athenæum* and the *Spectator* have taken up my *Morality* somewhat hostilely. I suppose this was to be expected; for they are both of them strong movement journals, and my speculations are, upon the whole, conservative. But I was not quite prepared for the amount of unfairness, and the evidence that neither critic had read the book so as to judge of it in any connected manner. I do not see that they have noticed anything which I could have put in any better form: and there are one or two blots in the book which they have not hit. I want to see a reasonable estimate of it as a system; but I suppose this is too much to expect from the present practice of our criticism.

*   *   *   *

Always truly yours, W. Whewell.

Hastings, *Aug.* 20, 1845.

My dear Herschel,

We are glad to find by Lady Herschel's letter received to-day, that your boy is going on well. I hope you will soon be freed from your anxiety about the rest. We got here on Monday, and were glad we had done so when yesterday's storm beat against our windows. Through the whole tempest we had three cutters sailing a race in the sea before us; but one of them was twice baffled before she could round the flag ship.

I did not send you the reprints of my letters on glaciers: I suppose Forbes did. I think you and I are coming near each other, except that I should call substances *viscous* when they are in the condition you describe; the parts being constantly detached and forming new attachments of equal value. That they do this by starts, as in snow (if it be so), makes no difference in the mechanical results of the case.

I should like much to have your opinion on a matter connected with the cultivation of mathematics in England. Cambridge is, and I hope will continue, the principal school of English mathematics; but the course of things there is not what one could wish it to be. I think this arises from our most active students being encouraged to study rather the last supposed improvements, contained in memoirs, journals, and pamphlets, than the standard works of mathematical literature. I am disposed to recommend, as the standard of our highest mathematics, a list of *classical* works, to be announced by authority from time to time; say at present,

Newton's *Principia*, Euler's *Mechanica*, and *Calculus Integralis*, Lagrange's *Théorie des Fonctions* and *Mécanique Analytique*, Laplace's *Mécanique Céleste*, Monge's *Application de l'Algèbre à la Geométrie*, Airy's *Tracts*. I exclude in this list works of great historical value, as those of the Bernoullis and Maclaurin, and on the other hand I exclude the most recent memoirs and the like, not admitting them into our scheme till some time has elapsed, and the mathematical world has given them its sanction. Such exclusions are, I think, necessary, in order that our best mathematical men may really know the great works of the science rather than employ their time and activity in hunting out the last novelty. (Of course I include in the examinations problems requiring original investigations.) The obvious objection to such a list as I have proposed is, that it narrows and fetters the examination, and that our best mathematicians ought to be acquainted with the *whole* of mathematics. To this I should reply, that men at the time of taking their degree have not in general had time to acquaint themselves with the whole of mathematics, and will be better taught if they have a limited task, provided it be large enough, as I think the above list is. The real question is, whether we shall take our selection from the general judgment of the mathematical world, or from certain conventions and ephemeral works of our own. This would be my answer, but I should like to know how it would strike you, and whether you incline in favour of an unlimited scheme for our highest men.

I have no direct purpose of obtaining the authority of the University for any scheme which I may propose, but the evils of our present condition are much complained of; and a scheme which had good reason in its favour would have a fair chance of producing its effect when published. I am going to publish a book about University matters, and this among the rest.

We stay here till Friday, when we go to Hare's at Hurstmonceux. Our kindest regards to Lady Herschel. God bless you.

Yours always, W. WHEWELL.

We have been to St Leonard's, and we have seen the smaller bodies of your system, who look very nice and globular.

My dear Hare,

It was kind of you to waft to me—not an "evil dream" like that which Zeus sent to the ships of the Greeks—but a very pleasant dream of our living near you, and bringing the Worsleys with us—so may it be at some time or other! and if not for years, at least for months. But I think this good fortune cannot be brought to us by our taking Hurstmonceux Place, which, with its rather large and naked rooms, chills even my wife's zeal for furnishing, and does not correspond to my ideal villa, which is a place where I may think, and write, and dose away a few summer months almost in solitude. But still I am glad and pleased at your combining my hopes with your neighbourhood. Perhaps I shall next summer ramble in some other part of England in the like manner: but I shall not find any place which has the charm of your "good dream."

Among my small projects is still one of which I have already spoken to you. It is to get a fair hearing for English Hexameters. Will you help me? I will tell you how. I should like to publish (without *printed* names I think) hexameter poems by several persons respected by the world—your translations of Göthe—Herschel's translation of Schiller's *Spaziergang* (a beautiful performance)—my own *Herman and Dorothea.* I have one or two other Hexametrists in my thoughts, but I will not speak of them till I know what you think. The measure is as suitable for our language as for the German, and would be a gain to our versification. You shall have no trouble except that of sending me your Göthe's Epistles. How say you to this scheme?

I have just been reading with great delight Bunsen's *Church of the Future.* I hope it is to be translated. The appearance of such a work at present would, in my opinion, be a great blessing to us Englishmen. It is very true that it would be hard to translate; but so is all philosophical German, for obvious reasons. Pray help to persuade Bunsen to give us an English edition of the work. As a speculator about such matters, which indeed we all are, I am quite charmed to have such views put in circulation.

* * * *

Always yours, W. WHEWELL.

TRIN. LODGE, *Dec.* 27, 1845.

MY DEAR FORBES,

I have been longer than usual in answering your last letter; but here come Christmas and New Year's day, bringing, among other blessings, reasons for ending such intervals of silence, with good wishes for a continuance of blessings. There are few persons whom I include in such wishes more constantly than I do yourself, and now of course I include in them Mrs Forbes. Your last letter brought me the intelligence that she had presented you with a second daughter. I hope both she and the baby are quite prosperous.

I have been reading the article in the *Quarterly* on the *Cosmos* (which is certainly the right English spelling). I am much pleased with it. It appears to me to contain a very good abstract of some of the most striking parts of the book, and, what I value much more, judgments pronounced on several of the controverted points, which suggest themselves, in a truly judicial spirit. I cannot say the book disappointed me, but I do not very highly admire it. Its value is great as a large and lively summary of many branches of science according to recent views, but it pretends to be something more—one does not precisely see *what;* and it is the more difficult to discover what its aim is, because it does not occupy the mind in any way except as a collection of facts. I think you have formed nearly the same judgment. I am much obliged to you for mentioning my book in the review. The fact is Humboldt has mentioned my book in a note, but in such a way as to shew, I think, that he had not read it, if even he had opened it. He joins me with Ampère, and with a certain American Professor Park, who has written a Pantology, blaming us, if I understand him, for too learned classifications of Sciences; but I do not well make out his drift in this part.

By the way, my books, the *History* and the *Philosophy*, are quite out of print, and I am going to reprint them. I shall of course correct errors, and make additions where any thing belonging to any of my *Epochs* has been done. I have not yet quite determined in what way I shall introduce these corrections and additions; but I shall be very glad of any suggestions from you as to defects which you have observed in the book. Some of your remarks I recollect, for instance, the imperfect justice done to Playfair's illustrations, and shall profit by them.

Faraday has on hand something very great, in his eyes surpassing the connection between electricity and polarization, which I think you refer to in your expressions about him. But I believe he does not wish his hopes to be talked of till his result comes out. According to what he wrote to me lately, this is not likely to be long.

My book which I sent you circulates rapidly, and, I suppose, will produce some effect here, but that must be slowly. However, we have already in progress measures in some degree founded upon it. In another term perhaps they may reach the form of a grace.

I am glad to see in the *Quarterly Review* an article tending to bring the episcopal churches of England and Scotland into closer union. Do you know whose writing it is? I am afraid Bishop Terrot must have been much disturbed by the controversy to which it refers.

I have not yet congratulated you upon the happy termination of the business of your pension. I am very glad your merits have been recognized in that public manner; glad for the credit of the ministry and the country, as well as on your own account.

*   *   *   *

With Mrs Whewell's kind regards and good wishes, I am, my dear Forbes,

Yours most truly, W. WHEWELL.

TRIN. LODGE, CAMBRIDGE, *Feb.* 23, 1846.

MY DEAR FORBES,

The author of the Article which you inquire about is Peacock; so you see your judgment of my style was right. I suppose you have been discussing the Cavendish and Watt question, and of course have seen Harcourt's decisive article in the *Philosophical Magazine*. I wish it were in some channel of more general diffusion, for the correction will reach few of those whom the misrepresentation has reached. I do not know what connection between Cavendishes and Harcourts there can be. I believe the Peerage shews none.

I agree with you about our Examinations. I have been complaining in council, as well as in print, of the neglect of Newton. We have some measures on the anvil which are intended to remedy this, among other evils; but we get on slowly with such work.

Pray what do you Scotch Episcopalians say of a Trinity College, which is establishing, I think, near Perth? I was applied to once or twice about it, and a little while ago the Bishop of Edinburgh wrote about a person to place at the head of it. If it is a good design, I hope it will prosper : but your bishops a little perplex us with their *very* high church notions.

I suppose you have the same marvellously vernal weather that we have here. Our rose trees are in leaf, and appear to have no suspicion that it is only February : but I am much afraid that March will come upon them like a harsh schoolmaster who finds the boys out of school before their lessons are ready.

Our kind regards to Mrs Forbes, and believe me, dear Forbes,

Yours most truly, W. WHEWELL.

TRIN. LODGE, CAMBRIDGE, *Ap.* 4, 1846.

MY DEAR HERSCHEL,

I have a project, in which if you will give me your countenance (which will cost you no trouble), I am persuaded we shall benefit the world. I have always been of opinion that English hexameters have not yet had justice done them. I want to print a collection which may win them some favour; and I want to insert in it your *Walk*, which appears to me superior in several of its parts to any example we have, and throughout excellent. I shall reprint my *Herman and Dorothea*: Hare will give me some translations of Göthe's *Epistles* made long ago: perhaps Hawtrey of Eton, who is an accomplished hexametrist, will also let me print some of his performances. I have not yet asked him: but, if you will join me, I shall hope to carry all before me. I do not contemplate printing the names, though of course the veil of *anonymy* will be a very loose one. And if you have any other hexameters and will let me have them, it will be better still. I see by indications in several quarters that there is a hexameter spirit alive; and I do not see why we should not be as prosperous hexametrists as the Germans.

I hope you are in better health and comfort than you were when I saw you. When are your southern stars to dawn on our eyes? Have you seen or repeated Faraday's experiments; and can you get any theoretical hold of them? We are still, it seems to me, thrown again and again on the same difficulty—the nature of the crystalline structure of bodies. I wish I could imagine to myself a quartz crystal.

I hope Lady Herschel, and all the rising generation, are well and happy. My kindest remembrances, in which my wife joins. I hope Lady H. will give her approval to my hexameter scheme.

Always truly yours, W. WHEWELL.

TRIN. LODGE, CAMBRIDGE, *May* 5, 1846.

MY DEAR FORBES,

It would have been very pleasant to have had your imagination realized, and to have fallen in with you in our mountain region during this summery May; but that is not to be at present. We had intended to be, at this time, on the coast of Suffolk, in a certain cottage-mansion[1], nestled in a wooded hollow of the earth-cliff close to the sea; for such a house we

---

[1] Cliff Cottage, Lowestoft: see Vol. I. page 154.

have taken, and hope to visit often by the help of the railroad. But Mrs Whewell has had an attack of serious illness, which has prevented this; and though she is now, I am thankful to say, recovering, this scheme is for the present at an end. I cannot of my own knowledge help you much in your glacier researches in the lake districts. James Marshall went in pursuit of that game, I think, in company with Buckland (and, I doubt, Sedgwick), and believe that he saw the traces very distinctly. I believe your host, Captain Wauchope, has often been on the hills with James Marshall; and, therefore, he can perhaps tell you the localities where the sight is to be seen. I recollect only one place, on the side of Coniston Old Man, going up to the mine; and there, certainly, the appearances are plausible enough, and I should think worth looking at by one who is familiar with glacier phenomena, which I am not. When you go to Keswick, call upon our brother and sister, Myers and Mrs Myers. He is at the parsonage close to the pretty church, and will be glad to see you, if he has not already been made known to you, which perhaps he has. I think he is the only member of our clan at present residing among the lakes. Mrs Marshall is in London.

I agree with you as to the overcharged importance of Faraday's view of his recent discoveries, *in their present form*. But if we had any one who could theorize well about them— I do not say, rightly (that is too much to expect for a while), but give us good theories which take up many of the facts, and of course help us to better—in that case the steps he has made might be very important. The difficulty is, as I conceive, to imagine any machinery in a crystal, and still more in a fluid, by which circular polarization shall be produced. To do this requires a clear mathematical head, and a complete mastery of all that has hitherto been done in mathematical optics. I am grieved not to see in any of our mathematicians any symptom of aptitude for such speculations. The worst attempt they could make, on this subject, would be of far more value than mathematical investigations in subjects quite unfit for mathematical handling, like geology.

I believe Lyell returns in June. I do not know whether he has yet seen my admonitions[1] to him on the very absurd part of his Travels in which he has chosen to attack me, or whether they are to form part of his welcome home. I have barely seen, not read at all, Murchison's Geology. These large books repel me. And, indeed, I do not know what I shall do with the subject, when I arrive at it in my second edition of my *History*. I think I have already told you that I am employed in preparing a second edition for the press, and that I shall be obliged to you for any corrections of errors which I have committed.

I regretted very much that we were in London when Lord Mackenzie and his lady were here. Several of our friends saw them and were much pleased with them, so that I hope they enjoyed their visit here. Mrs Whewell joins me in kind regards to Mrs Forbes, and I am always,

Yours most truly, W. WHEWELL.

TRIN. COLL. CAMBRIDGE, *June* 5, 1846.

MY DEAR QUETELET,

*   *   *   *

I have especially to thank you for a discourse on the subject of Stevinus, whom I am glad to see honoured as he deserves by his countrymen. I am now employed in editing again my *History of the Inductive Sciences,* and I shall use the information about Stevinus which you have given me. If you can tell me anything more about him or about any others of your Belgian men of science, I shall be much obliged by your assistance. I shall not make very large additions to my *History,* but if there is anything really important, which I have omitted, I shall be glad to supply the defect. I have not been able of late to give much time to matters of science, but I have not lost my love of the subject. One of the most interesting steps which has been made lately in this country is Faraday's discovery of the influence of magnetism upon light through the

[1] See Vol. I. page 161.

medium of transparent bodies. Perhaps you have seen Airy's article in the *Philosophical Magazine* for June, respecting the mode in which this property of light will affect the equations for the vibrations of the medium.

*   *   *   *

W. WHEWELL.

CLIFF COTTAGE, LOWESTOFT, *July* 3, 1846.

MY DEAR HERSCHEL,

I am greatly obliged by your information about double stars, and I will do my best to use it for my *History* in a proper historical spirit.

I am also very much obliged by your remarks on my *Morality*; for I am glad to know how the system strikes any candid thinker, and especially you. I will make a few remarks in reply.

I make no pretensions to construct a system which shall not *coincide with* expediency in its results; but I say that I cannot construct a system *upon* expediency, because expediency is too vague a term, and does not imply anything ultimate or fundamental. What is expedient must be expedient *for* some end; and therefore a system which professes to be founded upon expediency must really be founded upon that ulterior end. See *Morality*, 555.

If you say that this end is the general happiness of mankind, I say that happiness also is indefinite, and means *only* an ultimate end. To construct a system upon this basis, we must define, measure, or analyse happiness. Those who have founded systems of morality upon happiness *have* pretended to define happiness, and have, in my opinion, failed in doing this so as to supply any solid or coherent ground of a system. *Morality*, 550.

When I have taken the general sentiments of mankind, and the Rules and Laws they have established, as guides or suggestions for Moral Rules, I have *not* neglected the *motives* and *intentions* with which such Rules and Laws have been established. I have said that Rights must be established, in order that there may be at least a moderate equilibrium among man's conflicting desires; and I have attempted to shew *what* Rights must exist, by taking into account man's most general desires (*Morality*, 65, &c.). I have said that *Moral* Rules are determined by this consideration; that they must respect Rights, and must affect the sentiments and intentions, as well as the actions.

This is the whole of my system. As to plan, I cannot find any way of determining Moral Rules by considering them as means of getting some external object. I do find the means of determining such Rules by looking at the constitution of man; his Desires, his Reason which makes it possible for him to act by Rules, and by self-restraining, self-educating Rules. I found my Morality, not upon something which man is to get, but on something which he is to *be*. He is to be truly a *man*: he is to conform to Rules; to Rules which recognize a common Humanity in himself and others. I found my *Morality* upon this; that man must govern himself by *moral* Rules; and that this is not a vicious circular structure, I consider the result proves. We must conceive that man *is* a moral being, and then try to see *how* he can be so.

I do not reject *experience*; for it is by experience that we become acquainted with man's constitution, his desires, the relation of his reason to those, and the effect of rules upon his moral constitution. Experience enables me to understand man's constitution more and more clearly, so that I can see how he may be, and may be made, moral. Experience in all things operates to clear our ideas, and so, in morality.

The difference between me and those who pretend to set their systems by happiness, or by expediency, is not that we look at different ends of the telescope, but that they have given rules for putting the telescope *in focus*, which appear to me useless and necessarily ineffective. *My* way is to adjust the telescope so that I can see clearly through it. Or, if you like it better, you may take the image of the Rainbow which I have used in my Lectures, p. 135. (I think I sent you the book—in p. 152, you will find nearly what I have now said expressed in other ways.)

If you look at Articles 65, 66, 67, you will see that my system recognizes, as much as any, the way in which the necessity of Rules depends upon the pain, fear, and confusion which would arise from the want of them: and in Art. 68 I have stated why I do not think this is properly expressed by saying that *mutual Fear* is the foundation of Morality. And in Articles 536 to 555, I have given my reasons why none of the words *Pleasure, Interest, Happiness, Utility, Expedience,* seem to me to describe rightly the basis of Morality; also in my 6th Lecture. But I nowhere deny that systems based on these grounds, do, by the aid of *natural.assumptions,* or by compensation of errors, come to the same result as my system.

Do not imagine from this that I see no value or merit in any other system than my own. But I have tried to make a system which should have the merit of avoiding assumptions and errors which I saw to be such.

W. Whewell.

Trin. Lodge, 1 *Oct.* 1846.

My dear Forbes,

I send you, as I said I should, the Notes to the next of my Books which comes in the order of printing, though I do not know that there is anything which you have especially turned your attention to in this part. The next Book, about Magnetism, is a more special subject of yours.

I omitted to say in writing before that I have, like you, been led to a high estimate of Lambert's merits in giving to modern physics many of its most important features and views. I have been trying to get his books, for I do not think we do him justice in judging him from papers in Transactions. I believe I shall have a better opportunity of speaking of him in the *Philosophy,* when I have to discuss the modes of measuring Secondary Qualities, than I could easily make in the History.

I have now got to prepare the Book about Chemistry for

22—2

the press. Have you any clear opinion about Liebig's merits, and about Dumas's "types," and "substitutions"? I am not called upon to speak of either, but if I can see my way perhaps I may say a word on those subjects.

Yours always, W. WHEWELL.

TRIN. LODGE, CAMBRIDGE, *Dec.* 20, 1846.

MY DEAR MURCHISON,

I do not doubt that I shall have to modify the statement I have made about the relation of *Cambrian* and *Silurian* when I come to another edition, and I am not at all prepared to dissent from your statements, or to resist your arguments. But I have in all parts of my History avoided as much as possible taking a part in existing controversies; and have in several instances avoided to speak of the most recent steps in science on that account. The complete identification of Lower Silurian and Cambrian is an instance of this. I could not, however much I had been convinced myself, have made up my mind to declare Sedgwick's "Cambrian" equal to *zero* as a separate formation. If it be so, it is a very hard case, after all his laborious researches, and his courage and magnanimity in always taking the most stubborn portions of geological toil. I never thought him wise in doing this; but I cannot be one of the first to proclaim the main action of his geological life a failure. You will not suffer by my using still the language you would all of you have used a little while ago, and I am content to be a few months behind the time of day, if I can so avoid giving pain to scientific workmen. I hope you are too wise and longsighted to receive pain from what I have written, or left unwritten. You have better roads to fame, even if my book had all the authority I could ever desire for it.

*   *   *   *

I have tried to weed all offence out of the book, and I am persuaded you will not be the man to insist upon it that I have failed. I send back your paper, and am, dear Murchison,

Yours most truly, W. WHEWELL.

CLIFF COTTAGE, LOWESTOFT, *Aug.* 12, 1847.

MY DEAR JONES,

I have been reading Twiss's Lecture on the History of Political Economy, and have been wishing repeatedly that you would write a review of it for Empson. Surely he would be very glad of such a review, and it would be little trouble to you who have all the literature of the subject in your head. You might find in that way an easy opportunity of giving your opinion on the leading points of the subject, and might, I should think, make an amusing article. He has, so far as I can judge, taken hold of the main points of the history very well; though I think in some cases he has let the points slip too easily. For instance, he appears to me to have muddled away Adam Smith's great distinction of productive and unproductive labour, and to have given his assent too easily to some of Say's extreme opinions, of which the fallacy appears so very obvious. When they say that there can be no excess of production which is not capable of being regarded as a defect of responsive production, they forget how difficult it is to devise such production. To find out such is an act of discovery, and discoveries will not come at call, whatever they may say. If you are to have articles suited to men's desires, you must begin by considering what men's desires are, and how limited. Are not people perpetually inventing new articles of production which do not *take?* and who shall say that at any given moment this difficulty can be overcome? If I recollect, Malthus has well disposed of this fallacy.

We are enjoying our sea breezes in this glorious weather. Peacock has just left us. Kindest regards to Mrs Jones.

Always truly yours, W. WHEWELL.

TRIN. LODGE, CAMBRIDGE, *Dec.* 6, 1847.

MY DEAR FORBES,

It is some time since letters passed between us, as at some times happens; and now, as is not unfrequent also, I break the silence by sending you a book along with a letter. I have printed a few verses, for copies of some of which I was occasionally asked; and as they are of various kinds, grave and jocose, original and translated, I have several chances of pleasing you, though all of them together perhaps do not amount to more than a small fraction. However, be that as it may, I know you will receive my jingle kindly.

I want to know if you can give me any advice as an experienced visitor of Paris in recent times. It is very long since I was there, and I am intending to take Mrs Whewell there to spend a few weeks this Christmas vacation. I think most of my scientific friends, on whom I formerly depended for access to good circles, have disappeared. You have been there so recently, under circumstances in some respects similar, that it is likely you can give me information on this point, and perhaps means of using it. Cuvier and Poisson and the like must make great blanks in Parisian scientific society, but no doubt these blanks, so far as society is concerned, are in a great degree filled up.

*   *   *   *

If you can give me any hints of the present state of such matters at Paris, or tell me of any useful acquaintances whom I might otherwise miss, you will add to the many offices of friendship you have done me.

Is there anybody at Paris who has taken up your glacier theory? I am still discontented with the want of justice towards you which *our* geologues have shewn, as I sometimes tell them. Do you make any use of the Malta mud-slide[1] of which I sent you an account?

I have been much amused by the controversy between *your* Sir W. Hamilton and De Morgan about the *quantification of the predicate*; but I cannot help condemning Sir W. very strongly. It was a great defect in literary morality to make a charge of plagiarism in vehement and disparaging terms, on little foundation, and, as appeared afterwards, on none, and with scarcely a possibility of its being founded: and it was a still greater offence,

---

[1] This account was sent to Dr Whewell by Mr Milward, and afterwards communicated to the *British Association* at Swansea: see page 70 of the *Transactions of the Sections* in the Report for 1848.

when he was compelled to retract the charge, to do so in words, but to go on repeating it in substance by insinuations. It is curious to see our friends, the analytical mathematicians, seizing on one subject after another—glaciers, faults of strata, earthquakes, and now syllogisms. They have at least as good a right to the last as to the others.

*　　*　　*　　*

Believe me, my dear Forbes,

Yours most truly, W. WHEWELL.

TRIN. LODGE, *Mar.* 6, 1848.

MY DEAR FORBES,

I ought sooner to have written to you, if it were only to thank you for a very kind and useful letter, which you sent me before I went to Paris. Your notices of the men of science there were valuable as mementos, and I found most of them much as you described them. Elie de Beaumont called upon me, but the persons of whom I saw most were Arago, Humboldt, and Leverrier. Arago was very attentive and obliging; and even then seemed not without a presentiment that his labours at the observatory might soon be interrupted or terminated. His brother Etienne had then just written a play which was performing at the Français, and Arago gave us tickets to go and see it, which I did—my wife was kept away by indisposition. The play is *Les Aristocraties*—the three Aristocracies—that of the old *régime*, of the empire, and of the manufacturing world—and the last was represented as superseding the others: a tone which prevailed in many of the stage representations which I saw. Etienne Arago is now Postmaster of the *Republic*, and, I suppose, will no longer tolerate even a manufacturing aristocracy.

*　　*　　*　　*

I saw less of the men of science than I had done on former occasions, but I can hardly say that this was entirely owing to any backwardness of theirs, for we were a good deal engaged with other friends at our end of the town. We were at the Hôtel

Vantini, which is not far from the one where you were; having been recommended to go thither by the Peacocks. I think that probably yours would have been better, though ours was good. Certainly yours would have supplied a better view of the Revolution, if we had staid to see it. In that case we should at any rate have had enough of it, for there were several barricades close to the door of the hotel.

What are we to say of the Revolution? I am disgusted with the ignoble fall of the monarchy, and afraid that nothing but mischief can be looked for from a democracy so extravagantly democratic as they are going to try to construct. When I was there, all persons, even Guizot's friends, were blaming the stiffness of the way in which his master and he were dealing with reform; but nobody could conceive that the government of Louis Philippe was so entirely without any roots in the soil. I suppose, in fact, it is a double revolution; the Government overthrown by the Reformers, and the Liberals by the Labourers. I only hope that they will work out within their own frontiers the problem which they have set themselves, and not throw Europe into a flame. If they do, we have the world thrown back into the condition in which it was when we were born, and have been grinding on for so many years to no purpose. And then we shall have the continent closed to us;—if indeed we escape having England opened to them, which will no doubt be one of the attempts in case of a war:—though I must say I never heard anyone in Paris dwell upon such a project, or speak of the alarm, which has been expressed in England, as otherwise than absurd.

Are you doing anything in the way of science? I have been once more turning to the Tides, and trying to put the Tides of the Pacific in order; but there is really no more to be done without a Tide expedition: the greater part of the recent observations are of no value. Some of our young men have been publishing a monograph of the Aurora Borealis of last September, which I would send you if I thought it had not yet reached you.

I hope Mrs Forbes is well. Mrs Whewell joins me in kindest regards, and I am always, my dear Forbes,

Yours most truly, W. WHEWELL.

MY DEAR JONES,

I hope you prosper, or have prospered, at Bournemouth. When you reported to me that you were going thither, I did not even know on what part of the coast it was; but I have since had it described to me as a very pretty and pleasant place in Dorsetshire [Hampshire]. I hear too, through Empson, that you flourish there, and am glad to hear it.

Of course you will soon look at Mill's book on *Political Economy*. It is full of interesting discussions on all the great social and economical questions of the day, and there are arguments and views extremely well put throughout. Nor have I found anything which I quarrel with (although I think he will draw to himself much odium by some of his ultra-Malthusian doctrines)—except the injustice towards you of which I think he is guilty. He says (Vol. I. p. 295) that your book is "a copious repertory of valuable facts," &c. Now this is very disparaging praise, and, whether he means it so or not, is the way in which people speak of books, when they want to deny their originality and philosophical value. The criticism is extremely unreasonable with regard to your book, because its peculiar and distinctive character is its originality in the point of view in which the facts are regarded, and its philosophical classification. I do not see how Mill can be either ignorant of the novelty or the value of your classification of cultivations, for he makes it the basis of his own speculations—B. II. C. 6, Peasant Proprietors; C. 7, continued; C. 8, Metayers; C. 9, Cottiers. He spoils the classification indeed by lumping together Cottiers and Ryots, who surely are very broadly distinguished, and, I think, dependent on quite different laws for their progress in improvement. But having read books on such subjects very carefully, as he has, he must have known that the idea of making such a classification, and, what is a great deal more, the making it the basis of principles which regulate the distribution of wealth and the progress of society, is entirely yours. He has been much more considerate towards a Mr Rae, whom he represents

(I. p. 197) justly, for aught I know, as the author of original views on capital. Then again, I think he has you in his mind in p. 217, where he holds that the proportional produce of land necessarily decreases with the extension of agriculture. Here, however, it is not so much that he is unjust, as that his argument is good for nothing—or rather proves his own classification and corresponding principle to be false. How can we say, with any sound sense or use, that the produce of land increases universally in a diminishing ratio, when we have to allow that there is a principle, which we call "the progress of civilization," skill and the like, which may prevent this diminishing ratio for centuries, and during the whole life of a nation? I wish you could be here and we could discuss this book day by day, as we used to do Malthus and Ricardo. I mean, I wish this, that I might have your help in getting hold of the subject now, as I had then. Besides, it was very pleasant—and would be so again. Adieu.

Always yours, W. WHEWELL.

TRIN. LODGE, CAMBRIDGE, *Oct.* 14, 1848.

MY DEAR FORBES,

I was much obliged by your taking for granted that I should be willing to revise your MS.[1], and give you my best opinion upon it. That kind of service is often so useful, and may be so easily rendered by any one who is willing, that it appears to me one of the most essential parts of friendship. And so I will tell you how the matter strikes me, as plainly and simply as I can. I presume, from the interest which you take in the question, that Macaulay's speech made a great impression in Edinburgh, as from its clever, lively sophistry, and flattering doctrines for the populace, it was likely to do. This being so, it is very natural that you should wish to obviate the effect of such sophistry upon your pupils; and your lectures would, I have no doubt, produce a very beneficial effect on *their* minds. If the lectures were published, they would

---

[1] See Vol. I. page 172.

certainly do you credit, so far as I can pretend to conjecture the public judgment; but I think you must be prepared to find that you do not carry the general sympathy of the public with you, I do not mean as to the substance, but as to the mode and occasion, of the controversy. I will try to give my reasons for thinking this. Even without any special circumstances, the act of publishing an argumentative pamphlet in reply to a lively occasional speech, the reply coming many months after the occasion, and when most persons had forgotten the speech altogether, would, I think, be considered as treating the matter too gravely and controversially. But it would probably be considered further that Macaulay had a right to a good deal of indulgence on the occasion. He was speaking for a Mechanics' Institute, and was to select within reasonable limits such views as were likely to be agreeable to the members of such bodies. It might be considered ungracious to quarrel with him because his topics were effective. Some persons might say "Let the Professor, if he pleases, warn his pupils that the reading of Academic Students, if it is to be of any value, is to be something very different from that thus recommended to the members of Mechanics' Institutes; but do not let him expect those, who have to address such bodies, to talk to them on principles which would shew all such Institutions to be absurdities." Then again the very cleverness and rapidity of Macaulay's sophistry makes it difficult to answer him effectively; and I am not sure that you do not increase the difficulty by taking into account what you call *practical* knowledge. For in regard to the advocate, a talent which is really much admired among them is that of *getting up* a case very *rapidly*, even if it is forgotten very soon when it is over; and in the case of the statesman it is not at all generally agreed what value *knowledge* has in his character, especially what we should call *profound knowledge*. What we want in him is *wisdom*, which is something different from knowledge. I do not say you are not quite right on these points, but I do not think you would carry the world with you so entirely as to make this a good first stage of your argument.

I am, as you may well suppose, quite ready to assent to your arguments as regards speculative knowledge. But I am not sure that, if I had had to treat the subject, I should not have added a few others; which however you may perhaps think, and it may be with a good deal of justice, are merely other ways of stating those arguments which you have given. I should for instance have exposed, and, if I could, ridiculed, Macaulay's notion that the growth of knowledge from age to age is a homogeneous measurable growth, so that a man is a dwarf to one age and a giant to another; the growth being more like that of the boy to the youth or the man. I should have tried to illustrate very clearly—as clearly and brightly as Macaulay himself, if I could — that a little knowledge is a very different thing when it is all that the age possesses, and when it is a little scrap of a great existing body of knowledge. And holding, as you know I do, a great opinion of the influence of words in such cases, I should have tried to explain that a *little knowledge*, taken out of a great up-grown body of knowledge by persons who had not followed the growth, must needs be verbal, and therefore worthless. But I should have had great difficulty in satisfying myself in the exposition of these views, and I cannot but laugh to think what a puzzle Macaulay has set us by a few sentences of his clever rhetoric.

I have said more than I intended, but nothing that I do not intend; for such a plain exposition of my views may best help you to decide as to the extent to which you will publish your lectures. If you do, I have no doubt that they will be considered sound and right. I do not think that in my eyes they would add to the high literary reputation which the style, as well as other excellencies, of your *Travels* has secured for you; but they will certainly stand as a vigorous protest against a seductive fallacy. I have spoken with Parker about publishing them. He is willing to give you half profits, taking the payment of the expenses upon himself. I generally publish upon such terms in matters of which the sale is doubtful.

I wish I could have gone with you to see the tubular

bridges. I tried in vain to make room for them in the course of my summer's wanderings.

I keep your MS, till I receive your directions about it, and am always

Yours very truly, W. WHEWELL.

TRIN. LODGE, CAMBRIDGE, *Oct.* 24, 1848.

MY DEAR FORBES,

I send back your MS, but I should do so with great reluctance, if such a step were to put a negative on your publishing the lectures. The circumstance which you mention, in regard to which I was in error, namely, that they[1] were not addressed to a Mechanics' Institute, but to a Literary Association, in the presence of eminent men, makes a very great difference in my mind, as you may suppose from the arguments I used; and I now incline to think you would do well to print the lectures. Perhaps you will ask some better advice than mine. I have read the MS. over again, and am much struck with the argument beginning on page 37. I still hesitate as to the argument drawn from the practical sciences. Do you not lay too much stress upon the *sudden* acquisition of knowledge? and will not this call up a plea for quick abilities? Have you not rested much upon the word *empirical* without fixing its meaning exactly? When it implies a reference to experience, is it easy to distinguish empiricism from experience? But does not the word *empiric* generally imply some pretence to knowledge which the empiric does not possess, and is not this out of the range of your argument? You see I tell you all my scruples about this part of the argument, for all the rest appears to me excellent, and such that the publication of it, when the occasion on which it was delivered is taken into account, cannot fail to do you credit. I send the MS. back however, because, even if you do print it, you may perhaps look at one or two notes which I have made, besides what I have said. I doubt the

---

[1] Macaulay's remarks.

prudence of accepting Macaulay's antithesis of *shallow* and *profound* knowledge, as applied to early and late times. The knowledge of the later times is more advanced, more extended; but not more profound necessarily, except in more profound minds. I do not know, however, whether you can easily alter this. You may easily suppose that I did not without much reluctance say anything which was likely to give you even slight and transient annoyance. I should have been ungrateful, as well as unfriendly, if I had been careless on this head after the way in which you have quoted me in the lecture. I do not know any ancient passage which much resembles Pope's. There is the noted line about Margites ascribed to Homer:

$$\text{῝Os ἠπίστατο πολλὰ κακῶς δ᾽ ἠπίστατο πάντα}$$
Much did he know, and ill he knew it all,

which has always been accepted as good satire.

I send the MS. separately. I shall be glad to put it in Parker's hands, if you resolve to publish through him. We are looking forwards with some solicitude to the voting about our New Regulations[1], which is to take place on the 31st.

Believe me, my dear Forbes,

Yours very truly, W. WHEWELL.

MY DEAR M. QUETELET,

I have received your letter, in which you gave me the gratifying intelligence that the Class of Letters and of the Moral and Political Sciences of Belgium has done me the honour to inscribe my name among those of its Foreign Associates. I am extremely grateful for the recognition of my humble labours in those departments, which is thus shewn by the eminent men of whom your Academy is composed, and I shall think myself very fortunate, if any speculations on such subjects, which

---

[1] These Regulations established the Triposes for Moral Sciences and Natural Sciences, and exempted those who passed in these or in the Classical Tripos from the necessity of taking Mathematical honours.

I may be able to carry on, may be found worthy to be communicated to that enlightened body, and may tend to promote
the progress of Literature, Morals, or Politics. I beg to thank
you, my dear Mr Secretary, for the manner in which you have
communicated to me this agreeable information, a new act of
friendship, in addition to many already received; and I have
the honour to be, with sentiments of the deepest regard for
you, and of respect for my honoured colleagues,

Your very faithful servant, W. WHEWELL.

KREUZNACH, RHENISH PRUSSIA, CURHAUS, *Aug.* 26, 1849.

MY DEAR QUETELET,

*   *   *   *

I have seen little of men of science while I have been on
the Continent. Argelander, as you probably know, has now got
a spacious observatory at Bonn, which he is gradually furnishing
with instruments. I saw there also Dove of Berlin, who is
travelling to various places where meteorological observations
have been undertaken, and comparing the instruments used with
his own. I dare say you know much more about the project of
these combined observations than I do, for I have of late been
out of the way of persons who take an interest in such matters,
but I still look upon you meteorologists (for I may consider you
as one, though you are other and greater things besides) as
among the cultivators of science, who are most likely to arrive at
some large and interesting results. We saw a great number
of falling stars on the 10th of August, but I have not yet seen
any report which enables me to judge whether a greater number
of such phenomena than usual were observed about that time.
My chances of visiting the continent become, I fear, less and less,
as years go on, and the ties of my position at home become
more numerous, so that I would not willingly lose any chance
of seeing you when I must pass so near you. I cannot help
congratulating you on the admirable fortune of Belgium in the
middle of all the disturbances by which neighbouring states have

been shaken. The result appears to me to prove the wisdom which prevailed in the establishment of your constitution after your revolution, but I should be curious to have your views on this subject. With my affectionate regards to Mad. Quetelet,

Believe me, yours most truly, W. WHEWELL.

TRIN. LODGE, *Ap.* 23, 1849.

MY DEAR JONES,

I hope your new Commission prospers. When I was in town, I found Buckland much troubled at the claims put forth, or implied, by the Duke of Richmond and the other Lessees in their manifesto. I told him I thought he might depend upon you and Lyall for preventing any mischief. But I have since seen their advertisement, and I do think it *is* impudent and bullying. They say that of church property the value of 14 millions belongs to the lessors and 21 millions to the lessees, " according to recent calculations." Now either their "calculations" assume a right of renewal, or they do not. If they do, they assume what the lessors deny, and what has no foundation in law, and, in many cases, none in practice. If they do not assume this right, apparently the calculation is exaggerated ; but, if it be correct, there is no reason for their movement, for nobody wants to take their present leases from them. So much for their case ; but one can by no means afford to despise impudent bullying in these times. So pray take care of the lessors. And query, why ought they not to be heard by counsel, if the lessees are ?

To talk of other matters—of your book, which has never had justice done it. We have had an intelligent Irish member staying here, to whom I shewed your chapter on Cottier Rents, and he objected to it that, in fact, the Cottiers mostly pay their rents in labour, so that they are really serfs. I shewed him a passage in Bicheno's Evidence on Ireland in 1836, which confirms your account, and he would have it that Bicheno's views were coloured by the recent perusal of your book. However this may be, it is plain that the designation of such a class is requisite in order to trace its peculiar consequences. I told you how Mill had spoilt your great normal division of tenants by confounding Cottiers with Ryots, and had not had the grace to acknowledge where he got what he spoilt. I wish somebody would do justice in these matters. People will not mind me when I write Political Economy, except indeed I join to it mathematics, and make nonsense of it.

You know that we are looking for a new Downing Prof. of Law. Amos and James Stephen are the best candidates, and I had rather have Stephen in the History Professorship.

Always truly yours, W. Whewell.

TRIN. LODGE, *Oct.* 26, 1849.

MY DEAR HARE,

*　　　*　　　*　　　*

I was very glad to see Mrs Augustus Hare here a short time ago. I gave her for you a short critique of some of Hegel's vagaries[1]. There is nothing which so entirely deprives me of all respect for German heads in the matter of reasoning as the way in which they have allowed Hegel to domineer over them. It appears to me that on every subject he is equally fanciful and shallow, though he may not be so demonstrably wrong as in the matter of Newton. Sedgwick is mightily delighted and entertained with my paper. He is writing somewhat about the Germans, *apropos* of Strauss, in a new preface to his ancient Academical Sermon. This preface will be of monstrous length, as it well may be, for it has been growing for several years, but, I have no doubt, will contain many excellent things excellently well said; and I wish much it were done. I am much afraid his lectures will overlay it this term, and give rise to another procrastination.

*　　　*　　　*　　　*

Give our kindest regards to Mrs Hare.

Affectionately yours, W. Whewell.

[1] See Vol. I. pages 317 and 318.

TRIN. LODGE, *Jan.* 2, 1850.

MY DEAR MURCHISON,

Sedgwick's arm is broken, but the fracture is as simple as may be; the right upper arm bone broken just below the shoulder; so high up that it cannot be set with splints, and is bound tight against his side. He was attended to immediately after the accident, and is, I believe, going on as well as may be. He is subject to some oscillations in his progress towards convalescence, arising principally from things which are in themselves very good—high spirits and numerous friends. On Sunday he saw so many persons that he got excited, and passed the ensuing night sleepless and in distress, which made him much worse the next day; and indeed he has hardly yet got over that check. But Humphry, his surgeon, says he is going on well, and we take care to have him well nursed. The fall was in the frost, and when he was going at a canter, so that it is not surprising if he was a good deal bruised in various parts. The mammoth will play him no more such tricks, for his knees were so much damaged that all thought of riding him again was at an end. Our excellent friend has certainly been very unfortunate in his horsemanship. You know perhaps, as I do, how patiently he takes the confinement and pain which these calamities bring upon him, but all his patience cannot prevent great suffering and weariness. He was just upon the point of printing the new edition of his sermon, with a preface of above 400 pages! This accident will again delay a work which has already been promised, I think, for ten years.

I hope Lady Murchison is well. Pray offer her from Mrs Whewell and me our kind remembrances and best wishes that 1850 may be to both of you a year of health and happiness. Always, my dear Murchison,

Yours very truly, W. WHEWELL.

I shall not be able to send you any more bulletins after this, for we go to Lowestoft to-morrow.

MY DEAR FORBES,

I am afraid you must have thought me almost obstinately silent of late, so far as correspondence with you is concerned. In truth I have been more busy than usual, giving lectures on Moral Philosophy to a greater extent than I have hitherto done, in order to make our new examinations as effective as I can; and I think I see a promise of good working for the new system, if we can go on steadily and judiciously for a few years, till our new lines of study take something of a *set*. You will easily understand that this is a matter in which failure is possible, for we shall have examiners to educate as well as candidates; and if the Tutors do not take up the new scheme cordially, we shall have the enactments of the University evaded by some merely nominal conformity to them. However, as yet I see no prospect of any difficulties, except such as we may hope to get well over. Our new Professor of English Law has not yet made his lectures very instructive, but he will probably learn to do so as he comes more into contact with the students, and we expect great things from our new Professor of Modern History, Sir James Stephen, who is to lecture next term. He has been, as you probably know, one of the great luminaries of the Edinburgh Review in the historical line.

I have not yet turned my thoughts to our task of reporting to the British Association on the result of our former Report. Indeed I do not yet quite understand what we are desired to do. You, I think, have a view on this point derived from what passed at the meeting. I should be very glad if you would draw up a sketch of a Report for us to make. However slight and imperfect the sketch, it would shew me the nature of what is intended, and the form into which any special suggestions must be cast. With regard to your proposal that new experiments on the laws of conduction should be made, I should think it a great gain to science if you would undertake such; and a recommendation to that effect would be a very valuable part of our Report, if it were likely to lead to any such result. If I find that I have any plans which I want to propose, and which may be connected with our Report, I will try to put them into shape, but, in truth, I do not know that I have any which will bear that operation.

Our plans, as well as yours, are as yet unsettled for the summer. If I were at liberty, I should like well to come into your country and to travel there a little, for it is long since I have seen it, and it is pleasant visiting ancient haunts in company with one's wife. My Edinburgh friends are, I find, diminishing. Lord Jeffrey and Mrs Alison are lately taken away; both of them persons whom I much valued. But I have doubts whether we shall be at liberty to go northwards the ensuing summer. I think it likely that Mrs Whewell will be advised to go again to Kreuznach, near Bingen, for the waters, the place where we were last summer, and in that case I must content myself with excursions of which that point is the origin; these, however, by the aid of railroads may go to some distance. I was as far as Basle last summer, and should have gone into Switzerland further, if I had had time.

I do not know whether you have heard that Ellis[1] has been very ill in the South of France. He left Nice and took the road to Italy, but when he had gone a short way he became so ill that he was obliged to stop at a little village, Saint Remo I think, on the Cornice Road. At his request we had his case referred to in our chapel prayers. Since then I am glad to hear that he is much recovered. I do not know how he has finally modified his plans.

I do not know anything about the Royal Society and their pension[2] except what I saw in the newspapers. If they had taken counsel of the persons who have managed that part of the British Association business, they might have learnt both the difficulties and the advantages of having money for such a purpose.

*   *   *   *

Yours very truly, W. WHEWELL.

---

[1] Robert Leslie Ellis, of Trinity College.

[2] This refers to the Government grant of £1,000 a year placed at the disposal of the Royal Society.

TRIN. LODGE, CAMBRIDGE, *Mar.* 17, 1850.

MY DEAR MURCHISON,

I have been making my arrangements to come to London for a day so as to be present at the meeting of the Royal Society committee on Thursday I shall be very glad if plans can be devised by which the Government boon may be made a benefit to Science; but I think you will agree with me that to engineer it so as to produce this effect is not a very easy matter. We know the great difficulty which we had in managing undertakings supported by the money of the British Association; by no means mainly from the difficulty of procuring the money, but much more from the difficulty of avoiding the appearance and the reality of waste, caprice, partiality and jobbing. Some persons, I find, doubt whether the old practice of applying the screw of opinion in the scientific world to Government on each special occasion was not better than this perennial stream of bounty. But as the boon has been offered, an offer very creditable indeed to the spirit of the Government and the nation, it is very fit that we should try if it cannot be worked out to some good purpose, and I shall hope to attend the meeting on Thursday, and to hear what suggestions can be offered for such a purpose. I might perhaps make some myself, but I am not quite sure that I like the responsibility of handling, or directing the handling of parliament money.

I have so long ceased to take any part in the business of the Royal Society, that I shall probably find myself quite perplexed with the views and practices of the present managers. As to the question of the medals, the project of giving all or the principal of our prizes to merit, without any regard to the scientific achievements of our own Society and our own countrymen, appears to me a cosmopolitan claptrap, which, followed out, would make an English Royal Society an absurdity. The project of commuting the medal for a money payment I look upon as very base. I think (as I always thought) the cycle of subjects for the prize an inconvenient and pedantic restriction, so that I shall probably come very near to you in my conclusions.

*　　　*　　　*　　　*

Yours always, W. WHEWELL.

WESTMINSTER, *May* 6, 1850.

MY DEAR FORBES,

I write from Westminster school where I am placed in an
official chair for two days to listen to the examination of the
scholars for election to Christ Church and Trinity College.
Mrs Whewell wrote, I think, the other day to Mrs Forbes, and
expressed, I suppose, the regret, which we both very strongly feel,
that we cannot look forward to the pleasure of being with you at
Edinburgh in the beginning of August. It is doubtful whether
our sojourn at Kreuznach will be terminated at that time, for
though we set off early in June, and the doctors talk at present of
six weeks only, they are much in the habit of finding at the end
of the first period that a fortnight or so may be added with advantage; and when the discipline of the place is over, it leaves the
patient in a condition not very favourable to travelling or any
other exertion. Perhaps we have a better chance of seeing you at
Bonn. If you are there at the time that we pass up the Rhine,
we may be able to stay there a day or two for the pleasure of
seeing you. I hope this letter will reach you before you set out
upon your travels, and if so, pray let me know when you will be
at Bonn, and how you are to be found there. And when you
leave Bonn for watering places, why should Kreuznach not be one
of them? It is a place by no means devoid of attraction for a
little while. I believe you know where it is. To get to it you
must leave your Rhine boat at Bingen and go by omnibus (of
which there are abundance) or other carriage about eight miles up
the valley of the Nahe. There is a fine gorge, where the river
breaks through some porphyry rocks, and some interesting ruins.
And if you go elsewhere to other baths within a moderate distance of Kreuznach, I may again have the pleasure of seeing you
there; for though Mrs Whewell is fixed in her brine (like a sea
nymph), I make circuits of various forms and amplitudes around
her. Last year I roamed as far as Basle, besides various excursions to the Brunnen. Pray say whether you cannot take account
of these things in planning your summer movements. I hope
Ellis is better and mending, though slowly and doubtfully. Cope,

one of our Fellows, went to him at Nice, and he then intended
to set off very soon afterwards, and to travel by easy stages towards England. This sudden inspiration of Lord John's will,
I am afraid, disturb our progress very much. He means us no
immediate harm, and the young Whigs like the project out of a
pure and overwhelming love of Blue Books. But on a subject
where there is so much ignorance, there is no saying what turn a
Blue Book may give to the thoughts of reformers; especially if
we come to have a radical ministry, which does not appear to be
at all out of the sphere of possibilities. Sir J. Stephen is giving
admirable lectures, and gaining favourable opinions in the University rapidly, notwithstanding the misfortune of coming under
accusation as a heretic: they found out that he was a Gnostic,
much to his astonishment. I think your article on the Observatory at Greenwich must please all readers who like information
on such matters, by its fulness and clearness; and the subjects
are popularized as much as can be done without sacrificing essentials.

*   *   *   *

Believe me always truly yours, W. WHEWELL.

TRIN. LODGE, CAMBRIDGE, *June* 2, 1850.

MY DEAR HERSCHEL,

I send you all the papers which have been published by
the authorities here, including Lord John's letter, which is of
course the most important of them, so far as your determination
is concerned. I have no doubt I speak the sense of the whole of
the University, when I say that there is no one person whom they
would more willingly have as Commissioner than you, if Commissioners they must have. I think too I can venture to say, that
though it may be thought proper to publish some protest against
the justice and reasonableness of such an act on the part of the
Government at the present time, there will be no substantial opposition made to the application for information on the general
management of the University and of Colleges, except perhaps if

the details of the management of property are gone into, which it may be thought unbusiness-like to promulgate without necessity. I may say further, that I do not think it impossible that the Commission should result in some good: for instance, if they were to publish such an account of the crying evil of the university—*private tuition*—as might enable us, who have hitherto fought with it in vain, to get dominion over it. You will find Peacock quite as strongly opposed to this practice as I am. Even that part of the purpose of the Commission which alarms our Colleges the most—the conversion of some of the non-resident Fellowships into resident Professorships—if it could be done without invading the corporate rights of Colleges, as perhaps it might, would be a very material gain to our system.

If you are afraid of any disfavour which the office of Commissioner might bring you in your own College, you might write to your Master, and ask him whether he would be likely to share such a feeling, and, if not, I do not think that you need fear its existence on the part of any one else. Your Fellows supplied the strongest part of the opposition to the new system; and some of these opponents would not join us in opposing the Commission, but they opposed it on grounds of their own. Still, if it must be, I do not doubt that all would like to have you on the Commission. I shall be glad to know if you accept.

Yours always, W. WHEWELL.

[Sir J. F. W. Herschel became one of the Commissioners, under the condition that he should be at liberty to resign his post after a year's service if he wished to do so. He continued however to the end, and signed the *Report*, with the other Commissioners, on August 30th, 1852.]

BINGEN, *June* 25, 1850.

MY DEAR FORBES,

I have unwillingly put off writing to you so long, for I hoped to write that I was coming to Kissingen, or else to announce myself in person. But I went to Frankfort nearly a week ago, and found the weather so hot, that, after making an excursion to Homburg, I had not courage to undertake the fifteen hours' drive to your sojourn, with the necessary consequence of at least an equal journey back again. I say *at least*, for one of my plans was to go from Kissingen to Fulda, which is a great ecclesiastical name in middle age history, and where I suppose there must be some architectural remains of former greatness, though I believe much has been rebuilt in modern times. As the thermometer was, I gave up all such plans, and rushed away from the blazing Zeil into the valley of the Rhine. Here, at this bend of the river, and with the deep ravine near you, you can almost always find some shady corner; and as soon as I got here, I was delighted to get into a skiff and plunge into the afternoon gloom at the foot of Rheinstein. Besides, the river with its eddies brings cool airs, and with its rippling reminds you pleasantly that it does so, and you can go to innumerable points without dust, jolting, or hired horses. I have been at various neighbouring places within the last few days, and shall return to Kreuznach to-morrow. I think the Rhine brings us down a taste of your friends, the glaciers, out of which it flows. My boatman, who is a sort of philosopher, made the remark to me, and bade me dip my hands in it by way of proof. I admire the constancy which you shew to your ancient love, Mont Blanc. I shall be surprized if it does not appear that Mrs Forbes becomes jealous of her. I should like much to visit some of the glaciers under your guidance, but I cannot manage to execute such a plan at present. Later, if Mrs Whewell is well enough, we have thoughts of moving in that direction; but any such movement must be later, and indeed after your return to England. I am very sorry that I cannot be at the Edinburgh meeting. I have directed copies of some recent (printed) papers of mine to be sent thither; one on the Tides for the members of the Committee of Section A; and one on Political Economy for the Committee of that Section. Perhaps you may have time to enquire if they arrived and are distributed. I think the Political Economy will perhaps interest you. I grieve not to see you while we are describing our continental orbits. Kind regards to your sister.

Always truly yours, W. WHEWELL.

KREUZNACH, RHENISH PRUSSIA, *July* 5, 1850.

MY DEAR JONES,

*   *   *   *

I have heard nothing more of our University Commission, but I have little doubt that it will issue in some form or other; and I should suppose we may meet it in such a way as to incur no needless danger; but, I confess, I have a strong persuasion that Lord John will not be satisfied till his move has ended in something being done as to the distribution of funds, and I do not see how anything of that kind can be done without a most perilous infraction of our corporate rights. However, when we see the course that is taken, we must try to make the best of it. Tell me if you learn what is likely to be done in this matter.

Just before I left England, I printed a paper (in our Camb. Phil. Trans.) on certain algebraical ways of dealing with certain questions of Political Economy. I forget whether I directed a copy to be sent to you. Perhaps not, for I deemed it likely that you would "pooh pooh" the algebra. But the paper really does contain a refutation of certain vaunted theorems of John S. Mill on International Trade; and shews them to be true, even on their mathematical assumptions, within very narrow limits. Since I have been here, I have been pursuing the subject, and I think I have discovered the solution of a problem which I recollect telling you twenty years ago I wanted much to solve, namely, what is the cause and what is the measure of the different values of money in different countries. The main point of my solution comes to this, that the value of money is high in a country which has the (money) balance of trade steadily in its favour, and of course, low, when the balance is against the country. I do not know whether this is new: it is certainly very different from Senior's celebrated solution ("on the art of procuring money"), and it is not in John Mill; but it follows irresistibly from his principles. This conclusion also entirely overthrows the doctrine which appears to me to be current, that there can be no permanent balance against or in favour of a country. This is no more true, than it is true that, because water seeks its level, two neighbouring ponds connected by a narrow channel cannot be permanently at different levels. They may be at levels very different, if there be a machine which constantly pumps the water out of the lower into the higher. Such a machine is the power of producing articles for which there is a demand abroad; and this machine may work in one direction with a surplus of result to a very large amount. This is not a mere figure of speech; or, at least, it is a figure of speech which may be confirmed by figures of arithmetic in the most undeniable manner. I believe I shall be able to discuss several other of John Mill's doctrines in a similar way; which you will not wonder at, if you recollect that they are, as he gives them, little more than arithmetical examples of principles, assumed as if proved to be absolutely and mathematically true. We shall be here for a month longer, if your summer holiday-making leads you this way, as I hope it will.

Always yours, W. WHEWELL.

KREUZNACH, RHENISH PRUSSIA [1850].

[To Professor FARADAY.]

MY DEAR SIR,

I am always glad to hear of your wanting new words, because the want shews that you are pursuing new thoughts, and your new thoughts are worth something: but I always feel also how difficult it is for one, who has not pursued the train of thought, to suggest the right word. There are so many relations involved in a new discovery, and the word ought not glaringly to violate any of them. The purists would certainly object to the opposition, or co-ordination, of *terromagnetic* and *diamagnetic*, not only on account of the want of symmetry in the relation of *terro* and *dia*, but also because the one is Latin, and the other Greek. But these objections, being merely relative to the form of the words, would not be fatal, especially if the new word were considered as temporary only, to be superseded by a better when the relations of the phenomena are more clearly seen. But a more serious objection to *terromagnetic* seems to me to be, that diamagnetic bodies have also a relation to the earth, as well as the other class; namely, a tendency to place their length transverse to the lines of terrestrial magnetic force. Hence it would appear, that the two classes of magnetic bodies are those which place their length *parallel* or *according* to the terrestrial magnetic lines, and those which place their length *transverse* to such lines. Keeping the preposition *dia* for the latter, the preposition *para* or *ana* might be used for the former; perhaps *para* would be best, as the word *parallel*, in which it is involved, would be a technical memory for it. Thus we should have this distribution—

Magnetic {Paramagnetic: Iron, Nickel, Cobalt, &c.
{Diamagnetic: Bismuth, Phosphor, &c.

If you like *anamagnetic* better than *paramagnetic*, as meaning magnetic *according to* our standard, terrestrial magnetism, I see no objection. I had at one time thought of *orthomagnetic* and *diamagnetic*, directly magnetic and diametrically magnetic, but here the symmetry is not so complete as with two prepositions.

In considering whether I quite understand the present state of the subject, I have asked myself what would be the effect of a planet made up of bits of bismuth, phosphor, &c., of which the general mass had their lengths parallel to a certain axis of the planet. I suppose all *paramagnetic* bodies would arrange themselves transverse to its meridians, and all *diamagnetic* bodies in its meridians. Am I right?

I rejoice to hear that you have new views of discovery opening to you. I always rejoice to hail the light of such, when they dawn upon you.

I have been at the meeting of Swiss Naturalists at Aarau where I met Schönbein, who talked much of you, and told me you were going to explain his views of ozone.

I shall be in London in a few days, and shall perhaps try to see you when I am there. Letters sent to Cambridge always find me. Believe me, my dear Sir,

Yours most truly, W. WHEWELL.

[To Professor FARADAY.]

MY DEAR SIR,

Your discovery is a door into a very wide chamber, though, to my eye, not fully open; but I shall rejoice to see you turn it well round on its hinges, as I know you will.

As to a name for the antithetical classes of bodies, I consider thus. I suppose you called one class *dimagnetic* from analogy with *dielectric*. I think you ought to have said *diamagnetic*; for the bodies *through* (*dia*) which electricity goes would have been called *dia*electric, but that vowels in such cases coalesce. I think you may keep *diamagnetic* for this class, and give to the opposite class a name implying that they *rank along with* magnetic bodies. I propose *paramagnetic*. Will it not do to talk of iron, nickel, &c., as *paramagnetic*, and glass, phosphorus, &c., as *diamagnetic?* Then this new branch of science, for so, of course, it will soon become, will be *Paramagnetism*. I do not see any inconvenience in any of the obvious relations of the word. But it is difficult to judge of this without knowing the relations of the facts better than I do, and, if you find that the words I have suggested will not work well, (for that is the main thing), I will try if I can suggest better.

I give you joy at having such good reason to work hard—I hope you will not work *too* hard.

Yours very truly, W. WHEWELL.

MY DEAR FORBES,

I suppose by this time the bustle of your Edinburgh meeting must have pretty well subsided. I have not seen any detailed account of the proceedings, but from such hints as I gathered, I should suppose that it was carried on in an animated and interesting manner. When you have leisure, I shall be glad to have a word or two on the subject from you. I directed certain papers to be sent to the meeting for distribution. Perhaps you can tell me whether this was done. I came upon your track in Switzerland. Studer at Berne told me you had been there, and Von Buch, whom I met at Aarau, had seen you. I went to the meeting of Swiss naturalists at Aarau, which was not well attended, the Genevese people not making their appearance. Hugi the glacierist was there, but said nothing of glaciers, nor, I think, did any of them. I am surprized that the Swiss, with their opportunities, do not follow out the subject more vigorously, now that the road has been pointed out. I went to Grindelwald, and travelled some leagues upwards by the side of the lower glacier, so as to get to what they call the *mer de glace*, where the ice is less irregularly broken; but I wanted you to point out some of the leading features of the case. The ribboned structure of the ice was not so conspicuous as I expected to find it. Still I was very glad to see glaciers once more with the new eyes which you have given us for them. I found at Aarau a Prof. Mousson of Zurich, who gave a remarkably clear exposition of what he called the *Whewellische* lines[1], which I observed a long time ago, and which are shewn when you look at a *dusty* mirror from a distance, holding a candle near your head. He described their main peculiarities, but did not attempt to deduce them from the theory. I dare say George Stokes will be able to do this for us. I hope that your sister enjoyed your tour, and that you found Mrs Forbes and her child well on your return. Mrs Whewell is, I hope, improved by Kreuznach. We are going to set off on a journey to the north, but your letters here will be forwarded.

Always truly yours, W. WHEWELL.

TRIN. LODGE, *Sep.* 26, 1850.

MY DEAR FORBES,

I was very glad to have your account of the Edinburgh meeting, and quite as much so to receive your notice of your own travels. We are returned from our tour among the English

---

[1] See Vol. I. page 318.

lakes, where we have been visiting a nest of brothers and sisters, who, as you know, have almost divided the lake country among them. While I was at Coniston, I made an expedition to the top of Sca Fell Pikes, our highest mountain, and by much the most rugged and harsh which I have seen. It was new to me, though it is an ordinary point for mountain climbers, and was occupied by the Ordnance Surveyors for some weeks. The remains of their habitations are still there, and are very much like a High-land village with the roofs taken off. I returned by that which is, I believe, the oldest pass way across this knot of mountains, the pass of Hardknot and *Wrynose*, as it is commonly called: but Airy holds that it is rightly *Raynose*. Of course these mountains and passes are small matters compared with our Swiss friends, yet still they have their own interest and beauty, and even their own difficulty, especially if you take a wrong road, as we did. But I must ask you a question before I expend all my paper. It appears that a book has been published, entitled *The Theory of human progression, and natural probability of a reign of justice,* which, though anonymous, is known to be the work of a Mr Dowe of your good town. Do you know anything of the man, or of the book? I ask for the information of Victor Cousin, to whom the book is dedicated, and who desires to learn something of his admirer. He says the book is good but rather radical. Perhaps you can tell me something, though the book and the man are a good deal out of your usual way. I have received the Edinburgh Royal Society Transactions, which, I see, contains a paper by you on the Vivarais, which must have been the result of a good deal of hard work. You must be glad to have finished what has been so long on hand. I see too a paper by Piazzi Smith on Comets, which appears to me novel, but, so far as I yet see, plausible.

The vision which you summon up of a visit to Scotland is a pleasant one, but hard to embody in space and time. I am glad to hear that you are going to experiment on Heat.

Always most truly yours, W. WHEWELL.

My dear Hare,

I am very glad to hear from Mrs Hare that you are much better than you were in London, and able to take your usual work. I should like to see your charge on the present aspect of church affairs; but you are so far in arrears in the process of *discharging* these charges in a printed form, that I fear I must wait long for this.

You will think I am very idle if I try to call back your thoughts to etymologies, but I do not want you to do much, nor indeed to try much myself. You will still recollect with pleasure our old Etymological Society of (I think) 1832, though so many of the members have been called away from this world of words to the realities beyond the grave. I want to send a few memoranda respecting that society to the existing Philological Society of London. I have all the papers belonging to our old Society, and some of them may throw some light on similar labours which others are pursuing. You were always the great workman with us, and I have got some of your papers, and especially a printed paper on *Words corrupted by false analogy or false derivation*; and I want you to tell me if this paper was ever finished and ever published. A part of it was published in vol. 1 of the Philological Museum of that day, with the title *On English Orthography*, but a part only; and my copy ends in the middle of a sentence. I should think that the Philological Society would be glad of the unpublished part of it; for it is very much in the line of some of their work, and is indeed a very interesting contribution to the history of the language. There is also a paper *On Words derived from Names of Persons*, in like manner, so far as I can discern, unfinished and unpublished. This, too, it would be a pity to have lost, for, though Lodge and I began the subject in a superficial manner, you bestowed some good labour upon it.

I have been looking through the box which contains the ancient treasures of our society, which has brought these things to my mind; but indeed they are as pleasant matters of speculation as one can easily find, for there is small satisfaction in the aspect of public affairs, or the conduct of men in great offices.

*   *   *   *

Yours always, W. WHEWELL.

[I extract a few sentences from Mr Hare's reply, dated Oct. 22, 1851. "I too have sometimes had thoughts of sending our two papers to the Philological Society, but have refrained from doing so because I could not find any means of finishing them. To do so on the scale of the earlier part would be impracticable, without the help of public libraries; nor would my eyes now be patient of such researches: but I now and then indulge a hope that, if I can ever clear off my score of promised books, I may work up the materials, which I have, in the form of references, to wind up our two papers, so that the labour spent on them may not be utterly wasted, and that I may publish a little volume of Etymological and Philological Essays. This, however, is so very problematical, that I would not in any way repress any wish of yours to send any portion of the printed part, which has never been published in any way, to the Philological Society; only I do not see the slightest prospect of being able to help you in completing either paper for the next year or two." The two unfinished papers were published in 1873 under the title of *Fragments of two Essays in English Philology*, with an *Advertisement* by Professor John E. B. Mayor.]

T. L., Oct. 29, 1851.

MY DEAR JONES,

I am afraid that the hope from St Paul's is extinguished. I see in the newspaper that a Mr Champneys is appointed. I have no doubt, however, that the applications made from so many quarters to Lord John must produce an effect upon him which will shew itself in the long run; or, as I would willingly hope, in the short run. Certainly, I would rather hear of your having a pension than a stall, on account of your health, among other things, which would not, I think, well suit attendance on daily service in a cathedral.   We see in Peacock's case how trying to the health such a discipline is.

I have been reading your lectures, which I had not seen before, and am very glad to have them.   They are very good in themselves, but of course are better still as the ground work of your oral exposition.   I hope you will now have health and time to weave them, and your other speculations on the like subjects, into a connected scheme.   Such a scientific exposition of the subject may be of use when men become sane on such subjects; which, at present, they do not appear to me to be.   It seems to be an absurd and humiliating result of so many years spent by Englishmen in the speculations of political economy, that the governing body should carry on the business of the nation on the basis of fallacies so very palpable as those of "Free Trade," and should persist in their course in spite of that diminished national production which, according to the arguments both of advocates and opponents of the policy, must be, as it is, the result.

I want to give Mrs Jones (not you) a little trouble for a minute or two.   I saw, in the waste-paper basket in your study, several letters of mine, a few of which I picked out; and I find that they contain some memoranda of my former employments, especially at the time when I was writing my History and Philosophy, which I should be glad to have.   Pray let her pick me out any that readily offer themselves, and send them to me.   I hope you are none the worse for trying to lecture, and that your porter prospers with you.

Always yours, W. WHEWELL.

[In justice to Mr Jones, his reply to the preceding letter must be given.]

HAILEYBURY, *Thursday*, [*Oct.* 1851.]

MY DEAR WHEWELL,

I scrawl a few lines at once to tell you you made a great mistake about the contents of the baskets in my study.   I have been clearing out my receptacles for letters, &c., which were *full*.

It was a work I could do when I could do little else. Into one basket what I meant to destroy were torn and put away for the fire. Into another the papers were thrown which I meant to preserve. This last basket will be sorted, and some of them tied up together and put in a safe deposit. I have not any thing like got through the work. I should rather have liked to have kept your letters while I lived. Still, it is but reasonable you should have them, if you wish it, and when I have collected them I will send them, if you desire it; but pray dismiss from your mind any idea of your letters being treated as waste paper.

Mrs Jones had nothing to do with the grand operation, although much approving it, and I shall (D. V.) finish, as I began, the labour myself. Pray tell Mrs Whewell so.

12 o'clock—I have just come home from lecturing for an hour—very tired. I think very ill of my future, but I will not plague my friends more than I can help about it.

CLIFF COTTAGE, *Jan.* 20, 1852.

MY DEAR JONES,

As I have now, I fear, no chance of seeing you for some time, I want to hear something of you. I suppose the overturn of the Derby coach has not yet produced any change in your fortunes, whatever it may do hereafter. I do not see how the Derbyites can rally if the present ministry behave with moderate prudence, and abstain from snubbing the Church and frightening the agricultural people; which operations, though very consistent with their past practice, are not necessary evidences of their consistency. If people have any good hopes given them on these subjects, I should think they would be glad to help to make the ministry strong, to prevent them being bullied by the Irish Brigade. But then I fear they are engaged to some heavy blow at the Irish Church.

However there is no use in my speculating about such matters here. I want to know who is to be your new Law Professor at Haileybury. You will not easily get one who suits you so well as poor Empson did. I have been looking at his articles in the

24—2

Encyclopædia, and am glad to find there is so much of them. I find however that he goes a long way in his admiration of Bentham. I want much to get from some of my lawyer friends information where I shall find any lawyer's classification of offences to compare with Bentham's. *That* appears to me very hard, as I have tried to shew in my lectures: but where is anything better to be found?

Are you doing anything with your lectures, or are you quite busy preparing artillery to batter the cathedrals? I hope Mrs Jones is well. Mrs Whewell brought here a bad cold, but I am glad to say the sea breezes have swept it away. Our kind regards to Mrs Jones.

Always yours, W. WHEWELL.

[When application was made to Sir Henry Holland for the loan of letters written by Dr Whewell, he expressed his regret that he had not preserved them. Subsequently he found one, and sent it, saying that he should be glad of its admission into Dr Whewell's biography, "as illustrative of his high and ever active philosophic spirit."]

TRIN. LODGE, CAMBRIDGE, *Ap.* 13, 1852.

MY DEAR DR HOLLAND,

I have been reading with great interest the "Chapters on Mental Physiology" which your friendship sent me. I should like to discuss some points of it with you, or rather, to learn from you in conversation more on points on which I learn much from your book. I was not aware of Amoretti's experiments with the divining rod which you mention p. 24; but I myself have seen many such experiments, and made the same remarks on the mechanism of them which you make. About five and twenty years ago Airy spent a summer in France, and became acquainted with a Frenchman who had written a book[1] on this subject, and had put it into a scientific form. According to him, the move-

---

[1] The title of the book is *Recherches sur quelques Effluves Terrestres par le comte J. de Tristan*...Bachelier, 1826. Sir G. B. Airy did not become acquainted with the author; and Dr Whewell does not write about the book with strict verbal accuracy.

ment of the rod was not determined by the proximity of water, metals, or the like; but by certain "bandes bacillogyratoires" which exist with intervals on the earth. Airy and I made many experiments, and there was certainly often a motion which we did not consciously favour. I believe I have the book somewhere, and, if I can find it, I will tell you more about it. Your remarks on Mesmerism and Electrobiology touch, I am persuaded, the real secret of the matter; but there is so much of imposture and bad faith mixed up with the greater part of these exhibitions, that it is almost disgusting to analyse them so as to see what remainder of other elements there is. Your remarks on Time appear to me very valuable. They seem to me to point to a doctrine which I have always maintained, that the mind is active as well as passive at every moment of existence; and that Time is a Law of this activity—a *Form*, the Kantians say, and the word has some advantages. Your remark about the patterns, p. 63, is another exemplification of the mind's necessary activity and perception, and in that case of a power of using this activity so as to obtain varied results. I read too with great interest what you say about Sleep; and am ready to assent to your doctrine of its being a succession of states. Still I think that the whole of this succession is separated by a definite line from waking. There is such a thing as *falling asleep* : what is the physiological fact in this case ?

But there is no use in asking you such questions on paper. I am but too glad that you spare some time from your patients to give to philosophers, and delighted with the result. Perhaps I may come to London to-morrow and meet you at the Club.

Yours always very truly, W. WHEWELL.

[The next two letters were addressed to Mr Cochrane; see Vol. I. page 294.]

KREUZNACH, *Sept.* 9, 1852.

SIR,

I take the liberty of introducing myself to you as a person, like yourself, interested in the success of English hexameters, and

also as an admirer of your translation of *Louisa* so far as I have
yet seen it, which is only to the extent to which the *Athenœum* of
Sept. 4 has enabled me to do so. I am also especially desirous
that you should be aware of a very general dissatisfaction, which,
I am sure, exists with reference to the partial and perverse mode
in which that journal and other critics of the same school treat
the subject of English Hexameters. It appears by what the
Athenœum critic says, that you have referred to the *vox populi* as
being the obstacle to the general acceptance of this measure. But
the fact is that the common unperverted ear is well satisfied with
the English Hexameter when well constructed : and the persons
who are opposed to it are those who have had their habits formed
by classical studies. It is not the *vox populi*, but the *vox gram-
matici*, which cries down such attempts. They cannot get rid of
the habit of scanning by classical rules, and supposing that
spondees are needed; the fact being that classical rules of scan-
sion have no application, for the simple reason that the ear does
not recognize them. For instance, in the passage the critic
quotes, "And so thīnlȳ ăttired" : the syllable is not different
from what it would have been if written "*atired*." And spondees,
feet of two equal syllables, are inadmissible in all verse which
proceeds by accent, the only kind of verse which our ears can
accept as verse. For accented verse must proceed by *alternation*
of stress. Those who have written English hexameters have
mostly spoilt their verses by not being aware of this. It is easy
to get spondees, but they are much to be avoided in general.

The fact is, though these critics do not or will not see it, that
there is no difference whatever in principle between English
hexameters and English anapæsts or any other verse. And all
the objections of such critics to hexameters on the ground of
many monosyllables, and the like, apply just as much to the
most common kinds of verse. I have transformed some lines of
Beattie's into hexameters to show this :

> Oft at the close of the day when the hamlet is still in the twilight,
> And when mortals the sweets of forgetfulness prove in their dwellings,
>   &c.   &c.

It is easy to go on.

I may add that I have myself been a labourer in the same field with you. I published, a short time ago, in *Fraser's Magazine* a translation of the first book of the *Luise*; and, some time before, a translation of the *Herman and Dorothea*, in which, however, I was not sufficiently on my guard against the spondees. In the same volume, *English Hexameters*, are some much better specimens of the verse by others.

I shall, in a week or two, be in England, where my address is *Trinity College, Cambridge*: and I look forward to having the pleasure of reading the whole of your translations. I am, Sir,

Your faithful and obedient Servant, W. WHEWELL.

CLIFF COTTAGE, LOWESTOFT, *Jan.* 7, 1853.

MY DEAR SIR,

I am much obliged by your letter and your corrections of the commencing lines of my *Luise*. I agree with you in general, that many of our hexameters suffer from having too weak dissyllable feet : but I would not go to the length of making peremptory rules against such feet. English dactylic hexameters, like ordinary English anapæstic verses, must be allowed such licence as suits the poet's ear. Such rules as you propose would make them smoother, but the violation of such rules does not make them cease to be verse, nor necessarily to be good verse. Your rule that the last foot shall be a trochee I also assent to, as a rule of *prudence*, to avoid giving offence to English ears. For my own part, I really prefer an occasional solid spondee in the last place, to keep up the old Odyssean strain. But this is a fancy which I should only indulge when hexameters have obtained a fair hearing in England, which as yet they have not. I hope your translations will do much towards this : but an original poem like Evangeline does much more. With regard to my fellow labourers in the *English Hexameter Translations*, I suppose, like myself, they only give a few occasional minutes to such performances, and would by no means undertake such a task as the translation of the *Messias* : add to which, that I do not admire the poem sufficiently to translate it; and it has, I think, already had its climax of fortune in England, in a prose translation. There is a little poem lately published in Germany, called *Hannchen und die Küchlein*, which is very pretty, quite as pretty as *Luise*, and which might be worth translating. I have it, but at Cambridge. I should not, I think, undertake it; for the apathy shewn towards the *English Hexameter Translations*, and the absurd injustice of our grammarian critics have somewhat disgusted me with the public to which I should have to appeal. I shall, however, write a few pages more for the *North British Review* to convince, or at least, to confute them. I am much obliged by your offer of helping me with materials for this task. I think I cannot go into details, such as would introduce the passages from Göthe and Coleridge to which you refer: but I should be glad to have the references at your leisure. Klopstock's *Preface* I have always omitted to look at, and should be glad to see it. When I return to Cambridge, I shall probably find it in the University Library, but, if you can send it here without inconvenience, it will be very welcome. Why should not you, who have read, it would seem, all that has been written about hexameters, discuss the subject in some publication at greater length than you have done? Göthe's opinions on all subjects attract great notice in England, as well as in Germany, and would do something towards stemming the prejudices, and informing the minds, of our countrymen. By the way, one of the lines of your corrected version of the opening of my *Luise* wants a foot, as I read it:

Picked from the hánd which Chanticleer gäy with his wífe-train.

I think anything which I may write for the North British Review will probably take an historical character, rather than that of an estimate of the merits of the poems noticed, so that I shall perhaps need the indulgence which you promise me, in case I should not make any one writer very prominent. I am, my dear Sir,

Your very faithful servant, W. WHEWELL.

MY DEAR CHEVALIER BUNSEN,

Will you have the kindness to give me the name of the work in which the hexameters and pentameters which I translated[1] were to be inserted, and tell me whether it is yet published? If it is not, perhaps I might revise and improve the translation. I forgot to request that I might see it in proof while it was passing through the press, which I should have wished to do.

I have read through your *Hippolytus* with the greatest interest. I hope it will do my countrymen some good. It is already in many hands here, and I hope will soon be in more. I have proposed a subject for the Hulsean Essay, which will necessarily turn the attention of the candidates upon the book:—*The history and position of Christian Bishops, and especially of the Bishop of Rome, during the first three centuries.* In general, our Hulsean Essayists shew an amount of learning which you would hardly expect, considering that they must be young men under the degree of M.A. I believe Mrs Whewell will be ordered to Kreuznach again this summer. I do not know whether you are acquainted with the place. There is there, in ruins, a very pretty chancel of the *Paulus Kirche*, which belongs to the *Evangelische Gemeinde.* It is almost the only specimen of Gothic architecture in the neighbourhood. I suppose all the rest perished during the Mordbrenner Krieg. I have sometimes thought of trying to get it restored, and appropriated to the service of the English residents there. Do you think that your Government would look with favour upon such a project, and that your law would afford facilities for carrying it into effect?

We are just returned from Suffolk, where we have been spending our holidays, and are now here till Easter. Mrs Whewell unites with me in kind regards to Madame Bunsen, and I am always   Yours most truly, W. WHEWELL.

---

[1] See Vol. I. page 294.

DEAR SIR, [Mr COCHRANE.]

I am glad you are translating *Hannchen*. My copy has two *n*s in the first syllable, and I never doubted that it was the diminutive of *Hannah*. I should be disposed to make that name the title of the poem, for more than one reason. In the first place, *Hannah* is a good English name, common, yet not common-place; and, what is a great recommendation to me, it is a scriptural name. My notion of the style suitable to the hexameter epic is that no words or phrases should be admitted which would be out of place in our translation of the scriptures. I tried to bring the diction of *Herman* to this standard. All attempts to give a modern poetical tone to the poem by taking such a name as *Jessie* appear to me to give a *falsetto* character to a style which ought to be simple and self-justified. The Odyssee should be our model. Then, I would not take Hann*chen*, because the word to English eye and ear is harsh, and the diminutive has no grace for us. I do not know any English diminutive for Hannah. *Annie* is, I believe, a Scotch diminutive for *Ann*; but, as I have said, I do not think that the right key. And so you have my thoughts about your title.

You are very welcome to make any use you choose of the translation of the *Luise* which I published in Fraser. I have a little more in MS., but it is hardly worth while looking it out, for the value of any hints which it could give you. I am glad to hear they are going to publish *Louisa* in America. Longfellow wrote me word that they had published more than one edition of my *Herman and Dorothea*, without having the grace to say whose it was. But even that was better treatment than they gave me with regard to another book of mine, *Elements of Morality*, for they stereotyped the first edition, and would not take the improvements which I introduced in the second. I hope they will use you better.

I received the volume you were so kind as to send me, and shall read it with great interest; but I have had to go into Germany since it arrived, which has taken up my time. I beg you to accept my thanks for it, and to believe me

Yours very faithfully, W. WHEWELL.

[I have stated on page 205 of Volume I. that the sheets of the Essay on the *Plurality of Worlds* were read by Sir James Stephen, as they passed through the press. Dr Whewell's share in the correspondence will now be given in a series of letters extending over the period from Sep. 7 to Nov. 14, 1853].

LOWESTOFT, *Sept.* 7, 1853.

MY DEAR SIR JAMES,

I am much obliged by your remarks upon the sheet which I sent you. That the defect of construction in the Title struck you, is quite reason enough for reconsidering it. As to any more extended estimate of the book than I have given in my Preface, I intended to make such an estimate the subject of one or more of my Lectures, and, when it has been delivered in that form, probably to publish it, with other matter. As for a biographical picture of Grotius such as you would wish for, I, and many others, would feel the same wish: but I have always felt also, that the best manner in which such a wish could be gratified would be by a biographical sketch from your pen, such as the admirable ones which we already possess of other great men. I will not willingly give up the wish in this form.

I am much obliged by the interest you take in my cosmology, and by the confidential manner in which you have received my communication of my intention—a degree of reserve which I fully appreciate. I do not know whether you would be willing to look at any of my proof sheets, with a view of exercising your kindness by warning me when I have said anything likely to give offence to common readers. The public is so ready to take offence at any speculations on religion which are at all out of the common track, that I might have occasion for such warnings, without being aware of it. I have some speculations of the nature of those of Pascal, though I have not referred to him in them. I am not sure, however, that I derive so much satisfaction as you seem to have done from the new edition of Pascal.   By giving us what the author himself had rejected, the impression of the work appears to me to be confused.   Poets and philosophers may win praise by our knowing "what they discreetly blot," but I do not think they gain in effect.

I am greatly grateful for your kind sympathy in our sorrows. They are not single, for James Marshall has just lost a child. Well for us if we can find in our philosophy anything which can elevate and moralize our natural sorrows.

Believe me, my dear Sir James,

Yours very truly, W. WHEWELL.

LOWESTOFT, Sept. 13th, 1853.

MY DEAR SIR JAMES,

I have directed the first sheet of my *Plurality of Worlds* to be sent to you.   It is not that I expect there is in it anything to interest you, but I wish you to possess my argument from the beginning.   In my first three or four Chapters I must be employed in unrolling the argument from Chalmers's point of view; afterwards I go beyond him.   In the first Chapters I have dealt with the astronomical cosmologists as my friend Grotius does with belligerents, and have granted them, or seemed to grant, many things which I must afterwards take away.   I shall be glad of any remarks of yours on any part of my Essay; but of course you will not hesitate to send me the sheet without any remark, if nothing especial occurs to you.   I do not wish to occupy your time with writing on or about every sheet; but in a later period of the work passages may occur, on which, as I said, friendly monitions, by a person of religious and philosophical views, may be very valuable to me.

I shall look with great interest for the railway book which you promise us, and which is sufficiently needed as to its subject. I read Spedding's notes on Campbell's life of Bacon with the feelings which I always have when that subject is under discussion —indignation at the popular misrepresentations of Bacon's character, and the levity with which each succeeding writer aggravates them ; and grief at the difficulty of setting such popular misconceptions, or any popular misconceptions, right. Perhaps my disturbance on this subject is enhanced by feeling that I have on my hands a work of the same kind.

Believe me, my dear Sir James,

Yours very truly, W. WHEWELL.

LOWESTOFT, *Sept. 25th*, 1853.

MY DEAR SIR JAMES,

Your remarks continue to be of the highest interest and value to me. Some of them you will, I hope, find provided for in the future parts of the work. As to a summary of the argument, perhaps I may attempt something of the kind in the Preface, when I come to that. I hoped that the titles of my Chapters would in some degree point out the course of my argument. You may observe that the title of the Chapter on which you have last remarked is, "Fuller statement of the *Difficulty.*" I do not suppose that any arguments there have *nullified* the Jovians ; but only that they have shewn that they are assumptions attended with such difficulties, that it is worth while examining whether they are assumptions supported by strong analogy. This I shall do in the next six Chapters, on physical grounds, drawn in a great measure from recent scientific discoveries. I am afraid six physical Chapters will weary you, though I believe to many persons they will appear the most substantial part of the book ; at any rate, such arguments ought to be weighed. After those Chapters I have got as much metaphysics, at least, as our generation in this country will well bear. But I abstain, as much as I can, from employing arguments from Scripture. The doctrine of Good and Bad Angels belongs to a part of Religion which I fear to mix with Astronomy ; and I hope to give you good reasons why such beings cannot have their habitations in suns and planets, or, at least, not in planets.

My last sheet must have offended your taste by many faults of style, arising (some at least) from the errors of the press ; but

I wanted to gain time by having it sent directly to you. In the last page (48), which connects with the sheet which probably has since reached you, you must, if you please, in the 10th line read " firmly " for " piously," and in the 17th line, " events " for " wants." Also the end of paragraph 14 is to be altered. And the paragraph at the end of this chapter in page 61 is to be recast.

I think I am indebted to you for sending me a Shrewsbury paper containing a very agreeable and instructive lecture delivered by you. Am I mistaken in conjecturing that this was the little book which you described yourself as being employed upon a few days ago, and as probably destined for railway readers ? It is well suited to make their journey pleasant and profitable.

I am, my dear Sir James,

Yours very truly, W. WHEWELL.

LOWESTOFT, *Sept. 25th*, 1853.

MY DEAR SIR JAMES,

Your divination of the argument which I intend to found upon my exposition of geological doctrine, is not very far from the fact ; though perhaps I flatter myself that I shall bring the argument more to a point, and add to it some important elements. I might be mortified at finding that what I expound so much at length may be expressed so much more briefly ; but I am rather pleased to find how obvious the argument is ; for obvious as it is, it has never, so far as I am aware, been put forward with any adequate distinctness and force. As to the unsatisfactory impression produced by the spectacle of a universe void of intelligent life, *that* is of course one of the difficulties of maintaining such a thesis as mine. For myself, I can say that the result of my speculations has been reverence, not diminished but increased, for the greatness of the Creative plan ; and that this is a result of my views, I hope to make apparent in the course of my book. There are many things to be said, to remove the dissatisfaction which the view may at first excite, which I have tried to say there : as, that we are not to determine arbitrarily in what way the Creator's attributes shall be manifested—that the plan of the whole pre-existed in the Creative mind—and the like. And there is one thought tending the same way, which I have not put in the book, that all those who do not hold the world to be eternal *a parte ante* (as the old theologians said) must hold that there was a time when there were no inhabitants of this creation to adore the Creator. But I hope to learn your impression of this part of the subject when you have read more. I took the liberty of directing the printer to send you a revised proof of pages 35...41, that you may see that your advice has not been quite thrown away. I have tamed down the tone of eager assertion of one side of the alternative to a more balanced exposition of the difficulty. The argument comes, as I have said, in the succeeding Chapters.

Believe me, my dear Sir James,

Yours very truly, W. WHEWELL.

TRIN. LODGE, *Oct.* 1, 1853.

MY DEAR SIR JAMES,

I write a single line to say that, if you should have any remarks to bestow upon me respecting any sheets of my *Plurality*, from this time I shall be obliged to you to send them to this place, not to Lowestoft. Probably you have not with you the first volume of my edition of Grotius which you had from me. When you conveniently can send it to me here, I should be obliged if you would do so. I had marked in it some *errata* which I would use.

Believe me yours very truly, W. WHEWELL.

TRIN. LODGE, CAMBRIDGE, *Oct.* 7, 1853.

MY DEAR SIR JAMES,

It would be an ungrateful return for your valuable remarks on my proof sheets to attempt to involve you in an epistolary controversy, and I have no intention of doing so. But in order to account for my not despairing of the effect of my geological argument after what you have said of your views, I should like to make some brief remarks to show that I think I can see reason for still hoping that it will have weight with some persons.

1.   I allow all the difficulty of a beginning of creative activity after a period of inactivity and blank : but I did not invent the doctrine.   It is the one commonly current; therefore the one to be used in reasoning *ad populum*, which I am doing.

2.   The opposite doctrine of the past eternity of the world leads to equal difficulties.

3.   These difficulties affect space in the same way as time. It is as difficult to conceive a space *where* the creative agency does not work, as a time *when* it does not.

4.   Hence my argument, which is founded upon the relation or analogy of space and time, still has force.

5.   The eternity of the *human* world, as a fact, is, I conceive, disproved by geology, which shews that it had a beginning and a recent beginning.

6.   I am aware that there are arguments derived from Scripture, to prove that the doctrines of geology, as to praehuman series of animals, are false.   I will not meddle with these controversies, but take the received doctrines of geology as true.   This I have very distinctly said in the book.

7.   If, to fill up the infinite past time implied in the eternity of the world, you suppose cycles of praehuman rational beings, what I have to say is that I proceed upon physical evidence; and that in this I can find no fact pointing that way, and many pointing the opposite way.

8.   If therefore it be proposed to place the assumption of inhabited planets and systems on the same footing as such praehuman cycles of rational beings, I have no objection.   I think it will be placing them exactly on the right footing.

But I will not say a word more on this part of the subject, and so farewell to my geological argument; and then come my astronomical proofs, which I hope you will think more persuasive. You will have had a sheet of them by this time.   I have to observe in the way of explanation, that I shall have a frontispiece to my volume, on which will be represented two of the most remarkable of the *spiral nebulae.*

I am much obliged by the proof sheets of the Grotius which arrived by rail; but what I especially want is not the second volume which you have sent, but the *first volume*, which is bound up in drab boards, and which probably you will find near the place where the sheets came from.

Thanks many and deep for the thought you give to my speculations.

Yours very truly, W. WHEWELL.

TRIN. LODGE, CAMBRIDGE, *Oct.* 8th, 1853.

MY DEAR SIR JAMES,

You may very naturally and most reasonably expect that, when you have kindly stated to me any difficulties which my arguments against the popular multiplication of worlds and systems raise in your mind, you have done with that part of the subject, and are not to be asked to recall your thoughts to it again. And I have no wish to trouble you with any repetition of the discussion, which would be misplaced. When I have had your first impression, I have had that which is of great value to me. But for my own satisfaction I should like in some cases to say why I do not adopt your suggestions of further explanation. And I will do this the more unhesitatingly, because I will state my reasons very briefly, and you are not at all bound to read them. Why do I not tell my readers why Sirius or the stars of the Great Bear may not as well be called *dots*, as those into which nebulae are resolved by Lord Rosse? I have told them, so plainly, that I despair of making the reason plainer. It is because, in the nebulae, we see that the dots are connected by shining streaks of smaller dots, which shew that they are parts of a coherent mass, of which the dots are comparatively small elements; a condition which is quite different from the detached and insulated aspect of the obvious stars: and again, because the *Magellanic Clouds* have shewn Herschel and me following him, that nebulae and clustered stars do not differ merely in distance, but are different kinds of things, lying side by side in the same celestial space.

The other remark which you make arises, I think, from my speaking of the *resolution* of nebulae as astronomers use such language; and not in any looser way, in which common speakers may talk. The Zodiacal Light would not be *resolved*, in the astronomical sense of the term, by spectators who should see *one* dot, the Sun, in it; but only by those who should resolve the *whole* light into dots, or stars; which even we, so close to it, cannot do. According to the astronomers, we do not resolve the nebula by regarding it as a sun seen through a cloud; but by supposing the cloud to be made up of a number of suns, and to be really not a cloud at all. It is not that the fog *envelopes* a kernel of light, but the fog is *made up* of kernels of light; and that these kernels are suns, with systems revolving round them and inhabitants in the systems, and these inhabitants intellectual, appears to me a bold assumption. I am afraid even such explanations, involving no sort of call for being noticed on your part, may make you think your office of friendly reader of my speculations in their printed form a troublesome one. But I hope you will recollect what a valuable friendly office I hold it to be.

Believe me, yours very truly, W. WHEWELL.

In sheet 9, p. 134, if you read it, read *Sun* for *Earth* in line 22.

Oct. 9th, 1853.

MY DEAR SIR JAMES,

I will only add a word to thank you for your note received this morning. I am aware of the danger of taking too much the tone of an advocate, but not the less obliged to you for warning me of the danger. But when I think my arguments very strong, as I did in the case referred to, it is difficult for me not to speak as if I thought so. And it is to be recollected that no one, so far as I know, has been the advocate of one world against many worlds, I mean in recent times. I am not bent upon emptying the other planets of all life, as you will see, if your patience does not fail, which it very naturally may.

Believe me, yours very truly, W. WHEWELL.

TRIN LODGE *Oct* 10, 1853

MY DEAR SIR JAMES,

Have the kindness to send to Lowestoft any remarks which you may be kind enough to make about my Worlds. I am afraid you will think my double stars very tiresome.

Yours most truly, W. WHEWELL.

LOWESTOFT, *Oct.* 13, 1853.

MY DEAR SIR JAMES,

Your commendations are even more acceptable than your criticisms, and, I hope, not quite unprofitable. They make me feel as if I had already the best reward of an author's labours, the favourable interest of an enlightened reader, and they afford a strong reason for removing any blots which you notice. The one which you remark in page 151 I acknowledge to a certain extent. The successive revelation of visual facts at different distances would require us, if we were to speak rigorously, to modify our mode of asserting facts. But I believe we must be content in general to overlook this inaccuracy. It is only an inaccuracy, not an error; for since the *regression* of the past in time increases *gradually* as the scene of the fact is removed from nearer to more distant localities, the *order* of events is still the same as it appears; and this order is what we reason upon. To *translate* the assertion of facts into the language of successive visibility would, I think, throw a puzzle in the reader's way which does not really affect the argument. Did you see a little book called *The Stars and the Earth*, which was circulated anonymously a few years ago? That book shewed what startling conclusions follow from this doctrine. I do not want so to startle my readers: and the conclusions do not really affect me. But I will bear in mind your remark, as well as others which you have made, and may have an opportunity of explaining the difficulties which you notice in subsequent parts of the book.

You will think me obstinate about my *dots*, but I am disposed to defend the phraseology. I should not think of calling the stars of the Great Bear, *dots*; they are stars.   By calling the elements of resolved nebulæ *dots*, I mean to imply that they have no *prima facie* right to be reckoned stars.   Suppose that in examining some texture microscopically I discover dots which others hold to be animals, but which I do not.   I call them dots.   But I should not therefore call a lion or an eagle a dot, or even an oyster or a sponge.   This seems to me a parallel case.   With regard to my expression of shewing all possible favour to the proposed inhabitants of double star systems, I meant that any tone of pleasantry, which the expression contains, should be felt principally by Herschel, who made the proposed scheme.   I am struck by the expressions which you use on the supposition of the Unity of the World, as opposed to the Plurality of Worlds.   I hope you will find that the expressions in the further parts of my book harmonize with the sentiments which you express so well. I return to Cambridge on Saturday.

Believe me, dear Sir James,

Yours very truly, W. WHEWELL.

With regard to your promised illustrations of Grotius, you probably know Martens's *Causes célèbres du Droit des Gens.*

LOWESTOFT, *Oct. 14th*, 1853.

MY DEAR SIR JAMES,

Your theorist would speak as many theorists have spoken, and probably as many of my readers will speak; but a good deal of what he assumes is at variance with what we know, and I think I have shewn this in the pages which you have seen.   His assumption, that in the solar system each member is necessary to the sustentation of the rest, has not the smallest countenance from astronomy; and that which is the basis of his whole theory is the assumption that the stars are like the sun—may be classed with him—an assumption, as I conceive, clearly disproved in many instances, and proved in none; and, at any rate, *the* assumption against which I argue at some length.

As to the argument from the long time required to bring us

notices of the distant stars, that they may since that light was emitted have become orderly worlds, I place against it, as at least equally probable, the supposition that the distant stars were sparks and fragments thrown off in the formation of the solar system, that they are really extinct thousands of years ago, and survive in appearance only in consequence of the light which they then emitted. As to the possibility of inhabitants of Neptune or the moon who have no bodies, or bodies which do not require such conditions of light and heat as terrestrial creatures, I have to say that in that case I do not see that they require a planet to live upon. They may live in the Zodiacal light, or in the tails of comets, or in the blank spaces between the planetary orbits, just as well as on Mercury or Neptune. But my object is to trace the results of physical reasonings, to reason on the supposition that the vital powers in other regions are governed by the same laws as they are here. To make this supposition with regard to *mechanical* forces was Newton's great step.

I cannot persuade myself to lay down logical rules for my readers. So far as I have ever seen, to do so is only to add a new class of subjects of dispute to those which the matter introduces. If my readers are not persuaded by my arguments, put as clearly as I can put them, they will not be convinced by my proceeding as if I knew the rules of reasoning better than they do. Excuse my stubbornness; and do not punish me by discontinuing your remarks.

I am, my dear Sir James,

Yours truly, W. WHEWELL.

TRINITY LODGE, CAMBRIDGE, *Oct.* 16, 1853.

MY DEAR SIR JAMES,

I cannot deny the interesting and solemn character of the reflexions contained in your letter of yesterday; for they have often pressed on my own mind while I have been pursuing my speculations. I have no wish to deny their force, but I may mention very briefly some of the considerations which make it appear to me, that they do not really give any probability to *that kind* of plurality of worlds which has a place in popular belief or popular talk.

1.   I grant entirely that the aspect of this human world is, to the eye of reason, most unsatisfactory, to the eye of faith, very incomplete.   I am willing to allow that we should have a more consolatory spectacle for our thoughts, if we were to imagine another world of purer and better servants of God.   But allowing this, what single reason have we to connect this better world with stars or planets ?   Why must we find for it a place at all *in space ?* Is Heaven in space ? and if that better world be really in space, is it not most natural to suppose it in the regions about the earth below the Moon ?   Those regions may be occupied by glorious, pure, but to us invisible creatures ; and all the machinery of planets and stars may be to them, as it is to astronomy, so much dirt, so many clods.

2.   The better world, which is to console us for the foulness and misery of this, must, it would seem, have some connexion with this.   The worlds in the stars and planets can have, it would seem, no such connexion.   They are not what we want.

3.   If there be other worlds in which the subjects of the universal Lord have not fallen, such a belief exalts our opinion of the goodness of God to them ; but what view does it give of His goodness to man ?

4.   Science does, I think, in the way of analogy, give us a glimpse which is full of consolation.   Geology, if her analogy is worth anything, tells us that the Creator can produce a creature as much superior to man in intellectual, moral, spiritual characters as man is superior to brutes.

There are in the succeeding parts of my book other lines of thought in which I hope my readers will find something to elevate and console them, with reference to any trouble which my reasonings may cause them.   And I the more think that there must be such resource, since it does not appear that before the modern views of the universe were established Christian thinkers felt any need of other regions of God's servants in other realms of space to complete their view of Him, so far as it could be completed ; except indeed you say that Heaven, Hell, and

Purgatory are in space ; and of them Astronomy tells us nothing. Dante is much more trustworthy on such subjects. But do not suppose, I pray, that your warnings and reflexions are lost upon me ; and believe me,

Yours very truly, W. WHEWELL.

TRIN. LODGE, CAMBRIDGE, *Oct.* 22, 1853.

MY DEAR SIR JAMES,

The remark that there are large portions in the scheme of the universe in which we cannot see design, is intended to support the argument which I urge against those who ask, what the design of the stars and planets is, if they be not inhabited. Perhaps pages 215 and 216 may be considered as bearing pretty closely upon this argument ; but, if I recollect my own writing, it is more closely applied afterwards.

I did not answer what you wrote to me a few days ago, about theology having been so narrow before our astronomical discoveries, and so much enlarged by them, because I could not call to mind any portions of theological literature which at all supported this opinion, except perhaps the latter part of Chalmers's *Astronomical Discourses*, supposing all his *may be* to be turned into *is*. Nor could I think that the discovery of new spheres in the sky helped us out of any difficulty, in finding a place for better and higher creatures than man. I did not know that there was any such difficulty, and I thought that a spiritual body was best placed on a *spiritual sphere*. But I dare not at present venture upon the old question of the *ubi* of spirits. The history of the earth gives us perhaps the best glimpse which science can give of the possibilities which belong to this subject, by shewing us from analogy that the Creator can bring into the world beings as much superior to man as man is to brutes.

Believe me my dear Sir James.

Yours very truly, W WHEWELL.

TRIN. LODGE, *Oct. 25th,* 1853.

MY DEAR SIR JAMES,

I will occupy as little as possible of your time with my answer to your kind and valuable criticisms; but with regard to the relation between the Christian scheme and the Plurality of Worlds, I must remark that it is by no means introduced for the first time in pages 246 and 247. On the contrary, *the difficulties* belonging to that relation are the starting point of my whole essay and of Chalmers's speculations, which are my text. I will soften the expressions which you notice, but the topic cannot be excluded; for it is in fact *the* topic of my essay. I do not despair of finding that, if you should read again my argument from "flowers that blush unseen," you will find that it contains an answer to your view. I had once intended to write my essay in the form of a dialogue, and then I should have rejoiced to have such an expression as that of Mr Wilberforce which you quote[1], to grace my colloquy. I should then have made *Philonous* reply, "But as God's smiles *here* do not need spectators, why do they in Saturn?" Besides Saturn has spectators. It is precisely because *we* see him and admire him, that *we* are talking about him. It is very doubtful whether, if we were upon him, or rather *in* him, for I am afraid there is no footing there, he would look so well.

I will not go into the metaphysical argument. You may find plenty of metaphysics in sheet 17; I am afraid, too much.

Believe me, dear Sir James,

Yours very truly, W WHEWELL.

TRIN. LODGE, *Nov. 4th,* 1853.

MY DEAR SIR JAMES,

It must be unsatisfactory when a person giving good advice is told in reply that the person advised has previously weighed the considerations urged, and has thought other con-

---

[1] Sir James Stephen had written thus in his letter: "I remember Mr Wilberforce saying as he plucked a flower, 'Ay! these are the smiles of God'; perhaps a borrowed, but surely a very just expression."

siderations more weighty. But such is the case with regard to what you said in your letter of Nov. 1st on the unity of my subject. I am aware that the Plurality of Worlds, discussed on physical grounds alone, would be a simpler book and probably more popular; but it would not be what I have to say, or should care to say. If the relation of Man and God is the great feature in the universe, it concerns me to inquire what that relation is; and in truth, that, and not the Plurality of Worlds simply, was, and was declared to be, my subject from the beginning.

On the other point, the unsatisfactory aspect of the world regarded as God's best work, I cannot but say, on the other side, that I cannot see what consolation it can be for *us* that *our* world is selected as the special abode of sin and misery. And the futurity which Christianity has unveiled surely tends to remedy, not to aggravate, the dismay which the spectacle occasions. And at any rate, whether or no there be vice or virtue in other worlds, there are sin and misery in ours, which are not got rid of by imagining scenes of bliss and purity unconnected with us. I do not like the phrase "the origin of evil;" but some way of regarding sin and misery on earth a pious and thoughtful man must have. Mine may be quite wrong: but my essay as a *Théodicée* is, I hope, as allowable as others. The distinction between God's action with regard to animals and to man I have tried to point out—one guess among many, if you will; but at least a guess that is guided by obvious phenomena. Animals are not capable of vice or crime (except by a remote analogy with human actions) because they are not moral creatures.

With regard to your other difficulty, you almost prevent my speaking by making me feel it presumptuous to do so; but you may recollect that I do not condemn those who think it probable or certain, that the creation is thronged with beings whose modes of existence are unimaginable to us. I have said you may have thousands of circles of these beings, even below the Moon. But what I do say is, that all the attempts to connect the existence of such beings with the *planets* and *stars* have failed; and, as I try to shew, fail the more, the more scientific light we throw

upon them. This is my first thesis. You may go back to Ruling
Angels, or to any similar belief, but you must allow me to say
that there is no physical evidence in favour of such belief; nor,
as I also try to shew, any metaphysical necessity.

You will be glad, I am sure, to know that you are near the
end of your friendly and valuable, but I am afraid unsatisfactory,
labour.

Believe me, dear Sir James,

Yours most truly, W. WHEWELL.

TRINITY LODGE, CAMBRIDGE, *Nov. 9th*, 1853.

MY DEAR SIR JAMES,

I am much obliged by your suggestion of a protest in the
Preface against being regarded as merely a philosophical person;
and shall try to profit by it. Yet I can hardly think it necessary,
for, even if critics say to me what you suppose, it will agree with
the common, and, I presume, orthodox expressions which theologians commonly use, when they say that philosophy cannot explain revelation; which the philosophers on their side do not
pretend to do. I should have expected rather the accusation of
my scientific friends, that I warp science to make it conform with
religion. But, as the times go, it will not be wonderful if both
the opposite charges are urged. In the course of your remarks
on my sheets, you several times spoke of the Plurality of Worlds
as a doctrine which supplies consolation and comfort to a mind
oppressed with the aspect of the sin and misery of the Earth.
To me the effect would be the contrary. I should have no
consolation or comfort in thinking that our Earth is selected as
the especial abode of sin: and the consolation which Revealed
Religion offers for this sin and misery is, not that there are
other worlds in the stars sinless and happy, but that on the
Earth an atonement and reconciliation were effected. This
doctrine gives a peculiar place to the Earth in theology. It is,
or has been, in a peculiar manner the scene of God's agency and
presence. This was the view on which I worked, as I stated
emphatically in the early part of the book. But if any persons

find their comfort and consolation in an opposite quarter, I should be very much grieved to disturb the convictions which comfort them, and would at least express my regret at the possibility of doing so, *if I thought the feeling was extensively entertained.* Can you give me any means of judging of this? Can you tell me of any book, except Chalmers's, where this view is dwelt on? On reconsidering your objections to my metaphysical Chapters, their abstruseness and prolixity, I found your remarks so much an echo of my own calmer thoughts, that I have resolved to sacrifice those Chapters, at least for the present; and I am reprinting that part of the work in a very abridged form.

With a strong sense of the kindness and patience which you have shewn in your critical office,

I am, my dear Sir James,

Yours very truly, W. WHEWELL.

TRIN. LODGE, CAMBRIDGE, Nov. 14th, 1853.

MY DEAR SIR JAMES,

Herewith I send you my Title-page and Preface, that you may see the last of my Essay, whose progress into print you have followed so patiently and kindly. I have been much interested in the remarks which you have made in your last letter, and read them with the respectful attention which is due to reflections so grave coming from you. I am certainly surprised to find that you think the Plurality of Worlds a doctrine so essential to Natural Religion. If we were to consider this as established, we should wonder how the Christians of the first 1500 years could believe that "God is love," and should hold Copernicus and Galileo to be the true religious teachers of the world. And yet the doctrine that God is love was delivered without any reference to any worlds but that in which man's scene of life is; and was understood till the last 300 years in the same special sense. Indeed I should think, as I have before said, that the greater part of persons would derive small satisfaction or consolation from thinking that the Earth is the especial and selected abode of sin and misery. But there is no use in dwelling on this difference of view, which may prevail more extensively than I suspected. I, as I have said, make it my business to deal with difficulties which have been clearly stated, especially by Chalmers. I was greatly interested with the little book of my old pupil and friend, Birks, which you sent me. He is on the same track of reasoning with myself; and his book and Hugh Miller's, which he quotes, also serve to shew, I think, that men's minds are ready for a fuller discussion of the subject, and therefore that my book may be seasonable. I am now waiting for a picture of some nebulae as a frontispiece.

I regret extremely the manner in which Dr Jelf has forced matters to extremities in the case of Mr Maurice. I fear such measures will do nothing but harm to religious belief.

Believe me, yours very truly, W. WHEWELL.

TRINITY COLLEGE, CAMBRIDGE, *Oct. 16th*, 1853.

MY DEAR BUNSEN,

You have made for yourself so important a place in our English theological literature, and you know so well how contentious a field that is, that you will not be surprised to find that opinions which you have published on theological subjects have received a notice in this University, more or less controversial. I do not tell you this as a matter which is important to you, or indeed to anybody; but still as a matter which, in a place where you have so many admirers, may not be uninteresting to you; and which, at any rate, you ought to learn from a friend, rather than from any common and vague report, which would probably reach you in some distorted and exaggerated form. And I hope you will allow me to look upon myself, from long and to me very interesting intercourse, though I am sorry to say, from circumstances which I have often wished otherwise, far more broken than I have desired, as a friend who will not be deemed officious in giving you such information. Indeed you told me, both when I spoke with you about your *Hippolytus* before it was published, and when you kindly sent me the book, that you regarded your *aphorisms* as an appeal to the thoughts of religious men in this country; and therefore you may naturally wish to hear what is the response which such persons make to them. With regard to the aphorisms in general, I have nothing to say, except what I believe I said before, that they interested me deeply, and that I conceive they touch many of the points on which the minds of religious men are most earnestly working. But of course those who have to preach on the current speculations of the time, relative to religious views, are naturally led to fasten their attention on special points which are there propounded in a striking manner. Such is the position of our Hulsean Lecturer, who occupies the pulpit of the University Church during this month of October, and who is required, by the conditions of his foundation, to discuss any modern opinions which affect the evidences of Religion. He has thought it incumbent upon him to make his subject, for several Sundays, what you have said on the nature of miracles, and the way in which the evidence of them is to be contemplated. He deals with this subject very temperately, and makes your aphorism on that subject rather the starting point of his preaching, than the doctrine to which his discussion is opposed. There is nothing in what he has said or, I should think, is likely to say, which calls for any reply. Still I have thought that you may like to know, since the fact is so, that your published views are put in this position before us. Pray give my kind regards to Mad. Bunsen. I have left Mrs Whewell at Lowestoft, I hope, improved.

Yours very truly, W. WHEWELL.

TRIN. LODGE, CAMBRIDGE, Nov. 4, 1853.

MY DEAR FORBES,

I possess Arago's *Éloge* of Young; but, as my habit is, I have bound it with other quarto pamphlets, so that it belongs to a considerable volume. I will send it you with great pleasure in this state, if you think it worth the while; but in that case perhaps you had better give your exact address, as I find the parcel delivery of railroad offices is not very exact or diligent. If you receive my volume, you will find Arago in somewhat incongruous company; the fact being, that I bind up pamphlets with little attempt at classification, to save time and trouble. I was with my wife yesterday at Lowestoft, and found her going on well, as I hope, enjoying the air and sunshine and able to walk about.

*   *   *   *

I am printing a book which will, I am afraid, appear very heterodox to many of my friends—perhaps not to you. I try to shew that the evidence is rather against, than for, inhabitants in other planets and other systems than ours. The habitual declamation on the other side is so very unfounded, that a statement of the arguments on that side may serve to produce a reasonable equilibrium. I must publish the book without my name, in consequence of the heresies which it will thus contain; so I give you this information in a sort of confidence. I think the case on my side is much stronger than ordinary persons have any notion of; but you shall judge for yourself shortly. Poor Ellis grows weaker and weaker. I am told he is now so weak that he cannot turn himself in his bed. We shall lose much by his not executing the works which he had in hand, and which he would have executed so well; but it is an elevating though mournful spectacle to see the soul retain its power so independently of the body as it does in him. I have not seen him lately. I do not believe the Dean of Ely has anything to say about his *Young*[1], but I will ask him.

Always truly yours, W. WHEWELL.

TRIN. LODGE, Dec. 30, 1853.

MY DEAR JONES,

Your letter was a great pleasure to me. I am myself disposed to believe that my book is well written, because I wrote it with pleasure and facility. When you come to the *hard* chapters I expect you will be staggered. The more ought you to admire me for my self-denial in cutting them out after they were written—indeed after they were printed. That you may see how the book looks so fashioned, I have directed Parker to send

[1] See Vol. I. page 49.

you a copy at 20, Duke Street, Westminster, to which I address this—I am not quite sure whether rightly.

I hope the new year begins for you and Mrs Jones with good auspices—I cannot say that I much like the "auspices" as we see them here. This letter of Lord Palmerston's is an ugly looking thing.　　*　　*　　*　　*

Mrs W. goes on, I hope, well; but I wish for her sake it were not such ferocious weather. I am just going to try to make my way through 100 miles of snow to spend New Year's day with her. Give my love to Mrs Jones, and believe me,

Affectionately yours, W. WHEWELL.

LOWESTOFT, *Jan.* 3, 1854.

MY DEAR HERSCHEL,

Probably by this time an anonymous book has found its way to you, on "the Plurality of Worlds." I do not know whether you are likely to guess that I have anything to do with it; but if so, pray do not encourage anybody in the same opinion. I believe the doctrines there delivered will be deemed to some extent heterodox in science; as they may well be, being so much at variance with opinions which you have countenanced. But I am sure you will not wish that discussion on such matters should be suppressed; and the author seems to me to have discussed the question very fairly. Perhaps you would not take it much to heart if the inhabitants of Jupiter, or of the systems revolving about double stars which you have so carefully provided for, should be eliminated out of the universe. Indeed, if in this way we could obtain a more satisfactory view of the government and prospects of us, the dwellers on this Earth, many of us would deem the loss a gain. But, at any rate, I hope you astronomers will let us speculate on the one side as well as the other; which is all that my friend asks.

We are here in the middle of intense winter; the ground covered with snow to the water's edge, the wind howling, and the shore strewn with wrecks in various gradations of destruction. I hope that, whatever your outward world may look like at present, you have within doors the sunshine which arises from good health and good family news, as you are sure to have the sunshine that comes of love and good humour. Mrs Whewell has been here all the autumn, gaining, I hope, in health and strength. Pray give my affectionate good wishes to Lady Herschel and all your family circle, and especially to my godchild Amelia and my deputy niece Maria. I hope she has pleasant recollections of our sojourn at Kreuznach. When we went thither, it was almost as cold as it is now.

Always, my dear Herschel,

Yours most truly, W. WHEWELL.

TRIN. LODGE, Feb. 19, 1854.

MY DEAR FORBES,

I am very glad to hear that you are reviewing my *Plurality*. There is no one whose criticism upon the book I should more wish to see, and I shall be much interested to learn what you object to, as well as what you approve. I shall not wonder at all at your hesitating to admit some of the astronomical arguments, though to me those about the Nebulæ and about the Solar System appear very strong; I mean, as compared with the arguments on the other side. As to the Nebular Hypothesis, you may pull it to pieces as much as you like, with my perfect good will. I have never, I think, argued upon it, except as an hypothesis which might be alleged on the other side. I shall be surprised if any one thinks that I give any support to the man of the *Vestiges*. It seemed to me that my book might have some value as a strong case exactly opposed to his. I will reconsider the expressions to which you refer; but at first sight I do not perceive that an abortive flame, as I speak of it, is more inconsistent with providential design than an abortive stamen or an animal abortion. As to the note p. 190, it is a purposely wild hypothesis, offered to balance another wild hypothesis, and to shew how easy, and how entirely inconclusive, such hypotheses are. With regard to the objection p. 21, it seems to me to be the basis of the whole of Chalmers's speculations; and certainly I have found it very strongly operating on men's minds. It seems to me to differ greatly from the objection referred to in p. 27; as much as the history and destiny of man differ from the life of animals. And if the objection is of little weight, I should like to know the answer to it. Some of the other points which you remark on I know to be somewhat defective, in consequence of my having cut out portions of what I had written, which I thought too *metaphysical,*—awful word to an author hoping for popularity! I struck out 100 pages after they were printed, in which, among other things, were attempts to illustrate the universality of Divine action. "The sparrow falling to the ground" was not forgotten. I should like you to see these omitted parts, I do not mean with reference to your review, which will of course take the book as it is, but because I should like to know what you think of my views; and from what you say they may interest you. I will get copies of the cancelled sheets and send you them.

I forwarded to you a short time ago a letter from Studer of Berne, which he sent to me, not being certain where you were, and having written to you at Basle and at Paris in vain. I am sorry you cannot give a better account of Mrs Forbes. Mrs Whewell is, I hope, going on well, but has been a little thrown back by the exertion which, it appears, was too great of going out in a carriage.

I forgot to say that I did not suggest to Parker to apply to you, nor knew that he had done so; but it was a very good inspiration of his.

Always truly yours, W. WHEWELL.

TRIN. LODGE, *Feb.* 26 [1854].

MY DEAR FORBES,

I send you the omitted parts of the *Plurality of Worlds.* In the Chapter on the *Omnipresence of the Deity* you will, I think, find some of the thoughts which occurred to you as having been overlooked. If you take the trouble to compare these omitted parts with the Chapter on the *Unity of the World,* you will see that the paragraphs, which in that Chapter may seem to be

somewhat unconnected, were parts of a continued chain of speculation, as originally written. It still seems to me that they are
too metaphysical for common English readers. And, though to
condemn and expose German speculations may be a good thing,
it seemed to be going too far out of the way to do it in discussing
the *Plurality*, with so many other strong prejudices to contend
against as the assertor of the *Unity of the World* has before
him. Mrs Whewell is better than she has been. I should be
glad to think that she is making a permanent progress. I hope
that you do so.

Yours always, W. WHEWELL.

CLIFF COTTAGE, LOWESTOFT, *July* 28, 1854.

MY DEAR FORBES,

It was a great pleasure to me to hear from you, for I was
very desirous of knowing what progress you had made in the
recovery of health, and I am rejoiced to find that you can give
on the whole a more favourable account of yourself than I had
of late from you. I had heard that you had had an attack of
pleurisy, and that in its result it was supposed by your medical
attendant to have left you better rather than worse, and also that
you had been allowed to return to Scotland. I am very glad to
find these statements confirmed by yourself. I hope, as time
goes on, you will be able to report yourself a strong and sound
man again, as you ought to be. I am much obliged by your list
of the densities of the planets. You will have seen in the
Plurality of Worlds what use I intended to make of it. Close
accuracy is of no consequence in my argument, and indeed the
whole argument is far from conclusive; but we must reason as
we can on this dark subject. The points which now appear to
me to be most likely to throw light on the question are the
amount of light in the Fixed Stars compared with the Sun, and
the changes which take place in the forms of Nebulæ. The
latter subject is as yet untouched; and on the former the estimates are so different, that the argument inclines one way or
the other, as you follow this or that author. Yet the point itself

seems to be not incapable of considerable accuracy. We can compare the sun with the moon, and the moon with the stars; and, it would seem, in both cases with some security. If you find anything which appears to you trustworthy on this subject, I shall be obliged to you to point it out to me. I have in the second edition of the *Plurality of Worlds,* which I have had to publish, referred to Herschel's statements in his *Outlines of Astronomy.* Do you know anything better? As to the changes in the Nebulæ, we must, I suppose, wait for our knowledge, or rather, leave the hope of it to another generation. I am afraid that even Lord Rosse will not be able to tell us anything on the subject for many years. I have taken little interest in the Royal Society's proceedings of late, but I much regret losing Lord Rosse as President. I know little of the new councillors. I have of late years been so little in London that I have almost lost all knowledge of the new men of science there. Some of them certainly I do not think much of. This year I have scarcely been a day in London, though I have been persuaded to give two inaugural Lectures, as they call them. They contained my ordinary general views with a few illustrations, but they seemed to be accepted as answering their purpose.

I am glad your *Dissertation* goes on. When it is published and I have read it, I shall feel some confidence that I know what has been recently done in physics. From the account of ordinary narrators I can collect little that satisfies me.

Yours most truly, W. Whewell.

Trinity Lodge, Cambridge, *March* 10, 1855.

My dear Forbes,

I was very glad to receive a letter from you, for my letter which crossed yours did not prevent my feeling as if I were in your debt, or, at least, as if I should like to write to you again, which I was about to do. I am especially rejoiced that you can give so good an account of yourself, and hope that, now you have weathered an Edinburgh winter, you may emerge from the condition of an invalid. I grieve to say I cannot give you a good account of my wife, except so far as this : that she is better than she has been ; for she has had the influenza added to her other discomforts, and has now got rid of that part of her troubles. What you tell me about the progress of your *Dissertation* interests me much.   I have the most entire confidence in your justice and in your judgment, and therefore I highly value what you say of your habit of using my books.   I know that it requires some effort to act in such a case with perfect impartiality, when one knows that what one has to say will disappoint worthy and laborious people ; but the feeling of the value of historical truth supports one under such annoyances.   I cannot think otherwise than that your book will be well received ; though perhaps not by that class of popular readers who have been led to believe that they can judge of scientific discoveries without knowing anything of science—an impression studiously strengthened by various persons, some of whom know no better, and others ought to know better.   I shall be curious to see what you say on many points.   I think magnetism comes within your field.   Do you follow Faraday in his notion of *physical lines of force ?*   I have told him that the notion must be followed much more into detail than he has yet done, before we can compare it with other theories ; but that, as an expression of the facts, it will probably have great value.   The wrangle about the correction of magnetism in ships, carried on in the *Athenæum* between Airy and Scoresby, is probably too much a point of practice for your field.   I should have been better pleased, like you, if Airy's determination of the density of the earth had agreed with former results ; but I am much disposed to believe that his method is the most trustworthy. If the increase were very great, it might affect the argument about Plurality of Worlds ; but it does not go far enough for that. Have you seen a strange little paper of Airy's in the Proceedings of the Royal Society, in which he assumes the interior of the earth to be fluid ?   I can tell you nothing of the life of Young, as to the chance of its speedy appearance ; for I have almost ceased to question Peacock about it, thinking the subject must be a disagreeable one to him, as a work so long overdue would be to me.   I have not yet looked at the volume of Arago.   In fact, this term I am mainly employed with my Moral Philosophy, having to lecture and examine in that subject. My interest in the subject by no means abates; but I am meditating to ease myself of the employments which it brings, by resigning my Professorship. I find lecturing in addition to other matters rather a burden; and I have now got our Professorial scheme into a condition, which has, I hope, some coherence and stability, so that I may devolve the management of it into other hands.

*　　*　　*　　*

Yours most truly, W. WHEWELL.

TRINITY LODGE, CAMBRIDGE, April. 15, 1855.

MY DEAR SIR GEORGE [CORNEWALL LEWIS],

I am much gratified that, although you have now the empire of England to manage as well as the history of Rome, you have had so kind a recollection of me as to send me a copy of your book on the latter subject. I am reading it with great interest; the more so, inasmuch as our scholars here have been much in the habit of accepting Niebuhr's views of Roman history recently, as servilely as they did Livy before. I am glad to see that you are resolved to judge for yourself, and to enable us to do so.

I am amused with your comparing the speculations of the Niebuhrian school to the inquiries whether the stars are inhabited. And the comparison is very just; for when you have called a portion of history a mythe, the persons who figure in it are supposed to be as far out of our knowledge as the inhabitants of the Moon or of Saturn.

Believe me, my dear Sir George,

Yours very truly, W. WHEWELL.

TRIN. LODGE, CAMBRIDGE, Nov. 15, 1855.

MY DEAR AIRY,

I am Vice-chancellor, as perhaps you know, and must therefore regard matters in a Vice-cancellarian way. Here is one.

A little while ago one Dr Whewell offered the Observatory here
a set of Brooke's magnetic photographic apparatus such as you
have at Greenwich. Now the V. C. and Syndicate have to
consider what is to be done with this gift, whether it is to be put
in use for observing. For that purpose it will be necessary to
build a magnetic house, such as you have, and to employ an
additional observer. The expense will be considerable, and will
be, I fear, almost beyond the means of the University. But what
weighs with the V. C. still more is that it does not appear of what
use the observations will be when made. It appears to me that
those recorded observations which you have at your Observatory,
accumulated and accumulating, have as yet been of no use to
science ; and neither you nor your Visitors can see how they are
to be of use of that kind. This being the case, I do not see the
wisdom of setting up another expensive establishment, at the
cost of our poor body, and heaping up still more of these useless
measures. As we must make a report to the Senate on this
subject, I shall be glad if you will tell me—what is the size of
your magnetic house, and, if you can, what has been the cost of
it ; how far you agree with me that it is not of much value to
science, in its present state, to establish an additional magnetic
observatory of this kind ; and anything else which you think
bears upon the question. I have called a meeting of the Syndicate
on Saturday, and should be glad to hear by that time.

Always truly yours, W. WHEWELL.

TRIN. LODGE, CAMBRIDGE, *July* 18, 1856.

MY DEAR FORBES,

I have no doubt that you have long before this time with-
drawn from the city to the country, and wisely done so. I hope
you will thereby lay in a stock of health and strength which will
carry you onwards when the summer is done. I shall therefore
probably not see you when I come to Edinburgh, but I will not
visit your city without letting you know that I am going to do so.
There is a meeting of the Archæological Institute, which assembles
in your very archæological city on the 22nd, and holds its sittings

and makes its expeditions for about a week. This meeting I hope to join; having found some of my brother Heads who will take my Vicechancellor's duty for a few weeks. The liberty thus gained will enable me to visit my friends in Cumberland and Westmoreland, which I am very desirous of doing.

I hope you go on prosperously, in bodily health, and in intellectual activity. How does your history of the recent progress of science proceed? I should under any circumstances have asked this with great interest, as you will easily believe; but I have moreover an especial reason for wishing to know what you have done and are doing in this way, inasmuch as I am going to print a new edition of my *History*, and shall wish to give some indications of the progress made in science since the last edition was published. I do not think you will be afraid of my appropriating your stores; but I shall be glad to learn, by comparison with what you do, that I am in the right path of investigation, and to have my views confirmed by yours, as I have little doubt they will be. At present I am mainly employed about the old people, Plato and Aristotle, and the like. The books which have recently been published about them give occasion to a few remarks. By the way, who is to succeed your great philosopher of the Middle Ages, Sir Wm. Hamilton? I receive interminable collections of printed Testimonials with regard to the Candidates. I suppose Prof. Ferrier must have in his reputation and connections much to recommend him. I hope you will obtain for a colleague some one who does not hold that mathematical studies ruin the mind. I shall be here till the 22nd, and then at Edinburgh. With my limited time I fear I cannot go further north. My kind remembrances to Mrs Forbes.

Always truly yours, W. WHEWELL.

TRINITY LODGE, CAMBRIDGE, Oct. 3, 1856.

MY DEAR FORBES,

I do not know how it is that I have not written to you sooner, for I have often thought of doing so since I left Scotland. I hope you are returned to Edinburgh improved in health by the summer's rest and retirement. For I collect that you are returned from your communication to the *Athenæum*. I think you are right in your remarks there made. It grieves me to see labour thrown away in making indexes and works of reference on so large a scale that they cannot be used with any advantage. The persons who make them must be persons whose judgment can be trusted, or else their labour can be of little value; and if they are such persons, the results of the exercise of their judgment should take a permanent form, and modify the shape and extent of their records. I conceive that your own offer to give your labour to the cause is connected with the work you have had to go through in preparing your *Dissertation*. I hope it is in a state of forwardness; and I entertain this hope, not only from the interest which I shall of course feel in it on general grounds, but also because I expect it will keep me out of error and deficiency in the new edition of my *History*. Anything which I may insert with reference to events more recent than my last edition (1847) must of course be very slight and general; but still to know what is the result of a more detailed survey of the ground will be of use to me, as I have said, in keeping me right. When do you expect to go to press? I am here busy with the end of my Vicechancellor's year, and with getting our new constitution into working order, which the Act makes to be my business. I do not expect that the change will be very conspicuous, or that, if you come to Cambridge, you will see any great alteration in the aspect of things; and I am especially afraid that the new system will not do any of the good which needs to be done.

*   *   *   *

I am always, my dear Forbes,

Yours most truly, W. WHEWELL.

TRINITY LODGE, *Oct.* 23, 1856.

MY DEAR FORBES,

I have just cast my eye over your *Dissertation* (I have not yet had leisure to do more, for this is a very busy time with us), and I like very much both the plan and the execution. I shall return to it in a short time, when I can write the additions to my new edition of the *History*, and then I shall go carefully through it and will give my judgment on parts of it, if you care to have it. Our plan is a good deal different, so that I shall not have so much temptation to steal from you as might at first appear, and the leading topics of my additions are already sketched. I see that the Edinburgh Review, in the article on Arago, has at last adopted pretty decidedly the undulatory theory. Do you know who wrote the article? The part of your history where I cannot yet help having some misgivings about an alleged discovery, is that about the mechanical equivalent of heat. I believe it rather on Wm. Thomson's authority, than because I have satisfied myself. Are you quite satisfied? I send you copies of my memoirs on the Tides which you want, and which I have, the sixth, seventh, and fourteenth; I have not, I fear, the eighth and ninth, but they are not important ones. The *sixth* series I am glad to send you, for I look upon it as my great achievement in Tidology. I still want a Tidal expedition. I believe *now* that the oceanic tides cannot be represented by cotidal lines; they are forced vibrations, as you say, and I cannot make them out clearly, except an expedition be sent to *hunt* them. The British Association has authorized a new application to Government for such an expedition; but I fear there is no chance of the Admiralty taking the matter up.

*   *   *   *

Always truly yours, W. WHEWELL.

TRIN. LODGE, CAMBRIDGE, *Feb.* 8, 1857.

MY DEAR SIR [MR DE MORGAN],

Your letter has remained unanswered in consequence of my absence from England. I have been at Rome, where among other things I think I ascertained that the story of the reversal of the sentence of Galileo amounts only to this: that on a revision of the Index Expurgatorius in 1818, his works (along with others, and as a matter not unusual) were removed from the list, not without opposition.

With regard to Rothman, I should think the most exact and complete account of his family circumstances and cardinal dates will be heard from Pashley, now Police Magistrate, who was his executor, or from Rothman's sister, to whom Pashley can probably direct you. We can tell our usual story about him—that he was elected Scholar in 1820 (I think his Freshman's year), Fellow 1824. He was of Airy's year, but became Fellow a year later than Airy did. Senior in 1843. He filled at various times various college offices, which would not interest the world. You know his London life better than I do.

The Bromheads have, I believe, a brother living; who, I suppose, must be the present baronet, and must be Sir Edmund Bromhead, Thurlby Hall, Lincolnshire. Charles Ffrench Bromhead was of my year. He was very tenacious of the double *Ff*; also very desirous of having it recollected that an ancestor of his, Gonville, was one of the founders of Gonville and Caius College. We reckoned our year a great year. The Fellows whom it gave to the College were, besides myself, Sheepshanks, Julius Hare, Hamilton (Conic Sections, and Dean of Salisbury), Edward Elliott (Horæ Apocalypticæ), Higman, Stevenson, now vicar of Dickleburgh, and Bromhead. He was proud of the group to which he belonged. He had some mathematical talent, and took a share in the discussions and attempts which our *new* mathematics then occasioned. His elder brother, Sir Edward, was before me : he was an associate of Herschel, Babbage, and the like; perhaps Sir Ed. Ryan. I believe he was one of the writers in the very remarkable volume of *Analytical Memoirs*, published at Cambridge in 1813 by Herschel, Babbage, Maule. And he wrote on the *Fluents of Irrational Functions* in the *Philosophical Transactions*, 1816. Afterwards he was mainly employed in the life of a country gentleman ; but he wrote upon Botanical Classification, I think in the *Philosophical Magazine*. He belonged to the class who aim in all things at generalizations.

Yours very truly, W. WHEWELL.

LOWESTOFT, *Sept.* 19, 1857.

MY DEAR FORBES

I was very glad to find a letter from you on my return to Cambridge. I should have written to you, but did not know whither to direct it; which often prevents one's writing, though one knows that the letter, if sent wrong, will be forwarded right. I hope you are the better for your spring at Clifton, and am glad that you are likely to enjoy your summer retreat at Pitlochry. I shall be very glad if at some future time I am able to see you there.

It is true that I was at Dublin during the British Association meeting; but I fear I cannot tell you very much about what happened there. I did not arrive in the place till the middle of the meeting; and then, a very unusual thing for me, I was so ill that I was hardly able to attend any of the meetings, or to enter into what went on there. I saw, as you say, in Lloyd's Address, that the dynamical-heat men are running their scent very eagerly. But the principal paper which I heard of on that subject was George Rennie's, which was reasonable and definite enough, being mainly experiments upon the old question of the heat produced by friction. The theory of molecules with atmospheres, rotations, and the like, which Lloyd seemed disposed to countenance, is fantastical enough, and different enough from any theory which sound science has ever established. I have no objection to let theorists indulge in such fancies. In the case of heat, Laplace and Ampère have done it before; and, if there be any real explanation of phenomena in such theories, the hypotheses will simplify themselves as they are worked out and compared with facts. But Presidents of the British Association and Edinburgh Reviewers ought not lightly to adopt such fancies; for theirs is a judicial position, and they give ordinary readers a wrong notion of the progress of science. I should hardly have gone to the meeting if it had not been that I wanted a reason for going to Ireland, to see Lord Monteagle's house and Lord Rosse's telescope. In my visit to the latter I had not the benefit of a fine day, though I did catch a glimpse of a nebula.

But I saw the mechanism and handling of the instrument, which
are very good and well devised, and which I had not fully comprehended before, though I suppose I might have done so, if I had
attended to the descriptions which are everywhere given of it.

*   *   *   *

Believe me always truly yours, W. WHEWELL.

TRIN. LODGE, *Feb.* 28, [1858.]

MY DEAR FORBES,

I did not reply immediately to the kind invitation which
your last letter contained, because I knew that you would interpret my silence to mean, what was the truth, that I could not
then come to you.   I shall be most glad to come into your regions,
and to you, when an occasion of leisure offers, and, I hope, in
Scotland, or in England, the year will not pass away without
our meeting.   I left the North, as I told you, to go into Kent
to marry my nephew to Herschel's daughter, which was very
happily performed.   The young couple are now at Rome, mainly
aided in their sight-seeing by our friend Pentland, as every one
must be who is willing to take his guidance.

But the occasion which sets me upon writing to you at
present is to tell you of the pleasure which your article in
Fraser upon the new edition of the *History of the Inductive
Sciences* has given me.   Your praise is worth having, because
though I know that I may reckon upon your friendship, I know
too that that feeling is incapable of influencing your judgment;
you judge both with more knowledge and more thought than
any other reviewer of the history of science.   I am much interested in the *Lessons* which you draw from your survey.   I hope
the present generation will profit by your telling them that
great discoveries are now as rare as ever.   Their self-complacency
will make this a strange announcement to many of them.   I have
been gathering some hints (small ones) about the history of
science in Cambridge from reading Barrow's works.   We are
going to publish a new edition of Barrow.   I think I am now
in a condition to prove that, though the Cartesian philosophy, as

a system of Mechanics, was accepted here (very wisely), yet that the notion of the Vortices, the subtile matter, and the like, was rejected from the first. So little basis is there for the jeers which your countrymen, led by Playfair and Stewart, used to cast upon us. I do not know whether we shall edit Barrow's mathematical works; that is, whether it will be worth while, as a trading speculation: for we are obliged to look upon matters in that point of view. We are in the hands of our Commissioners at present, and do not know what will come of it. No good, I fear.

Give my kind regards to Mrs Forbes, and believe me always most truly yours, W. WHEWELL.

WINDERMERE HOTEL, *Sept.* 1, 1858.

MY DEAR FORBES,

Your letter of the 28th reached me yesterday, just when we were undecided what course to take next, partly because we had concluded a round of visits, and partly because the weather seemed to be breaking up. To-day, however, it seems as if this glorious summer had not yet exhausted its bountiful spirit, and we have some heart to proceed further northwards; the more so, that your friend, my dear brother-in-law, Ellis, after having had a bad attack lately is now much better, at least for a time. I may mention that I am sure a letter from you would give him great pleasure. Though blind and sadly afflicted in many ways, his mind is as active as ever; he retains his interest in old subjects of speculation, and has always something bright and original to suggest on every subject that is presented to him. The clearness of his mind and the kindness of his heart under his circumstances are a striking and touching sight. We have been enjoying exceedingly, Lady Affleck[1] and I, our northern tour. All the north is new to her, and delights her much. And, as I have said, we have some thoughts of extending our travels still further north. And I want you to tell me whether we could for a few days have good accommodation near you,

[1] See Vol. I. page 222.

that my wife may have the privilege of becoming acquainted better with you and Mrs Forbes. I think you must have a good hotel near you.

*   *   *   *

Lady Affleck desires her kind regards and will be very happy to renew her acquaintance with you and Mrs Forbes.

Believe me, dear Forbes, yours most truly,

W. WHEWELL.

TRIN. LODGE, Oct. 10, 1858.

MY DEAR SIR [MR DE MORGAN],

In the *Athenæum* of Sept. 18 is an article about Bacon, about which I am sure you will feel an interest. I am nearly sure that you wrote it. I will therefore send to you a notice which will interest the writer thereof, relative to a point there mentioned—Barrow's acquaintance with Bacon's works. I happen to be able to speak very distinctly on that point. Among Barrow's *Latin* works is a speech (of 1652) on the *Cartesian hypothesis,* not meaning thereby *vortices*—I doubt whether any Cambridge man ever thought seriously about the vortices—but the hypothesis that all natural phenomena may be accounted for by matter and motion. Barrow argues against this hypothesis, and, after adducing various other arguments, says, "Committo autem cum eo (Cartesio) *Platonem,* &c., at imprimis ac præ omnibus tanquam πρόμαχον hujusce prœlii, *Verulamium* nostrum, &c." He quotes him so as to shew he had read him well—generally and particularly—his warnings, his images. And again in a later part of the same speech he goes still more into the detail of Bacon's philosophy, quotes his *motus libertatis, nexus, continuitatis, ad lucrum, fugæ unionis, congregationis,* and instances, after Bacon, the electrical operations of Gilbert as examples of *motus ad lucrum.*

I have no doubt Bacon's works were familiar to all Cambridge men of Newton's time. Have you overlooked Cowley's Ode *To the Royal Society?* Macvey Napier wrote a Dissertation on the influence of Bacon's works in the Edinburgh Transactions, and

I think, since published in his Essays. Gassendi has a very copious analysis of the *Novum Organum*.

I am only recently returned from a summer tour with my wife. In the course of it I received a letter from you containing some kind expressions to both of us, which were duly valued. Lady Affleck and I are now here for the rest of the year.

Believe me, yours very truly, W. WHEWELL.

We had the Newton celebration at Grantham, when of course we did not tell our hearers that his discoveries were according to your anagram, *Not New* : but I told them something from Barrow about our Cambridge philosophy of that time. I have said what I had to say about the new edition of Bacon in the Edinburgh Review for October last, I believe.

TRINITY COLLEGE, CAMBRIDGE, Nov. 26, 1858.

MY DEAR M. QUETELET,

It is so long since I wrote to you that I almost seem to myself to have forfeited your friendship, but yet I will not believe that it is so. I have been prevented from writing to you of late, by knowing of the heavy domestic calamity which had fallen upon you, and feeling my inability to offer you any consolation. I hope that, notwithstanding all that you have lost, (and I could appreciate the greatness of your loss), you are still rich in family treasures, and happy in the happiness of your children. I do not know if you have heard that I have found a dear and sweet companion to solace my solitude. I was married to Lady Affleck last July; she is the sister of one of our good Cambridge Mathematicians, whom I think you know, Robert Leslie Ellis, one of 'the editors of the new edition of Bacon's Works. I think he once passed an evening at your house at Brussels ; he is still a person of admirable clearness of mind, but has been confined to his bed for several years, and afflicted by the severest bodily sufferings, including loss of sight.

Though I have no children as you have, I have young relatives in whom I take a paternal interest.

*　　*　　*　　*

One young nephew is married to a daughter of Herschel, so
that I have the satisfaction of being connected by family ties
with him, one of my earliest friends. I gladly talk to you of
these domestic matters, for I persuade myself that you will not
be sorry to hear of me, and these things interest me more than
scientific news, if there were any good news stirring, but at present
I hear of none. Our mathematician Adams is to succeed Dr
Peacock as one of our mathematical Professors here. You knew
Sheepshanks, his love for Astronomy, and perhaps his love for this
College. His sister, mindful of his feelings, is about to give
£10,000 sterling to Trinity, for the promotion of Astronomy.

A few days ago I was pleasantly reminded of you by receiving
a parcel of books, including some volumes of your *Bulletins de
l'Académie Royale*. This led me to consider how large and how
valuable the series is, and I have resolved to give my series to our
College Library. I do not know whether it would be possible for
you to put the name of the College, instead of mine, on the list of
persons to whom the *Bulletins* are sent. If you could, the series
would go on in future to benefit the Library. I may mention too
that at present I cannot find some of the volumes, namely,
Tome 1 ; Tome 2 ; Tome 9, Pt. 1 ; Tome 13, Pts. 1, 2, and 3 ;
Tome 14, Pt. 1 : perhaps I may find them hereafter among my
books. Once more excuse my long silence, and believe me yours
with unaltered regard, and let me have the pleasure of hearing
from you and hearing how you are.

Yours faithfully, W. WHEWELL.

LOWESTOFT, Jan. 18, 1859.

MY DEAR SIR [MR DE MORGAN],

After having invited your criticism on my *Novum Organon*,
I ought not to omit to thank you for it. I have been, as
I expected that I should be, instructed and interested by it. Nor
can I deny that it is in a great measure just. My object was to
analyse, as far as I could, the method by which scientific discoveries
have really been made ; and I called this method *Induction*,
because all the world seemed to have agreed to call it so, and

because the name is not a bad name after all.   That it is not exactly the Induction of Aristotle, I know; nor is it that described by Bacon, though he hit very cleverly on some of its characters, erring much as to others.   I am disposed to call it *Discoverers' Induction*; but I dare not venture on such a novelty, except in the indirect way in which I have done.   With such a phraseology I think my formulæ are pretty near the mark, and my *Inductive Tables* a good invention.   But I do not wonder at your denying these devices a place in Logic; and you will think me heretical and profane, if I say, *so much the worse for Logic*.

I shall be amused if you succeed in persuading the world that Bacon had little to do with the modern progress of science.   The cry in his favour has been so strong and universal during the last 150 years that the task will be difficult; but perseverance may do much.   Others of my friends are bent upon re-establishing his moral character, and with them too I agree in a great measure. It will be odd if I live to see Bacon no longer reckoned a great philosopher and a bad man, but a poor philosopher and a good man; but yet this is not far from my thought, and may become a popular opinion.   'The greatest, brightest, meanest of mankind' may become 'ambitious theorist, but honest man.'

I can express every whole number from 1 to 15 (I think) by means of four 9s .   Thus $2 = \dfrac{9}{9} + \dfrac{9}{9}$.   Is it worth while working this out further ?

Yours very truly, W. Whewell.

Trin. Lodge, *Feb.* 14, 1859.

My dear Sir [Mr De Morgan],

I go on drawing my Barrow upside down, according to the imagination of our examiners; for I am now, last of all, printing his mathematical works, which are, I think, his best performance.

---

[1] Two more examples taken from a fragment in Dr Whewell's handwriting will illustrate the meaning of this :

$$6 = \sqrt{9 + 9 + 9 + 9}, \quad 10 = \frac{99 - 9}{9}.$$

In doing so, I have found a passage[1] which will perhaps amuse you, and which you will think a good thing frankly said. He asserts that Algebra is not a part of *Mathematics* but of *Logic*. I have copied out the passage and send it inclosed herewith. I still think, notwithstanding your startling illustrations, that Logic has got her name up by being supposed to have something to do with discovering truth, and that it is so much the worse for her, if she cannot keep this part of her character. You make her the old maid of the family of Truth, who nurses the children of all the sister sciences, but has none herself. Certainly in the old time she used to flirt with a large body of lovers, and gave a hope of a numerous progeny. According to your account, she does not really help us to sharpen scythes, which may reap a really valuable harvest, but only to whet razors, which may scrape away what is superfluous from the surface. If you will not let me treat of the Art of Discovery as a kind of Logic, I must take a new name for it, *Heuristic*, for example, only that, as you know, I do not assert such an art to exist. I am not sure, however, that it is not possible; or rather, if I had £20,000 a year which might be devoted to the making of discoveries, I am sure that some might be made. The projects of Solomon's House, and Cowley's College, and the like, are not quite visionary, as the British Association has shewn. And though such machinery can only collect facts at first, collected facts will suggest discoveries; especially now, that we know in a good degree the way of extracting laws from facts. Your Astronomical Society is a valuable instrument of *my* Logic; so I wish you success in the composition of your Report.

Yours very truly, W. WHEWELL.

TRIN. LODGE, *Feb.* 29, 1860.

MY DEAR FORBES,

Your kind letter has remained too long unanswered; partly that I was waiting till you were likely to receive the volume which I have just published, the end of my Philosophy. I am

---

[1] Barrow's *Lectiones Mathematicae*, Lect. II.; page 45 of Dr Whewell's edition.

afraid that I publish so many books, that they must be a burthen to the friends to whom I give them; but this, though in a great part only a new edition, has in it so much new matter, that I could not help asking for it a place on your shelves. I have ventured to follow out my views as they lead into religious philosophy, being convinced, as I have there said, that men will not be satisfied with any comprehensive philosophy which does not do this. The want of this is that which to my mind makes Humboldt's *Cosmos* so unsatisfactory as a whole, and deprives it of any real principle of unity. I am aware, of course, how ticklish a matter religious philosophy is; and do not expect you or any one to agree exactly in my views, though I shall be much interested to know in what light they strike you. I am led by my course of reasoning to the field of a hot controversy which is now going on between Mansel of Oxford and F. Maurice; but I think no one can accuse me of being controversial in what I say on that subject. I am also led to my controversies with Sir W. Hamilton, whose metaphysics I always thought as worthless as he was subtle and learned. And now I hope I have done with my History and Philosophy for a long time, for a new edition will not soon be wanted.

Your notice of dear R. Ellis in the *Athenæum* has given great pleasure to many persons as well as to us. I have considered what he has said about F. Bacon, in my additional Chapter about Bacon in the volume of *Chapters*[1] which, as I have said, I hope has reached you by this time. And now I am going to proceed with my translation of Plato. Acting on your Scottish proverb, I have taken care to be off with the old love before I am on with the new; and I shall now finish a second volume, as the public have taken very kindly to the first. In your letter you noticed what I had said in my sermon about public schools. You are, however, to understand it was said, not on the ground of what I know of public schools, but with reference to certain books which appeared lately....It seemed to me that these works, as well as the behaviour of our young officers in India, and the language

---

[1] The title of the book is *On the Philosophy of Discovery, Chapters Historical and Critical.*

which they used in their narratives, must give foreigners a great disgust for us as a cruel and savage people; and so I took the opportunity of saying. When you send any of your children to school, I hope you will find a school very different from those thus described.......

When do you move to St Andrew's? I am sorry that you have difficulty in getting a suitable house there, but something may turn up unforeseen. Our kindest regards to Mrs Forbes. Lady A. begs her cordial remembrances.

Always truly yours, W. WHEWELL.

BAYONNE, *July* 24, 1860.

MY DEAR FORBES,

I think I did not tell you that Lady A. and I had made up our minds to go to Spain to see the eclipse of the 18th in its totality. We did so, and saw it very well from a range of hills near Orduña, south of Bilbao. The effect was very striking, above all things the corona, which shone out as soon as the eclipse was total. I had not imagined any thing so sudden and luminous. You will probably hear and read a good deal about it, for there were many English observers scattered about the country, and Leverrier with bodies of French and Spanish ones further south. To me, I may say I think to all at our station, the corona appeared to consist of white light with *radial* rays at various points; but other observers whom I have seen speak of the diverging rays being curved in various ways. Galton who was, I believe, at Miranda has made a wonderful picture with curved and tangential rays. The red prominences I could not see *distinctly* with the naked eye, and I used no telescope. We did not see the four planets, as Capt. Jacob at another station did, but we saw stars and an extraordinary saffron dawn in the horizon, when all was very dark about us. Altogether the phenomena were quite a thing to see in one's lifetime, if possible. We were the more pleased, as the previous days and the morning till eleven or twelve were most unpromising with cloud and rain; and, I believe, the same was the case at all the other stations, and at almost all the sky cleared most benignantly just in time. We have purchased this enjoyment at the cost of some discomfort in Spanish inns, which, however, may be borne, even by ladies, when you have learnt the ways of them; and Lady A. has enjoyed her Spanish tour greatly; a little longer than our Scotch tour, for that was only nine days, and this was eleven.

*   *   *   *

Yours affectionately, W. WHEWELL.

CLIFF COTTAGE, LOWESTOFT, July 10, 1861.

MY DEAR FORBES,

I ought sooner to have answered your letter, but I have been in circumstances unfavourable to letter writing ever since I received it. We went to London early in June and remained there till a few days ago, a good deal involved in engagements such as belong to London at that season. I am not without a feeling of regret that you should have been in the south of England without coming to us; and still more grieved to find that it was the insecurity of your health which made you return to Scotland in so premature a manner as to pass us by on the other side of the island. I am glad, however, that you are discovering so much to enjoy in the climate of St Andrew's. As to your project of a common hall for your students, I should think it likely to be very beneficial; the more so, if you have any traditional habits connected with it, which may secure good order and good manners. I suppose some of your Professors will generally dine there. We have had to make some alterations in our arrangements in our hall, in consequence of our great numbers, which render it necessary to have two dinners every day. And our students and their tutors have become so fastidious about the details of the dinner-table that we have made several changes to please them : for example, the introduction of silver forks, silver castors, and the like, and the establishment of a hall butler to keep the servants in order; and I believe that now, in spite of our numbers, we have a very comfortable and civilized dinner-table. But there is among young men so herding together a great disposition to complain about such matters, if it be only to seem fine in one another's eyes. Talking of university arrangements, I have a letter from Matteucci at Turin, asking advice as to how they may best organize their University there.   I suppose he has written to many persons on the same subject; perhaps to you.   I do not know that I can refer you to any book which gives any account of any recent changes at Cambridge. Indeed there have been very few; for we do what we can to make the new statutes conform to the old usages. The main changes are in the tenure of fellowships at the smaller Colleges, and in the establishment of *minor scholarships*, to be obtained by competitive examination on beginning residence   an improvement which we introduced in anticipation of the Commissioners.   Our moral and natural sciences triposes, corresponding to the new *Schools* at Oxford, have not yet become very effective ; but we have made a good organization for them.

I did not, when I was in London, hear anything definite about Graham's colloid[1] condition.   It is too absurd to see how the Londoners wrangle against your word *viscous*, as if there were some deadly heresy involved in it.   But you need not fear those wranglers. If they establish a *colloid* condition, nobody except persons very obstinately prejudiced will be able to deny a viscous condition.

*   *   *   *

We shall stay here till the beginning of August, and then, I think, go to Spa for a short time, and perhaps to Vienna, mainly to see the new Alpine Railway to Trieste, which, I am told, is a wonderful piece of engineering.

I am sorry to hear that you have had ground for uneasiness about your daughter's health.   I hope that your whole family is now quite well and prosperous.   Lady Affleck joins me in kindest regards to you and Mrs Forbes, and I am always,

Yours most truly, W. WHEWELL.

---

[1] See the obituary notice of Thomas Graham in the *Proceedings of the Royal Society of London*, Vol. 18.

SPA, *Sept.* 23, 1861.

MY DEAR QUETELET,

I expect to be at Brussels on Wednesday and Thursday next, the 25th and 26th, and do not like to be so near you without trying to see you. I have been a very bad correspondent to you, but have never ceased to retain a lively recollection of your friendship and a strong wish to see you again. I am travelling with my wife Lady Affleck, to whom I was married three years ago, and we shall, I expect, arrive at the Hôtel de l'Europe about 3 o'clock on Wednesday. If you will let me know there when we can call upon you, it will give me great pleasure to introduce my wife to you. We have been travelling in Switzerland, and are on our way home. I hope that you and your family are well and prosperous.

Believe me, my dear Quetelet, yours with great regard,

W. WHEWELL.

CLIFF COTTAGE, LOWESTOFT, *Jan.* 4, 1862.

DEAR GWATKIN,

I am always glad to hear from so old and valued a friend as you; and am glad too that you approve of the omission of *Bel and the Dragon*. You must not however lay too much stress upon this omission, as if it was a deliberate act of the College. The fact is that both *Susannah and the Elders*, and *Bel* have for some years been exchanged for other lessons by the mere authority of the Dean for the time being; he thinking that these, and especially the former, are not edifying reading in a large body of young men. But the practice has not been very steady; and I believe *Bel* was read this year. I think it very awkward for us, who have the daily lessons read all the year round, that the Apocrypha holds so large a place among the lessons. I think that we should have the power to change the lessons, if necessary, and, as I have told you, our officers act as if they had this power. And in other ways I should be glad that our officiating ministers or our bishops had power to vary other parts of our liturgy. But I would not join any association for making changes; for any general scheme of doing so, if successful, would, I think, lead to a number of extravagant and unwise proposals. I had rather, as I have said, have power given to vary the liturgy according to circumstances. There is a portion of such power granted at present; and some of the bishops seem not unwilling to extend it.

I presume the book, which you mention as sent to me, is at Cambridge, where I shall see it when I return to College. I am at present at the sea side in Suffolk. Many happy new years to you from yours very faithfully, W. WHEWELL.

TRIN. LODGE, CAMBRIDGE, Jan. 29, 1862.

MY DEAR SIR GEORGE [CORNEWALL LEWIS],

I have received your History of Astronomy, and I return you my best thanks for your kindness in sending it me. I can truly say that I am astonished at the quantity of materials which you have brought together, and at the skill with which you have used them. It seems to me that if I had had to read all the books which you have read, I should never have had the courage to write my History of Astronomy, of which you are so kind as to speak. I am going through your book with great interest.

Believe me, my dear Sir George,

Yours very faithfully, W. WHEWELL.

TRIN. LODGE, Feb. 12, 1862.

MY DEAR SIR GEORGE [CORNEWALL LEWIS],

I am much obliged to you for sending me Mr Brown's learned and convincing dissertation[1]. I sent for Robinson who is of my College, and who is really a man of an enquiring mind; and I requested him to get his Tutor to look with him at the books you refer to about the Oscan and Iguvian inscriptions. He informs me that his friend Brown has followed exactly the

---

[1] This refers to a *jeu d'esprit* by Sir G. C. Lewis with respect to some modern restorations of extinct languages. The title is *Inscriptio Antiqua, in agro Bruttio nuper reperta, edidit et interpretatus est Johannes Brownius, A.M. Ædis Christi quondam Alumnus. Oxonii* 1862. See the *Letters of...Sir G. C. Lewis*, page 403.

same methods as the learned authors of those books, and that the result is no less certain.

I have given the copies which you sent to those who, I think, will appreciate them.

Believe me, yours very truly, W. Whewell.

Lowestoft, Aug. 2, 1862.

My dear Forbes,

I ought sooner to have replied to your last letter; but it invited me to speak of our summer plans, and this I could not do with any definiteness till now. Now I fear I must say that I do not think there is this year any prospect of our crossing to your side of the border. Indeed it is not certain that we shall even visit the north of England. My wife has a great desire for a sniff of Swiss air, and a look at your friend Mont Blanc once again. Last year we barely had a glimpse of him, sleeping at his foot for one night only. So I think when we quit this place, which we shall do in a few days, we shall go to the continent for about three weeks. It cannot be much longer, for I must be in Cambridge in the middle of September. The meeting of the British Association there on Oct. 1 shortens our vacation sadly. I wish there were any hope of our seeing you at the meeting. We should be most glad to have you for our guest then, and your being there would add much to my enjoyment of the meeting. The great pleasure of such occasions is the presence of old friends at them. The business of the meetings has fallen so much into the hands of comparatively new men, of whom I know little, that I do not take any great interest in them, except on the ground of such old companionships. I believe we shall have several foreigners there, and, I should hope, something which may interest you. I do not know if the subject of glaciers is likely to turn up. I cannot but regret that you did not print the paper which you say you prepared for the *Athenæum* on that subject. It would be a useful contribution to the history of the speculations about glaciers, and might save some future historian of science from trouble and error. Might it not be still read at the meeting? Even if you do not come,

I should be glad to take charge of it, and to do the subject such justice as I can. The present race of glacier speculators began in great ignorance of what had been done before, and are slow in learning, but truth has her claims, and is not to be given up.

I have been interested by the process of laying a telegraph wire almost from my door to Holland. The point where it leaves the shore here is only a quarter of a mile hence. I found the operator at the station employed in testing, as he told me, the *conductivity* of the wire. I was glad, for the sake of scientific propriety of language, to find that the word which I recommended years ago has made its way and established itself in the working vocabulary of galvanism as an art.

I am very glad to hear that your plans for the improvement of St Andrew's answer so well. I should much like to see your College there, and especially your hall and your painted glass; and also I may say, to see you in a position in which you find so much to do and to interest you; which means, with you at least, that it is done well.

Lady A. bids me say that she has been much pleased with the little printed paper about glaciers, but I believe she will put in my envelope a note in which she will give you her own thanks for it. Pray give my kindest regards to Mrs Forbes.

You have quoted a pretty fancy of Longfellow's about the glaciers. I suppose you know Byron's description, which is remarkable for being more true physically than most of the prose descriptions by men of science at that time.

> The glacier's *slow* resistless mass
> Glides onwards *day by day*,
> But I am he that bids it pass
> Or on its course delay[1].

It is in 'Manfred' I think spoken by the Spirit of the Alps.

CHAMOUNIX, *Aug.* 21, 1862.

MY DEAR FORBES,

I forget whether I told you that we intended to come here; but so it is that here we are. We came, without any purpose of

---

[1] The quotation is not quite accurate; see *Manfred*, Act I. Scene 1.

either extensive or scientific travelling, to gratify some recollections of the place which Lady Affleck wanted to revive on the
spot, and shall return to Geneva on Saturday, having come here
on Monday. We have had glorious weather, Mont Blanc and his
companions having been before us in broad sunshine almost the
whole of the time. As we came here with no special purpose of
studying your friends the glaciers, I did not bring your book with
me, which would have added greatly to my pleasure in being here,
nor even restored my recollections of the subject by reading anything which you have written before I came. But being here,
and looking at two or three glaciers, many of the points of discussion have come back to me; and, though I shall by my questions let you see how dim and unstable my recollection of your
lessons is, I will not hesitate to ask you some questions, that your
answers may revive and rectify my notions, such as they are. If
the subject should come under discussion in October at Cambridge,
I should like to be tolerably accurate in my history and philosophy
on the subject.

First, as to the form and direction of the crevasses. As seen
at the *Mer de Glace*, they are, predominantly, convex to the upper
part of the glacier. I think you have noticed this and explained
it. Hopkins is in the habit of saying that the direction of the
glaciers, according to his views and yours, is quite different. Is
this true? Have any of the recent glacierists given any observations *on a large scale* as to the direction which the crevasses really
follow? I have seen none. It seems to me that the direction of
the crevasses, perpendicular to the curve which a line on the ice
transverse to the glacier assumes by the glacier motion, follows at
once from the nature of the motion, without any discussion of
forces such as Hopkins introduces, which appear to me quite inapplicable to the case.

Next, as to the ribboned structure. I have not seen it any
where clearly; in some degree it appears in the fragments in the
lower part of the *Mer de Glace*. Have any of the new glacierists
made any observations of this phenomenon on a *large scale?*
I have seen none. And is there any difference in the direction of
the stripes on their theory and on yours? You will see my igno-

rance by the nature of my questions; but a few words from you will, I think, set me right.

There is a pamphlet giving an account of the discovery of the remains of the guides who perished in 1820, and which were found last year, as you predicted. You have probably seen this pamphlet; but, as it is possible that you have not, my wife is going to send it you. Some additional remains have been found this year. I saw Dr Hamel in London a few weeks ago, the principal author of the catastrophe.

We shall be in Cambridge in ten days or thereabouts, so write to me there, if you think that my questions deserve any answer.

Kindest regards to Mrs Forbes, in which Lady A. joins, and to you.

Believe me, yours very truly, W. WHEWELL.

Do you hold that there is any other leading direction of the crevasses besides the one transverse to the glacier? It seems to me that something of the kind is shown by the aspect of the terminal part of a glacier, and by the *pyramids* into which the ice runs.

C. C. LOWESTOFT, Jan. 28, 1863.

MY DEAR FORBES,

It was a great pleasure to me to receive your letter. I too had been, ever since the new year began, meditating a letter to you to convey good wishes and kind recollections, such as are suitable to such a season. As years go on, old friends become more and more valued, and it becomes more and more natural to express our value for them. I was pleased and interested to hear of your family interests. I too, though with no sons of my own, have some experience of the difficulty of finding a career, as the French call it, for young men.

*   *   *   *

But I am quite seriously of opinion that an English University Education is of little or no use to a man who has that only to depend on, except he is clever enough to obtain a Fellowship.

*   *   *   *

I am very glad that your College affairs are going on to your liking. I do not think that our Commissioners have done us *much* harm; but they have done us *some;* and they have failed altogether of the good they might have done. I dare say I shall find your Lecture at Cambridge when I go there, which will be next Tuesday. I meant my Lectures to be a sort of new-year's offering to you, and intended, as I have said, to write and tell you so. They were delivered to the Prince of Wales at his father's request, as I have said in the Preface. The last two Lectures, I think, are important and new to the literature of the subject as generally read—though their novelty is not mine, but my dear friend Jones's. The doctrines seem to me demonstrably right.

TRIN. LODGE, CAMBRIDGE, *Feb.* 28, 1863.

MY DEAR SIR CHARLES [LYELL],

I am reading your book on the Antiquity of Man, as all the world has done or is doing; and, like most of your readers, am deriving great pleasure from the perusal. I am especially delighted with your Chapter 23, *On the Comparison of the Development of Languages and of Species.* The comparison appears to me admirably adapted to explain the difficulties and the solutions of the one theory by the other. You may, perhaps, recollect, that I had used the relation of the two sciences, geology and glossology, to explain my notions of the structure and progress of my "palætiological sciences," though I do not mean that I have in any degree anticipated your arguments.

So far as I collect from your expressions, you would not disagree with me in the main conclusion which I then drew, and which still seems to me to stand: that in all these sciences no natural *beginning* is discoverable, no origin homogeneous with the known course of events. We can in many of these sciences go far back, and exclude many suppositions, but we cannot find an origin to which all these lines converge, though we always seem to be approaching to it.

I am much interested too in what you say about the mental development of man. Perhaps it might be pursued further, by tracing in a definite way some of the steps of which the secular mental development of man consists : for, according to my philosophy, the formation of new sciences has been accompanied by the development of new ideas in modern times ; and those can never fully appreciate man's mental development who go spinning round and round the ancient starting places of geometrical and arithmetical truth.   This, however, I have tried to work out in my *Philosophy of Discovery*, and may perhaps have something more to say upon it.   How strikingly do your speculations illustrate and extend Cuvier's remark, that the geologist is an antiquary of a new order !   Believe me, dear Sir Charles,

Yours very faithfully, W. WHEWELL.

THE ATHENÆUM, *May* 14, 1863.

MY DEAR SIR CHARLES [LYELL],

I find that I did not give you a reference to the part of my *Philosophy* to which I especially wished to direct your attention— the part about *languages*.   You will find it in Book X. of the *Philosophy of the Inductive Sciences*, and in the corresponding part of the *History of Scientific Ideas*, a subsequent form of the same work[1].

In Chapter II. Arts. 7 and 15 of that Book, I have spoken of the Classification of Languages, and of the causes by which Languages are affected as they are derived from each other : and in Chapter IV. Art. 7, I have tried to shew that Science tells us nothing about creation.   I have also urged there the duty of mutual forbearance between the cultivators of science and the upholders of tradition ; but this is a more general topic.   Believe me, dear Sir Charles,

Yours faithfully, W. WHEWELL.

---

[1] *Philosophy of the Inductive Sciences,* second edition, Vol. II. page 661 ; *History of Scientific Ideas,* Vol. II. page 280.

MY DEAR SIR CHARLES [LYELL],

I was interested by your remarks on the passage of my works to which I had referred you. I may say at once that I did not conceive that I had in any way anticipated you in arguments drawn from Glossology and Geology. Only I was struck and pleased with the way in which you had illustrated the arguments in the one subject by those in the other. Many persons will see the opposition of theoretical views more clearly in the case of *languages* than of *strata*. But I conceded that these parallel arguments, though they may be useful and striking in *stating* the question, are not of much weight in *deciding* it. There still remains the opposition between the catastrophist and uniformitarian view as I formerly stated it. I was aware that you inclined to uniformitarian views; I do not see that much has been added lately in defence of them. The past history of man has been made more extensive. Be it so. But we have always the whole of time at our disposal, and therefore nothing is gained this way. You say you can conceive all the changes which have produced the existing languages to have proceeded by slow gradations. But there has been nothing like uniformity in this matter. The English language was formed in less than a century from the Conquest, and has undergone little essential change since. I concede that it has lasted 1000 years. The Greek has lasted much longer, including its present revival. You say that I consider the past changes which have altered languages to have been of an order superior to those which operate at present. I do so, because my masters in the science of languages, Pritchard and the like, do so.

It seems to me that all that has been recently urged in favour of the uniformitarian scheme merely amounts to this : that we can interpolate steps so as to make a gradual transition from any one condition to any other—either of *strata* or of *speech*; and when you have got any change, however small, you say multiply this by 10000 cycles and you have the observed change. It does not appear to me that it is so. I think that no numerical multiplication of changes now in progress could produce the past changes; and I think I have all the philologers with me. But, as I have said, I rather regard this as an illustration than an argument. Only, as there must have been an origin, I do not see why we should avoid *Indications* of it in the facts. I have a great deal more to say, but I fear we should not agree. Thanks for your attention.

Yours faithfully, W. WHEWELL.

TRIN. LODGE, *Ap.* 10, 1863.

MY DEAR SIR [MR DE MORGAN],

I agree with you about the state of Aristotle's text: perhaps I go further than you do. In all cardinal passages it requires emendation: as in the passage about Induction I have shewn. And I hold that any emendation, which makes sense, is better than the reading of 100 MSS., which makes nonsense. I wonder what sense the translators and commentators, who accepted *invisible voice* as an illustration of the Infinite, attached to the words[1]. Your ἀόριστον is much better, though even so, I do not see how the *indefinite voice* is an example of the In-finite. However, it is an example of a thing indefinite, and so far gives the passage a meaning. It seems to me that endless confusion on this subject arises from using the adjective *Infinite* without a substantive. You ask—that is, *they* ask us—to define the Infinite. I reply, asking again Infinite *what?* Infinite space time, number, &c. I can define, but I do not know anything about *the* Infinite, without such application. But it is plain that Aristotle and his modern followers are aiming at some indefi-nite Infinite, and so make confusion. Still greater confusion arises from talking of the *Absolute*: any assertion about it must be foolish nonsense, must amount to this: "I assert that of that of which nothing can be asserted or denied." I do not think we have any Aristotelian metaphysicians here, so I cannot help you to any Cambridge opinion on the subject. I have various

---

[1] Mr De Morgan referred to a passage in the *Metaphysica*, Lib. x. Cap. x ; see also the *Physic. Auscult.*, Lib. III. Cap. IV.

translations and commentators, French and German; but they do not accept my first postulate—that the duty of a translator is to make sense at any rate, the author's sense if he can.

In the passage of Aristotle everything depends upon the word διέναι, and I do not see that it throws much light on the ἄπειρον. Do *you* think that it helps our conceptions much ?

Believe me, yours infinitely and indefinitely,

W. Whewell.

[The next letter was addressed to the Rev. Dr D. Brown, Professor of Theology in the Free Church College, Aberdeen.]

Trinity Lodge, *Oct.* 26, 1863.

Dear Sir,

I had the pleasure of receiving your letter from my nephew, Julius Elliott, on the subject of my little volume of extracts entitled "Indications of the Creator," which you kindly inform me you refer to in your Lectures on Natural Theology. I am glad that you think the book fit to be adduced in reference to that great argument. It was not intended as a treatise on that subject, as you see from its composition; and to my mind the effect of the views there taken has always been more impressive from being collateral results of other speculations, than they would have been, if they had been the direct purpose of my speculations. And I may say that the recent discussions which have taken place in geology and zoology do not appear to me to have materially affected the force of the arguments there delivered. It still appears to me that in tracing the history of the world backwards, so far as the palætiological sciences enable us to do so, all the lines of connexion stop short of a beginning explicable by natural causes; and the absence of any conceivable natural beginning leaves room for, and requires, a supernatural origin. Nor do Mr Darwin's speculations alter this result. For when he has accumulated a vast array of hypotheses, still there is an inexplicable gap at the beginning of his series. To which is to be added, that most of his hypotheses are quite unproved by fact. We can no more adduce an example of a new species, generated in the way which his hypotheses suppose, than Cuvier could. He is still obliged to allow that the existing species of domestic animals are the same as they were at the time of man's earliest history. And though the advocates of uniformitarian doctrines in geology go on repeating their assertions, and trying to explain all difficulties by the assumption of additional myriads of ages, I find that the best and most temperate geologists still hold the belief that great catastrophes must have taken place; and I do not think that the state of the controversy on that subject is really affected permanently. I still think that what I have written is a just representation of the question between the two doctrines.

I do not think therefore that I could be of any service to the cause by any new writing on the subject at present. One merit of what I have written is, if I may venture to say so, that it was not written with a controversial purpose. And it still seems to me to apply to most of the questions now under discussion. A person who ventures into the controversies which are at present agitated ought to have a great deal of specific knowledge, which I do not possess. Any attempt on my part to give a summary of the argument would be to me very unsatisfactory, because, as I have said, the impressiveness of those views to my mind arises from their being not the propositions of a system of doctrines, but the corollaries of the history and philosophy of science.

I may mention that, in the last edition of my History of the Inductive Sciences (1857), Vol. III., I have made some additional remarks on Morphology, and on Final Causes, which I should insert in the "Indications," if an opportunity should occur: and that in the "Philosophy of Discovery" (1860) I have given a Chapter on Man's knowledge of God, where I have given such theological views as appear to me to result from the study of science. The manner in which you refer to what I have written on such subjects will, I hope, excuse me for thus noticing my own performances.

Again thanking you for your favourable judgment of what I have written, I am, dear Sir,

Your very faithful Servant, W. WHEWELL.

CLIFF COTTAGE, LOWESTOFT, *Jan.* 4, 1864.

MY DEAR FORBES,

It was a great pleasure to me to receive your Christmas greeting. In return allow me to wish you and Mrs Forbes and all your family a happy new year and many more after this. I think with much gratification of the long continuance of our friendship: I feel very strongly that we cling to old friends the more closely, the fewer they become. From you besides I always learn so much. I read your article on the antiquity of man in the Edinburgh without knowing that it was yours; and I have since read it over again, and I think still more strongly, what I thought before, that it is the best critique of Lyell's book which I have seen. The evidence of the facts recently under discussion is most clearly given, and is completed by independent reading, and the case is summed up in a calm, impartial and judicial manner: in *your* judicial manner I am disposed to say; for I always admire your surveys of scientific questions for that quality. I have myself taken no share in the discussions on the antiquity of man; but I will not conceal from you that the course of speculation on this point has somewhat troubled me. I cannot see without some regrets the clear definite line, which used to mark the commencement of the human period of the earth's history, made obscure and doubtful. There was something in the aspect of the subject, as Cuvier left it, which was very satisfactory to those who wished to reconcile the providential with the scientific history of the world, and this aspect is now no longer so universally acknowledged. It is true that a reconciliation of the scientific with the religious view is still possible, but it is not so clear and striking as it was. But it is weakness to regret this; and no doubt another generation will find some way of looking at the matter which will satisfy religious men.

I should be glad to see my way to this view, and am hoping to do so soon.

We went abroad during the summer to various parts of Germany, Italy, and Switzerland. Our farthest point was Venice; and I think the most remarkable thing we saw was the railroad over the Sömmering Pass, a beautiful piece of engineering through a beautiful country. The greater part of our travels was employed in my wife and me seeing together what we had before seen separate; but Venice was new to me, and I rejoiced much to see it. We had however very hot weather there and very sanguinary mosquitoes. We ended our tour by a visit to Switzerland; not, however, trying to make a near acquaintance with any of your friends, the glaciers. Our greatest achievement (a very cockney exploit!) was our going to the top of the Rigi for a night; for which we were rewarded with a beautiful sunrise. On the top of the Pilatus opposite to us we saw a crowd of sightseers as wise as ourselves, who issued from their hotel to worship the sun-rise, as we did from ours. Since our return to Cambridge we have had the Airys with us for a week. They told us of their Scottish tour and of their disappointment in not seeing you. Airy is, as you say, an excellent mathematical workman, and his memoir on the strains of a beam is a good example of his skill. Do you recollect a suggestion of Brewster's, when he discovered the dipolarizing structure of glass under strain—that we might in this way see the distribution of strains in a beam, by making a glass model?

*   *   *   *

I am, my dear Forbes,

Always truly and affectionately yours, W. WHEWELL.

TRIN. LODGE, Dec. 12, 1864.

MY DEAR SIR [MR COCHRANE],

I was glad to hear that your Homer was still making progress, and obliged to you for sending me a specimen of it. I return it to you, and hope I have not kept it too long. I have no remarks to make upon it, except some of the same kind as

I made on the earlier part. In some cases I think still that the accent is forced, which is a bad fault in hexameters, worse than in any other kind of verse, because it affords food and pretext to the popular prejudice against hexameters. You may see in Lord Derby's preface that he has taken up this prejudice.

* * * *

As to this version of his, I agree with you entirely. It is a good blank verse version, but on the whole certainly not better than Cowper's, and in some parts very decidedly worse. I do not think in that verse they will be able to produce a better translation than Cowper's. Sir J. Herschel is going on with his, and has completed the twelfth Book. I do not know if he has got any further. His translation both as to metre, phraseology, and spirit, seems to me excellent. He has subjected the fidelity of his version to a severe test, by printing in Italics every word which is not found in the original. It is curious to see how few he has managed to make these.

As I have said, I fear that there is still a general and very stupid prejudice against English hexameters, but I hope such examples of them as yours and Herschel's translations of Homer will finally prevail over this impatient bigotry.

I am, my dear Sir, yours very faithfully, W. WHEWELL.

TRIN. LODGE, *Jan.* 2, 1865.

MY DEAR FORBES,

It is long since I heard of you, so I gladly take the occasion of the new year to ask news of you, and to give you my best good wishes as suits the time. May the year and many years more bring to you all blessings! I hope that St Andrew's continues to agree with you, and that you have had comfort and success in introducing the improvements which you contemplated a little while ago. We have been going on here much as usual. Our Commissioners who gave us our new code of statutes tried to provide an annual event for us. They appointed an annual college meeting of all the Fellows, at which any one might propose any change in the mode of conducting the College.

As might be expected, some of our younger Fellows could not resist the temptation to make experiments by changing our habits. Last year it was proposed to alter the mode of providing for our Sizars. This year it was proposed to abolish the order of Fellow Commoners. This latter proposal was negatived by a very large majority, which was a great satisfaction to me, and will, I hope, keep us quiet for some time to come. The proposition about Sizars was negatived by a sufficient majority; but our Seniors thought it better to do what was asked for, as it was only a matter of expense, not of principle. The Oxford people are more prone to change than we are. They have abolished their Gentlemen Commoners at all their Colleges except two; but then these two are Christ Church and St John's. We have a great deal of building going on here. I wish you could see our New Museums and Lecture rooms, which supply a want long felt here, and will, I hope, give new vigour to our professorial system. Pray give my best wishes to Mrs Forbes for herself and her family. I hope your children are all prospering to your heart's content. My wife joins me in every good wish and kind regard, and I am always

Most truly yours, W. WHEWELL.

TRIN. LODGE, Nov. 7, 1865.

MY DEAR SIR [MR DE MORGAN],

I believe you are quite right in your notions as to the ancient practice of this College, and of other Colleges, as I suppose. Pupils were admitted under many of the Fellows, perhaps under all, the consent of the Master being always had. The admissions were first limited to two sides in the last century, Jones of Trinity being by his popularity and his extensive connexions the cause of the limitation. This I was repeatedly told by the late Bishop Monk, who had examined this point of College history. But in the Admission Book I find that, though the great majority of admissions are under Jones from 1787, they are almost in as great a majority under Collier, and under Atwood *and* Postlethwaite, as under Jones afterwards. The relation, as you say, was rather parental than tutorial. Indeed we must recollect that there were no examinations to prepare for at that time, and a general superintendence of the pupil's reading was probably all that was needed. Still the Master did sometimes take pupils in some especial way. When it was intended that the Duke of Gloucester, Queen Anne's son, should go to Trinity College, he was to lodge with Dr Bentley. So says Monk[1]. I am not certain that the *late* Duke of Gloucester, when an undergraduate at the college, had rooms at the Lodge, but I think it is not unlikely.

You will find in my life of Barrow that Barrow had pupils; I suppose most of the resident Fellows had. Believe me

Yours very truly, W. WHEWELL.

TRIN. LODGE, *Jan.* 7, 1866.

DEAR SIR [MR COCHRANE],

I was glad to hear from you again, and to learn that you have completed your translation of the Iliad; for it is something to have accomplished so great a task. I am afraid I cannot suggest any course which will find a publisher for it. Sir John Herschel has finished his Iliad, and is in the same difficulty as yourself, for the publishers are very shy both of Iliads and of hexameters at present.

I will not conceal from you that I think Herschel's the best translation which I have seen, and that I hope it will be published sooner or later.

I have written nothing of late. The past year has been darkened to me by the greatest of domestic calamities—the loss of a beloved wife. I hope no heavy sorrow has fallen upon you.

I am, dear Sir, yours faithfully, W. WHEWELL.

[1] See Monk's *Life of Bentley*, Vol. I. pages 145 and 200.

THE END.

CAMBRIDGE: PRINTED BY C. J. CLAY, M.A. AT THE UNIVERSITY PRESS.